# Eau Canada

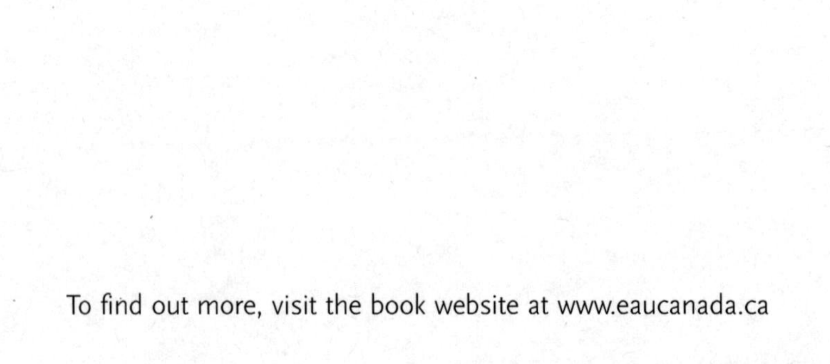

To find out more, visit the book website at www.eaucanada.ca

# Eau Canada: The Future of Canada's Water

Edited by Karen Bakker

UBCPress · Vancouver · Toronto

15 5 4

Printed in Canada on ancient-forest-free paper (100 percent post-consumer recycled) that is processed chlorine- and acid-free, with vegetable-based inks.

**Library and Archives Canada Cataloguing in Publication**

Eau Canada: the future of Canada's water / edited by Karen Bakker.

Includes bibliographical references and index.
ISBN-13: 978-0-7748-1339-6 (bound); 978-0-7748-1340-2 (pbk.)

1. Water – Canada. 2. Water – Government policy – Canada. 3. Water use – Canada. 4. Water conservation – Canada. 5. Water-supply – Canada. I. Bakker, Karen

HD1696.C2E28 2006 333.9100971 C2006-905375-8

Canadä

UBC Press gratefully acknowledges the financial support for our publishing program of the Government of Canada through the Book Publishing Industry Development Program (BPIDP), and of the Canada Council for the Arts, and the British Columbia Arts Council.

This book has been published with the help of a grant from the Canadian Federation for the Humanities and Social Sciences, through the Aid to Scholarly Publications Program, using funds provided by the Social Sciences and Humanities Research Council of Canada, and with the help of the K.D. Srivastava Fund.

Royalties from the sale of this book will be donated to the Canadian branch of the Waterkeeper Alliance.

UBC Press
The University of British Columbia
2029 West Mall
Vancouver, BC V6T 1Z2
604-822-5959 / Fax, 604-822-6083
**www.ubcpress.ca**

# Contents

# Figures, Boxes, and Tables

FIGURES



BOXES

TABLES

# Foreword

*David Schindler*

Canadians feel very strongly about water governance. A 2004 Ipsos-Reid poll conducted on behalf of the Council of Canadians found that 97 percent of Canadians agreed with the statement: "Canada should adopt a comprehensive national water policy that recognizes clean drinking water as a basic human right." A high proportion of Canadians blame politicians for the water crises at Walkerton, Ontario, and North Battleford, Saskatchewan. Therefore, this book should be of widespread interest to Canadians since, in language that is intelligible to the public, it summarizes many of the problems of water governance in Canada.

The libraries of academic institutions contain many papers on important facets of water governance in Canada. They are diffused through a broad literature and written in the professional jargon of scientists, constitutional lawyers, policy analysts, and other professionals. As a result, the public, and even most politicians, is unaware of the increasing problems that our country faces in securing its water supplies for the future. Bakker and her colleagues do a remarkable job of sifting through this diffuse mountain of academic and legal work, summarizing what is relevant to contemporary water governance. One hopes that politicians will read the book, people will demand action, and the frustrating decades of inaction by politicians on national Canadian water policy will end in the very near future.

The Rawson Academy of Aquatic Sciences once attempted a similar synthesis; unfortunately, however, government bulletins are poorly advertised (see M.C. Healey and R.R. Wallace 1987, *Canadian Aquatic Resources Bulletin 215*, Canadian Bulletin of Fisheries and Aquatic Sciences, Ottawa). The academy's book received only limited use by academics and attracted little attention from politicians or the public. It was written in the heady days following the Pearse Federal Inquiry on Water, and most academics hoped the inquiry would result in a strong federal water policy that would form the basis for international and interprovincial water management, to be enforced by Environment Canada's Inland Waters Directorate. This did not happen. Many of the reasons why are outlined in *Eau Canada*. Today, we still have no strong Canadian federal water policy, and the Inland Waters Directorate has been disbanded. The Rawson Academy of Aquatic Sciences, comprised of a collection of prominent academics interested in translating water research into strong policy, no longer exists, and federal water policies have become increasingly weak. The federal government is largely ignored by the provinces when they make their decisions about water. As a result, throughout the country, we have a mish-mash of water policies that are inconsistent with respect to the precautionary protection of the environment and to protecting the rights of Canadians. Shades of the Balkans!

The threats to Canadian water have changed since the Rawson Academy's summary was written eighteen years ago. At that time, much was made of the threat of large-scale diversions of northern rivers to support water-hungry American developments of the Midwest and Southwest. Fortunately, those threats have faded somewhat, thanks to the increased efficiency of desalinization, some modest conservation initiatives, and a better understanding of the huge costs and enormous social and environmental impacts of such large-scale diversions. Nevertheless, a few misinformed individuals continue to press for such massive engineering feats. Cost seems not to be a factor in other expensive undertakings (such as wars), and there is some danger that, at a time of water scarcity, little consideration might be given to the costs or ecological consequences of massive diversions. At a time when good will between the United States and Canada is at a minimum, some threat to our water clearly remains.

Modern threats to water are smaller but more abundant than were older threats. Examples include legislation to sneak small but numerous aliquots of groundwater from the Great Lakes Basin to US counties that straddle the watershed boundaries, or the consistent exceedance of the amount diverted from the Great Lakes to the Mississippi via the Chicago River – a diversion that was unilaterally approved by the United States before the Boundary Waters Treaty of 1909. There have also been worrisome and unprecedented violations of the Boundary Waters Treaty over the past few years, and Canadians should be aware of these. The diversion of water from North Dakota's Devils Lake into Lake Winnipeg, one of our most productive freshwater fisheries, is one such violation. There is also talk in the US midwest of taking water from

Lake of the Woods to replenish or replace the dwindling, unsustainably used aquifers of the western United States. In all cases, there is no reference to the International Joint Commission (IJC), which has adjudicated cross-border water decisions between the two countries for almost 100 years. Canada is also guilty of IJC violations. Pollutants from smelters in Trail, BC, to US waters, along with the threat of pollution from new Canadian mines in the headwaters of the Flathead River, have Montanans riled. It is time that mechanisms for our transboundary governance of water are reviewed and updated.

*Eau Canada* should make Canadians aware of the hypocritical nature of Canadian policies with respect to water governance. Diversions that would have created an international political row had they taken place on cross-border waters (Devils Lake being an excellent example) have taken place quietly and without fanfare within Canada. Cases include the combining of many rivers entering James Bay and the diversion of the Churchill River into the Nelson River in Manitoba. Much of the resulting hydroelectric power is sold to the United States. These have been destructive to ecosystems and Aboriginal cultures but are tolerated by Canadians because they occur within our boundaries. More of such projects are planned. We already impound and divert more water than any other country, and there are plans afoot to double the amount.

Canadians should also keep a sharp eye on what is happening to the Athabasca River in northern Alberta as a result of oil sands development to fuel our increasing fleet of gas-guzzlers. As in Quebec and Manitoba, key ecological resources and Aboriginal cultures are clearly threatened, and there is no sign that our federal government will intervene on behalf of Canadians to protect water resources and cultures that are an integral part of our country. As with hydropower, multinational companies, who plan to export most of the oil and gas for profit, are perpetrating much of this massive ecological destruction. Further downstream, it now seems virtually certain that a pipeline will be built in the Mackenzie Valley to transport northern hydrocarbons to southern markets, despite irreparable ecological damage to ecosystems and social systems in northern communities.

*Eau Canada* also spells out, in clear and concise language, the real facts concerning water abundance in Canada. The glib assurances of water abundance that we frequently hear from the media, politicians, and even environmental groups are lies, unsupported by meteorological or hydrological data. Typically, a square metre of average Canadian terrain receives no more water from precipitation each year than does the same amount of terrain in many other countries. The apparent abundance of water on the Canadian landscape is partly a function of the large number of depressions that hold water (i.e., lakes) left by retreating glaciers and partly the result of the fact that we are a cold country, so that losses to evaporation and transpiration are low. In other words, while Canada has a large freshwater "bank account," the interest rate is very low. As in finance, it is the interest that we have to use if we want to sustain our water capital. Southern Canada, where 85 percent of Canadians live, is much drier than the US average. Even the enormous Great Lakes have only 1 percent of their massive volume renewed by precipitation and runoff each year. Those who would send our water south, or even transport it by tanker to countries that have lost their water to unsustainable practices, either ignore or do not understand these facts.

As I write this, the Canadian election campaign of 2005-2006 has just finished. It has been bizarre to see debate by those who are supposed to be our leaders limited to issues that are secondary to Canada and Canadians in the overall scheme of things. The Conservative Party has been elected on its promise to ensure that federal politics are cleaned up. The party has declared that the environment is not a major factor in its agenda and has promised to replace Canada's commitment to the Kyoto Accord with a more industry-friendly pace of reductions in carbon emissions. It does not seem to grasp the fact that political scandals are much less important to the future of Canada and Canadians than are freshwater supplies, the ecosystems that sustain them, or many other environmental issues that are critically connected to our accelerating pace of "development." Warning signs of our unsustainable water policies are everywhere: Walkerton, North Battleford, and Kashechewan are examples. So are the prolonged Prairie droughts between 1998 and 2004, and the dwindling quantities of water reaching the Prairies from glaciers and snowpacks assailed by a warming climate. We should be asking political leaders: "What will your policies be on water issues, climate change, and sustainable societies?" Instead, we tolerate weeks of insubstantial nonsense as the basis for making our choices at the polls. It is to be dearly hoped that *Eau Canada* will be the start of a new dialogue between academics, the public, and politicians – a dialogue directed at ensuring that strong and sustainable policies underpin our future treatment of water and other natural resources.

# Preface

*Eau Canada* analyzes aspects of contemporary water management and governance practices in Canada, including decision-making procedures; business models; legal and jurisdictional frameworks; political cultures; water rights regimes; and questions of participation, transparency, and accountability in water management. Many other equally important areas of water management and academic inquiry – such as aquatic science and engineering – are excluded or dealt with only in passing.

Any analysis is an exercise in the drawing of boundaries, and the boundaries for some subjects – particularly environment-related issues – are more porous and artificial than others. Most experts working on water issues – including the majority of contributors to this book – would likely agree that an integrated approach, although difficult, is absolutely necessary, given that water is a multi-purpose resource that is essential for life, transcending jurisdictional and geopolitical boundaries and linking complex social and technical systems.

The relatively narrow focus on water governance and management is justified insofar as many Canadian water experts feel that water governance in Canada is in a state of *crisis* (in the true sense of the word): a turning point in which weaknesses are exposed, challenges are confronted, and opportunities for innovation arise. "Crisis" is an over-used and tired term, but it is appropriate here for two reasons: (1) the potentially dramatic consequences of the systemic weaknesses in contemporary Canadian water governance, and (2) the opportunity that this offers for innovation and renewal in our relationship to water.

This, in turn, lent impetus to the decision to target this book at a general audience rather than at specialists. Reflecting the diversity of stakeholders in Canadian water debates, contributors to this book have been drawn from universities, think tanks, NGOs, and government. They have written chapters aimed at a similarly diverse audience in the hope that *Eau Canada* will serve to spark debate about the nature, causes of, and solutions to persistent problems in our relationship with water.

# Acknowledgments

The support of many individuals was critical to the success of this project. Randy Schmidt and Melissa Pitts at UBC Press gave unfailingly helpful advice. Some of the text boxes in the book were researched and written in substantial part by Suzanne Moccia, Emma Norman, and Alicia Tong. Suzanne Moccia provided masterful editorial support; her assistance during all stages of editing the book was particularly appreciated. Carey Hill, Kathryn Furlong, and Alice Cohen provided research assistance on the appendix, while Hans Schreier, David Boyd, Oliver Brandes, David Brooks, Suzanne Moccia, Kathryn Furlong, Emma Norman, Philippe Le Billon, and Graeme Wynn provided comments on various portions of the manuscript.

Financial support from the Walter and Duncan Gordon Foundation's Freshwater Policy Program for the promotion and dissemination of this book is gratefully acknowledged.

Most important, the twenty-seven contributors to this book took time away from busy schedules to craft chapters that seek to speak beyond the confines of disciplinary boundaries; their time, energy, and creativity are much appreciated.

All royalties from *Eau Canada* will be donated to the Canadian branch of the not-for-profit Waterkeeper Alliance, an organization dedicated to grassroots action to prevent water pollution and to enforce water legislation. (For more information on the Waterkeeper Alliance, see http://www.waterkeepers.ca, and Appendix 3 of this volume.)

# Abbreviations

| | |
|---|---|
| AIA | Alberta Institute of Agrologists |
| BWT | Boundary Waters Treaty |
| CCME | Canadian Council of Ministers of the Environment |
| CEAA | Canadian Environmental Assessment Agency |
| CEPA | Canadian Environmental Protection Act |
| CERCLA | Comprehensive Environmental Response, Compensation and Liability Act |
| CGLG | Council of Great Lakes States Governors |
| COAG | Council of Australian Governments |
| CUPE | Canadian Union of Public Employees |
| CWWA | Canadian Water and Wastewater Association |
| DDT | dichlorodiphenyltrichloroethane |
| EAB | Environmental Appeal Board |
| EAGLE | Effects on Aboriginals in the Great Lakes Environment |
| EDS | endocrine disrupting substances |
| EPA | Environmental Protection Agency |
| FAO | Food and Agriculture Organization (United Nations) |
| FQRSC | Fonds québécois de la recherche sur la société et la culture |
| GMO | genetically modified organisms |
| IJC | International Joint Commission |
| IWRM | Integrated Water Resources Management |
| KW | kilowatt |
| KWh | kilowatt hour |
| lcd | litres per capita per day |
| MCM | million cubic metres |
| NAFTA | North American Free Trade Agreement |
| NAWAPA | North American Water and Power Alliance |
| NCE | Networks of Centres of Excellence |

| | |
|---|---|
| NGO | Non-governmental organization |
| NRC | National Research Council |
| NRTEE | National Round Table on the Environment and Economy |
| NWRI | National Water Resources Institute |
| OECD | Organization for Economic Co-operation and Development |
| P3 | public private partnership |
| P7 (P8) | group of world's poorest 7 (now 8) countries; counterpart to G7 (now 8) |
| PCB | polychlorinated biphenyls |
| POGG | peace, order and good government |
| POP | persistent organic pollutant |
| SCEE | Standing Committee on the Environment and Sustainable Development |
| SSRB | South Saskatchewan River Basin |
| UBCIC | Union of British Columbia Indian Chiefs |
| UNESCO | United Nations Educational, Scientific and Cultural Organization |
| WCD | World Commission on Dams |
| WRDA | Water Resources Development Act |
| WRI | World Resources Institute |

# Eau Canada

# 1
# Introduction

*Karen Bakker*

Canadians' relationship with water is rife with contradictions. We are fiercely protective of our water, yet hugely wasteful with it, using more water per capita than any nation in the world, except the United States (Boyd 2001). Images of pristine water are Canadian icons, yet we are one of the very few developed countries not to have legally enforceable water quality standards (see Appendix 1). Canadians are highly resistant to the notion of exporting water, yet Canada is one of the largest diverters of water in the world for hydropower (Day and Quinn 1992).

These contradictions in our approach to water stem from many sources: a mistaken belief in water's unlimited abundance; an assumption that water resources can be diverted to suit human purposes, with little regard for environmental consequences; our failure to comprehensively address threats to public and environmental health that arise from water contamination, poor environmental monitoring, and a lack of data and enforcement; and an inability to transcend provincial-federal turf wars over resource management. In a recent report on water management by the Senate Standing Committee on Energy, Environment and Natural Resources, the state of water management at the federal level was described as "shocking" (Senate 2005) and "unacceptable," an assessment with which many of the contributors to this volume would agree.

Over the past decade, these contradictions in our approach to water management have become increasingly difficult to sustain, and the "flush-and-forget" mentality that characterized our relationship to water for much of the twentieth century is giving way to increased concern. Canadian news coverage of water issues doubled between 1999 and 2001, and it has remained high since 2001.[1] Well-publicized water contamination incidents in Kashechewan (Ontario), Walkerton (Ontario), and North Battleford (Saskatchewan) have alerted Canadians to public health issues related to water quality (Laing 2002; O'Connor 2002; Parr 2005; Prudham 2004; Woo and Vicente 2003). Reports on increased threats to water quality and quantity from Environment Canada, the National Water Resources Institute, and the Senate have attracted renewed attention to water issues (NWRI 2001; Senate 2005).

At the local level, many communities are engaging in debates over water management. Droughts and floods in different regions are raising concerns about water supply and flood control infrastructure, and about the possible effects of climate change. Responding to the underfunding of public infrastructure, many cities – including Toronto, Montreal, and Vancouver – have recently debated the involvement of private companies in water supply (Bakker and Cameron 2005). Ongoing attempts to export water from Canada – whether by tanker from BC lakes or directly from the Great Lakes – are also generating heated discussion. In many instances, such as during the recent Great Lakes Annex debate (see Chapters 7 and 8), citizens' coalitions and non-governmental organizations are much more active than in the past; indeed, they are often at the forefront of debates.

The contributors to *Eau Canada* explore these debates, focusing on five themes: water governance; transboundary water management; water privatization; pathways to better water management; and changing water worldviews.

## Part 1 – Muddy Waters: How Well Are We Governing Canada's Waters?

At the provincial level, water governance and management in Canada have been undergoing a period of rapid change and intense debate over the past decade. Provincial governments have revised legislation and introduced innovations in water management, such as Alberta's water markets, Quebec's new citizen-run participatory "watershed organizations," and Ontario's requirements for full-cost pricing and accounting for water supply infrastructure. Several provinces, including Manitoba, Ontario, and Quebec, have revamped water quality standards and monitoring. Alberta is dramatically delegating citizen participation in water management (Alberta Environment 2003). Manitoba has created a new Ministry of Water Stewardship (the only ministry devoted to water issues in Canada). Leading not-for-profit organizations (Council of Canadians) and environmental think tanks (Canadian Environmental Law Association, Friends of the Earth, Sierra Legal Defence Fund) have launched high-profile water campaigns, as has the country's largest union (Canadian Union of Public Employees). Growing corporate interest is reflected in the Conference Board of Canada's new Water Research Forum initiative.

This activity is occurring within the context of what many of the contributors to *Eau Canada* portray as a diminished (and in some instances ineffective) federal government focus on water issues over the past two decades. Noting the absence of federal leadership on water policy, as evidenced by

BOX 1.1

**CANADIAN RESEARCH ON FRESHWATER PROBLEMS: FROM "BEST IN THE WORLD" TO "DOWN THE DRAIN"?**

David Schindler is one of Canada's top water scientists. In 1991, Schindler was awarded the first Stockholm Water Prize, water science's equivalent to the Nobel Prize. Schindler became well known for his groundbreaking work proving the link between water pollution and acid rain. He has warned repeatedly of a freshwater crisis in Canada, criticizing the cavalier attitude of Canadians to fresh water, the failure to deal with water pollution and to safeguard water quality, and the lack of a national water strategy and under-investment in water research on the part of the federal government. According to Schindler:

> In the mid-1960s, many aquatic scientists, myself included, immigrated to Canada because of new and exciting approaches to water research. Large freshwater laboratories were formed by the federal government ... The Experimental Lakes Area (ELA) was formed, becoming one of the few sites in the world where whole-ecosystem experiments could be done to investigate and solve pollutant and fisheries problems. This foresight caused great excitement in the global water science community: Canadian federal freshwater programs were envied throughout the world ... Many Canadian university and provincial programs also became strong ... Best of all, there was excellent interaction between federal, provincial, and university scientists, who worked together with contagious enthusiasm to develop the most powerful freshwater research teams anywhere in the world ... Unfortunately, these programs have been slowly strangled by a shortage of funds, poor salaries and the lack of replacement of departing staff. Politicians have stated the need to balance federal and provincial budgets as an excuse to reduce spending for environmental research and to decrease the size of the civil service ... It is well known that Canada's funds for research are a much smaller proportion of its national budget than in most First World countries. This must change if we are to adequately protect Canadian resources from degradation. (Schindler 2001, 24-25)

the failure to update the federal government's water policy since the last attempt in 1987 (Environment Canada 1987), some prominent Canadian water scientists have voiced concern about the underfunding of water research and the disbanding, in the 1990s, of federal departments devoted to water resources. They argue that this has dramatically reduced Canadians' capacity to assess the state of, and to safeguard, water resources (Box 1.1). As noted by Rob de Loë and Reid Kreutzwiser (Chapter 5), the ensuing fragmentation of water-related activities was so severe that the federal government had to assemble a "Where's Water" team in the mid-1990s to determine whether or not the federal government was meeting its water-related responsibilities.

One reason for the lack of federal attention stems from the division of constitutional responsibility for water between provinces and territories and the federal government. Fisheries, navigation, and international waters are federal responsibilities, yet water resources and water supply are provincial and territorial responsibilities. Water supply is, in turn, usually municipally managed. The fact that water is managed by multiple orders of government further complicates an already complex debate over how best to meet water quality, environmental protection, and public health goals in an era of public sector fiscal constraints. In the context of strained federal-provincial relations and provincial assertions of sovereignty, some stakeholders do not perceive increased federal involvement in water management to be either appropriate or desirable. Yet, this desire for a limited federal role does not resolve the question of responsibility for pan-Canadian or transboundary water issues.

Another reason for the lack of priority accorded to water issues may be, as aquatic scientist John Sprague explores in Chapter 2, a persistent "myth" of water abundance. Sprague argues that this myth is held by a majority of Canadians, and he illustrates his argument with statements made by Canadian politicians and the media. As Sprague demonstrates, the assumption that Canada is significantly more water-abundant than other nations (or the notion that "Canada is to water as Kuwait is to oil") is simply false. We are not the "Kuwait of water": Canada has under 7 percent of the global renewable water supply, and much of that supply flows north to areas relatively remote from population centres in southern Canada. These arguments are backed up by the findings of a recent report of the Senate Standing Committee on Energy, Resources and the Environment, which notes that:

> As Canadians, we generally don't spend much time thinking about water because we assume that there is plenty of it in this country to which we have ready access. [But] the fact is that certain regions of Canada, notably in the prairies, face important water challenges. Some parts of the prairies are semi-arid. In certain areas water consumption now matches or possibly exceeds what is renewed every year. (Senate 2005, 1)

Nonetheless, the myth of water abundance remains widespread in the Canadian media and public policy debates, and it is one of the reasons why, Sprague argues, both ground and surface water fail to receive the attention they deserve. Environmental lawyer Linda Nowlan, writing on Canada's approach to groundwater management in Chapter 4, agrees. She documents the relative lack of information about Canada's groundwater resources, despite the fact that one in four Canadians relies on groundwater for drinking water (Nowlan 2005).

As geographers Dan Shrubsole and Dianne Draper point out in their survey of water use and management in Canada (Chapter 3), the myth of water abundance is becoming increasingly difficult to sustain at the local level in many communities. Important and increasing stresses on water sources have been documented by Environment Canada. These stresses, in turn, have significant implications for water users. For example, Shrubsole and Draper report that Environment Canada surveys indicate that approximately 25 percent of municipalities experienced water shortages due to increased consumption, drought, or infrastructure constraints between 1994 and 1999 (Environment Canada 2002).

As subsequent chapters in this book point out, recognition of the stresses on water supply has been one trigger for innovation in water governance and management in Canada over the past decade, particularly at the regional, provincial, and municipal levels (see, for example, de Loë, Kreutzwiser, and Neufeld 2005; Kreutzwiser 1998; Plummer et al. 2005; Sproule-Jones 2002). As discussed in Chapters 4 and 5, enforcement and regulation is one area in which Canada has shown little innovation in comparison to other OECD countries. Canada contrasts dramatically to countries such as the United States and the United Kingdom, where large and well-funded environmental protection agencies (distinct from departments of the environment) fulfill key roles such as enforcement and monitoring. In Canada, these key tasks often fall to independent, often under-resourced NGOs. Drinking water quality, for example, is not nationally enforced or monitored; the federal government merely sets guidelines – which only a few provinces follow in their entirety (see Appendix 1). In some provinces, the best source for comparative data on water quality is an environmental legal not-for-profit organization, the Sierra Legal Defence Fund, which has produced a well-publicized series of "National Sewage Report Cards" and reports on drinking water quality (SLDF 2001, 2004). In contrast, in countries such as the the United Kingdom or United States, water quality reporting is undertaken by well-resourced public sector watchdogs, backed up by legislation; this results in more comprehensive, reliable, and accessible data for consumers than is the case in Canada.

Moreover, several contributors to this book argue that Canada falls behind other OECD countries in accounting for water quantity and quality. In the US, for example, stream flow and hydrology data are publicly available via the Internet. In Canada, however, data are less widely collected, and the number of hydrometric stations has been dramatically reduced in recent years, thus straining information-gathering capacity and raising concerns that network density in some provinces does not meet international standards (Lilley 2004; Scot, Yuzyk, and Whitney 1999). For example, about half of the hydrometric stations in the Okanagan Basin – the driest watershed in Canada – have been discontinued since 1973; this comes at a time when the need for long-term data is critical, given rapid population growth, development pressures, and the effects of climate change in the Okanagan. As indicated by the experience of the Sierra Legal Defence Fund (Box 4.1, Chapter 4), data collection does not automatically imply easy public access. And the limited data that is produced is not widely available or easily accessible.

The contrast between Canada and other developed countries is significant. In the UK, for example, water quality standards are legally enforced, extensively monitored, and the results published annually in an easily accessible form on the Internet site of the national regulator (Drinking Water Inspectorate); in Canada, water quality is not as systematically monitored, nor are data as easily available. For example, due to funding issues, Health Canada does not inspect water on planes to determine its safety (CESD 2005). As a result, the federal government "cannot assure the millions of Canadian travellers that potable water on aircraft is safe" (ibid.), despite its acknowledged responsibility. In comparison, the US government currently has agreements with the majority of airlines transporting the public and found that, in 2004, water was contaminated with coliform bacteria in about 15 percent of the aircraft tested (ibid.).

## Part 2 – Whose Water? Jurisdictional Fragmentation and Transboundary Management

### Jurisdictional Fragmentation

Another persistent problem in Canadian water governance, as discussed by lawyers Owen Saunders and Michael Wenig in Chapter 6, is the jurisdictional fragmentation that characterizes water management. Water legislation is a patchwork of provincial and federal laws, with inconsistencies and gaps in important areas of responsibility and oversight. This is not unusual. Water continually crosses political boundaries as it circulates through the hydrological cycle and invariably raises difficult questions of jurisdiction, in part because water is a multiple-use resource, critical for energy, agriculture, tourism, environmental health, and human water supply.

Delegating these responsibilities to different departments and scales of governance may make sense in theory. Yet, many of the contributors to *Eau Canada* argue that our approach to distributing governance between different jurisdictions has reduced Canada's effectiveness in dealing with challenges to water governance. For example, drinking water guidelines are established through a process that involves federal, provincial, and territorial governments on a joint committee (Federal-Provincial-Territorial Committee on Drinking Water [CDW]), which establishes guidelines for Canadian drinking water quality. Canada's Commissioner of the Environment and Sustainable Development (in the Office of the Auditor General) has audited the process the federal government uses to develop these guidelines and found a "significant backlog" (of approximately ten years) in updating them, despite Health Canada's recommendation that they should take no more than two to three years to develop or review (CESD 2005). The Commissioner found that many known contaminants are not even listed in the guidelines because of the time lag in updating them.

As contributors to this volume conclude, the federal government is "floundering" with respect to water policy (Chapter 12), and "our water protection capabilities are adrift" (Chapter 8). The federal government's performance with respect to water policy was strongly criticized in a 2001 report by the federal Commissioner of the Environment and Sustainable Development (2001). The commissioner's report argued that the federal government was not doing enough to protect the environment of the Great Lakes and St. Lawrence River Basin. It also noted that the federal government lacked a consistent and clear strategy for water policy in Canada, and voiced concern that the Federal Water Policy had not been updated since 1987. These concerns were reiterated by the commissioner's annual report for 2005 (CESD 2005). The Commissioner commented favourably on the renewed federal interest in water but noted that the future of the Federal Water Policy remained "unclear" and "uncertain" (ibid.). The Senate Standing Committee on Energy, the Environment and Natural Resources also criticized the federal government's failure to collect adequate basic data on Canada's water resources, stating that "this information gap is ... unacceptable [and] stems in large part from the Government of Canada's retreat from water management issues and from funding relevant research" (Senate 2005, 5).

In addition to impacting our management of water quality and resources, jurisdicitonal fragmentation affects policies and politics of transboundary water management and exports. Given the potentially significant consequences for the integrity of Canadian waters, this book devotes a chapter to each of these issues.

### Water Exports and Transboundary Management

Are water exports to be feared? In the past, Canadian water experts tended to argue that Canadians spent too much time worrying about the possibility of water exports and not enough time worrying about the damage caused by their own diversions for hydroelectricity (Day and Quinn 1992). As geographer Frédéric Lasserre argues in Chapter 7, the ecological effects of water diversions are significant and generally negative. Fears of water exports also sometimes underestimate the extensive cooperative mechanisms that have evolved to deal with waters shared between Canada and the United States. The most important of these is the International Joint Commission (IJC), a binational panel that oversees the Boundary Waters Treaty (the international agreement governing cooperation over shared watercourses between Canada and the United States). As Ralph Pentland and Adéle Hurley point out in Chapter 8, the Boundary Waters Treaty is regarded internationally as a model of bilateral cooperation over shared water resources, although its adequacy and the commitment of both nations to the treaty is increasingly under question. Finally, the fact that water exports to the United States already occur (although in very small volumes) and that the "floodgates" have not opened as a result has been interpreted as providing reassurance that

Canada and its southern neighbour are able to cooperate over shared waters (Boyd 2003).

Yet, in the past two decades, the situation with respect to transboundary waters has undergone significant change. First, the North American Free Trade Agreement (NAFTA) has affected Canadian ability to control domestic water policy and to pass legislation controlling or prohibiting water exports (Boyd 2003). The question of how to deal with Canadian water ignited fierce controversy during NAFTA negotiations in the 1980s. The federal government maintained that Canadian water (except bottled water) was exempt from NAFTA. Opposition leaders and critics called for a clause that would specifically exclude water from NAFTA. Instead, the Canadian, Mexican, and American governments issued a joint statement in 1993 to the effect that water was excluded from NAFTA (DFAIT 1999). Moreover, Canada's NAFTA Implementation Act explicitly states that nothing in NAFTA applies to water in its natural state.[2] Nonetheless, legal experts continue to debate whether NAFTA applies to water in its natural state and, thus, to bulk water exports. The closed nature of NAFTA tribunals and the lack of transparency in the decision-making process have further contributed to ongoing uncertainty over the legal status of water under NAFTA.

Second, in response to lingering domestic concerns over NAFTA and to public opposition to three controversial water export proposals, federal and provincial governments have passed new legislation banning interbasin transfers. In 1998, an Ontario-based company (Nova Corp) applied for a permit to take water from the Great Lakes and export it to Asia. In the same year, a Newfoundland-based company (McCurdy Group) applied for a permit to export bulk water from Newfoundland's Gisborne Lake. Neither proposal went ahead. A year later, a British-Columbia based company (Snow Cap) was granted a licence to provide water for export to a California-based company (Sun Belt); the licence was subsequently revoked, although Sun Belt threatened to contest the decision under NAFTA's provisions for equal treatment (see Box 8.1). Although these proposals have been dropped, new ones regularly appear, such as the proposal by a local real estate developer to abstract and bottle one millions gallons of water per day from Adams Lake (in BC's Okanagan Basin) for export to the Middle East.

Following the public debate engendered by these proposals, and an IJC study calling for a moratorium on bulk water exports (IJC 2000),[3] the federal government amended the Boundary Waters Treaty Act,[4] banning the abstraction of bulk water from boundary waters that fall under the treaty.[5] Yet this amendment did not address the issue of exports of non-boundary waters, responsibility for which is claimed by both provincial and federal governments (the former responsible for water resources, but the latter responsible for international trade). Accordingly, the federal government also sought to establish a cross-Canada consensus on water exports and encouraged Canadian provinces to introduce legislation to ban or limit the export of bulk water. All Canadian provinces, except New Brunswick, have passed legislation pertaining to water exports and/or water diversions.

However, as discussed by Frédéric Lasserre in Chapter 7, much of this legislation does not explicitly prohibit water exports but, rather, bans out-of-basin water diversions on environmental grounds. Canadian legislators may have relied on the fact that, with a few important exceptions, most of the major watersheds in Canada fall completely within Canadian territory; banning trans-basin diversions thus implicitly prohibits water exports. As Lasserre notes, the legislative protection accorded to water exports is incomplete: it may not provide protection in the case of transboundary watersheds, and it may be vulnerable to challenge under NAFTA, under which any exemptions must be in proportion to the objective being served. Provincial laws that prohibit out-of-basin diversions may be open to dispute as being disproportionate with regard to environmental protection – particularly as Canada already diverts significant amounts of water from one basin to another.

Increasing unilateralism on the part of the United States is a third important change in transboundary management. The recent case of Devils Lake is illustrative of these changes (Box 8.2 and Chapter 8). The export of polluted waters by the state of North Dakota into Manitoba's Red River system, via a pipeline constructed in the summer of 2005, clearly violates Article IV of the Boundary Waters Treaty, which states that "waters flowing across the boundary shall not be polluted on either side to the injury of health or property on the other side." After pressure from Manitoba's provincial government, the Canadian government asked Washington to join it in referring the Devils Lake case to the IJC, after failing to respond to an earlier request by the US to refer the matter. At the time of writing, the US government has yet to do this. Reasons include the relative decline in the Canadian-American relationship, the power of North Dakota lobbyists in Washington, the commitment of the current White House administration to "states' rights," and increasing American unilateralism on a number of issues (such as softwood lumber). Devils Lake sets a precedent for unilateral action on transboundary waters on the part of states and provinces – a situation, as several contributors to this volume argue, to which Canada is ill-equipped to respond.[6]

Why would Canada be ill-equipped to deal with these and other governance challenges? As several contributors to this volume argue, at the heart of Canada's inability to exert control over shared water resources is the ongoing "turf war" between federal and provincial politicians over water resources and the resulting federal timidity in exercising its full jurisdiction over water management issues that are, in fact, federal responsibilities. In the case of the debate over bulk water exports in the late 1990s, for example, David Anderson (then federal environment minister) argued that it was beyond the federal government's "jurisdiction to make decisions about provincial resources" (Boyd 2003, 58), even though provinces had repeatedly referred questions over bulk water exports to the federal government's jurisdiction over international trade. As explored in Chapters 7 and 8, this jurisdictional stalemate was evident during the recent debate over "Annex 2001" proposals for a new governance regime allowing bulk water exports by American states bordering the Great Lakes (CGLG 2001).

## Part 3 – Blue Gold: Privatization, Water Rights, and Water Markets

Another controversial debate regarding water governance in Canada in recent years has involved the shift toward market-based institutions in water resources and supply management. Historically, water management in Canada has been closely controlled by provincial, territorial, and federal governments. In Canada, as in most industrialized countries, water supply was mobilized during the twentieth century as a strategic resource to support industrialization, urbanization, and agricultural intensification (Bakker 2004). Canada's approach to water management was state-led and emphasized engineering-intensive hydraulic works as a means of satisfying water demands. This "supply-led" approach was based on a desire to provide sufficient quantities of water, where and when needed, such that economic growth could proceed unconstrained. Given high capital costs and long infrastructure lifetimes, public financing was believed to be critical for the development of water supply and resources. Another justification for state involvement was the "public good" characteristics of water: given that water is a partially non-substitutable resource essential for life, and of critical importance to public health, state involvement was thought to be necessary. Accordingly, Canadian governments financed, built, and managed hydraulic works and water supply infrastructure throughout the country.

Yet, in the past decade in Canada, as elsewhere, the consensus about the need for and nature of state involvement in water supply has broken down (Gleick 2000). This has occurred for many reasons: costly infrastructure investment needs that have outstripped public financing capacity; declining transfers to municipalities from higher levels of government; changing political views on appropriate roles of states and markets in services provision; and increasing support for water conservation and efficiency initiatives, some of which are thought to be more amenable to market-based rather than to state-based regulation (Water Strategy Expert Panel 2005).

As a result, new experiments with market-based water provision and regulation have emerged in Canada. One set of approaches focuses on "commercializing" water management: managing water in line with what are commonly identified as principles and economic instruments associated with the private sector (see, for example, Renzetti and Dupont 1999). Water pricing is a good example (Cantin, Shrubsole, and Aït-Ouyahia 2005). Over the past decade, many Canadian municipalities have begun or have extended the process of metering residential water supply, which implies charging by volumes consumed. As explored in Chapters 13 and 14, various benefits, both economic and environmental, are cited as reasons for metering, which is generally associated with increased efficiency and a reduction in demand for water (National Round Table 1996).

A second, and more controversial, set of approaches to water governance involves the introduction of private companies and private property rights into water supply management (Bakker 2004; Horbulyk 2005). These market-based approaches can be grouped into four distinct categories (which are often confused in public debate): (1) private sector participation in water supply; (2) commercialization of water management; (3) water markets for "raw," or bulk, water (usually for irrigation purposes); and (4) water rights trading. Private sector participation involves private, usually for-profit corporations in the management (and, more rarely in Canada, ownership) of water and wastewater infrastructure. Commercialization is more widespread and less controversial than privatization: it entails the introduction of commercial principles such as profit-seeking and full-cost pricing into water supply management. As documented by Karen Bakker in Chapter 9, several Canadian municipalities have recently signed contracts with private companies for water supply and sewerage management, although most water supply systems in Canada remain publicly owned and operated. Many more water supply utilities have commercialized operations, adopting a variety of approaches, including the conversion of municipal utilities into publicly owned for-profit corporations (e.g., in Edmonton), the introduction of metering (e.g., in West Vancouver), and market-based pricing mechanisms (e.g., irrigation water in Alberta). Commercialization has occurred in many Canadian provinces; in some cases, such as Ontario, recent legislation requires such commercial approaches to water management as full-cost pricing. As Bakker argues, private sector involvement in water supply is not a panacea: it does not provide cheap financing or solve water governance problems. Bakker recommends that Canadian municipalities need to improve water governance first before considering private sector involvement, keeping in mind the disadvantages as well as the potential advantages of such involvement (Aït-Ouyahia 2006).

This message is reiterated in the examination of water markets and water rights trading in Chapters 10 and 11. Water markets, as discussed by economist Ted Horbulyk (Chapter 10), are relatively new to Canada. Historically, water rights in Canada's Western provinces were associated with land ownership: a water right "attached" to a piece of land could not be exchanged or sold. In water-scarce agricultural areas, the resulting inflexibility in water use (with under-use of water on some farms and shortages of water on others) led, in the late 1990s, to proposals to create water markets in Alberta. In other jurisdictions, as explained by lawyers Randy Christensen and Anastasia Lintner in Chapter 11, although water markets have not been created, governments are creating new ways for users to exchange or trade water rights. Christensen and Lintner examine Canadian and international examples (including Chile and California) and argue that Canadian water governance is not strong enough to allow for widespread markets or water rights trading. They suggest that potential positive outcomes of market mechanisms would be severely hampered by Canada's fragmented jurisdiction, weak regulatory structures, a lack of basic data on water resources, and poor accountability mechanisms. Canada, in other words, needs to improve water governance before it begins experimenting with some of the potential benefits that some market-based mechanisms might offer.

## Part 4 – Waterwise: Pathways to Better Water Management

How could we manage water more wisely? This section of *Eau Canada* documents innovative water governance approaches that are being applied in

Canada and that many water experts argue should form the core of a "new water paradigm" (Gleick 2000). In Chapter 12, lawyers Paul Muldoon and Theresa McClenaghan argue that the provincial-federal "turf wars" – and the resulting stalemate in water policy at the national level – should be addressed through a new governance framework that enables provincial and federal levels of government to work together to streamline and enforce existing water legislation. In Chapter 14, political ecologist Oliver Brandes, natural resources economist David Brooks, and lawyer and environmental activist Michael M'Gonigle make the case for a sustainable approach to water management, an approach based on conservation as part of a broader strategy for maintaining environmental health and preserving ecosystem functions.

This strategy of "ecological governance" begins with emphasizing reductions in water demand (to conserve resources, save money, and reduce environmental impacts), with the long-term goal of "soft path" management. This soft path requires a new form of water governance – one that focuses on sustainability, breaks the link between increasing water consumption and economic growth, and radically changes our water use and disposal practices by focusing on the delivery of *services* rather than water.

As economist Steven Renzetti argues in Chapter 13, a key component of this new approach to water management is water pricing. Pricing, according to Renzetti, should balance four criteria: efficiency, fairness, economic equity, and sustainability. Getting the prices "right" should help to encourage efficient water allocation, improve water quality, provide adequate revenues to water suppliers, and encourage innovation and conservation (although in practice it may not always be possible to balance these goals). However, there are significant barriers to "getting the prices right," and better pricing cannot solve more fundamental questions about governance and sustainability (Brandes et al. 2005). How, then, are we to adjudicate between different goals and strike the right balance? The final section of *Eau Canada* attempts to answer this question.

## Part 5 – Water Worldviews: Politics, Culture, and Ethics

The final three chapters turn to questions of new water worldviews and their implications for water management. In Chapter 15, lawyer Ardith Walkem documents the increasing legal recognition that Canadian courts have granted to Aboriginal water rights. This has implications, she argues, well beyond water rights and Aboriginal land claims: there is a strong case for integrating

Aboriginal water management norms and ethics into Canadian resource management practices as a matter of environmental justice. The implications would be far-reaching not only for Aboriginal peoples' water sources and water rights but also for the ethical bases of water management practices in Canada.

Changing our water worldviews is, however, far from an easy task. As political scientist Andrew Biro explores in Chapter 16, our ethical bases for water management practices are deeply rooted in Canadian culture: images of pristine, or "natural," water are central to concepts of Canadian identity, which are often implicitly defined against the "modern hydrologically engineered (nature-dominating, imperialist) society to the south." These characterizations, however, break down upon closer inspection. As geographer Frédéric Lasserre demonstrates in Chapter 7, Canada is a much larger diverter of water than is the United States: in the search for hydropower we have, in fact, dammed and diverted more water per capita than has almost any other nation in the world. Nonetheless, the "myth of water abundance" persists, in part, Biro argues, because of its fit with the Canadian national imaginary: "Northern, vast and open, rugged, wild and, of course, cold and wet" (Chapter 16).

Walkem's and Biro's arguments contain important parallels to the call for a "new water ethic," which is articulated by geographers Bruce Mitchell and Cushla Matthews and political scientist Bob Gibson in Chapter 17. In this final chapter of the book, the authors survey emerging approaches to water management in Canada and internationally, and they note that several new (and sometimes controversial) principles and goals are increasingly being advocated (and, in some cases, adopted) around the world. These include ecosystem integrity, source protection, user participation, efficiency and conservation, precautionary management, and legal rights – for both humans and the environment. As the authors point out, these approaches are not always consistent, nor are they always equally applicable. Rather, they point to a new underlying approach to water governance and management that acknowledges the limits of the conventional approach (i.e., hierarchal management) and emphasizes the need for dialogue regarding how we govern not only our relationships with one another but also with the ecologies within which we live. In other words, we need to move away from the "environmental management" of resources and toward the "ecological governance" of human-water relationships, managing *ourselves* as well as the environment.

## The Focus of *Eau Canada*

This book focuses on two aspects of our relationship with water: water governance and water management. The difference between the two is subtle, but important. Simply put, "water governance" refers to the decision-making process we follow, whereas "water management" refers to the operational approaches we adopt. Governance refers to how we make decisions and who gets to decide; management refers to the models, principles, and information we use to make those decisions. Obviously, the two are interrelated; however, management is often the focus of debate, whereas governance is often overlooked.

The premise of this book is that Canadian water governance and management are at a crossroads. There is increasing recognition of systemic weaknesses in our approach to water governance, and increased debate regarding necessary reforms to our water management models. This has, in turn, lent new urgency to calls to reform our approach to governance and has also led to a period of fascinating experimentation with new approaches to management.

*Eau Canada* analyzes some of these weaknesses, challenges, and innovations. Specifically, the book documents recent changes in water governance and management in Canada; analyzes current challenges in Canadian water governance; and explores the different solutions being advocated by different Canadian water experts. This focus will, of necessity, exclude (or acknowledge only in passing) equally important debates in related fields, such as aquatic science and engineering. This is unfortunate insofar as most experts working on water issues – including the majority of contributors to this book – would likely agree that an integrated approach, although difficult, is absolutely necessary given that water is a multipurpose resource that is essential for life, transcending jurisdictional and geopolitical boundaries and linking complex social and technical systems. Accordingly, an appendix with suggestions for further reading has been included (see Appendix 2).

Significant changes are going to be made to water governance and management in Canada over the next decade. One of the premises of this book is that informed public input is essential to good policy outcomes. Accordingly, the goal of this book is to introduce these issues to the broader Canadian public in the hope that people will continue to engage with, and build upon, the ideas and debates presented here. Our water is too important to do otherwise.

## NOTES

1 A search of the Canadian Business and Current Affairs (CBCA) database was conducted on 30 August 2005. CBCA indexes articles published in and about Canada, taken from more than 1,500 popular and scholarly publications, including newspapers, magazines, and academic journals. News coverage of the keyword "water supply" was surveyed for the period 1987-2005.

2 North American Free Trade Implementation Act, S.C. 1993, c. 44, s. 7. The relevant clause states that nothing in the act (except Article 302) applies to water, where water is defined as natural surface and ground water in liquid, gaseous, or solid state, but does not include water packaged as a beverage or in tanks. Article 302 refers to the progressive elimination of tariffs (customs duties) between NAFTA signatories.

3 The IJC report stated that Canada and the United States "should not permit any new proposal for removal of water from the Great Lakes Basin to proceed unless the proponent can demonstrate that the removal not endanger the integrity of the ecosystem."

4 Bill C-15, An Act to Amend the International Boundary Treaty Act, S.C. 2001, 40.

5 Bill C-15 specifically "(a) prohibits the bulk removal of boundary waters from the water basins in which they are located; (b) requires persons to obtain licences from the Minister of Foreign Affairs for water-related projects in boundary or transboundary waters that would affect the natural level or flow of waters on the United States side of the border; and (c) provides clear sanctions and penalties for violation."

6 The proposed North Dakota Garrison Diversion (also known as the Red River Supply Project) is an example. Amendments by the US Congress to the Garrison Reformulation Act in 2000 removed the previous requirement for consultations with Canada, strongly opposed by the Manitoba and Canadian governments (Manitoba Water Stewardship 2006).

## REFERENCES

Aït-Ouyahia, M. 2006. *Public-Private Partnerships for Funding Municipal Drinking Water Infrastructure: What Are the Challenges?* Ottawa: Policy Research Initiative.

Alberta Environment. 2003. *Water for Life: Alberta's Strategy for Sustainability*. Edmonton: Alberta Environment. http://www.waterforlife.gov.ab.ca.

Bakker, K. 2004. *An Uncooperative Commodity: Privatizing Water in England and Wales.* Oxford: Oxford University Press.

Bakker, K., and D. Cameron. 2005. Changing Patterns of Water Governance: Liberalization and De-Regulation in Ontario, Canada. *Water Policy* 7 (5): 485-508.

Boyd, D. 2001. *Canada versus the OECD: An Environmental Comparison.* University of Victoria: Eco-Research Chair of Environmental Law and Policy.

—. 2003. *Unnatural Law: Rethinking Canadian Environmental Law and Policy.* Vancouver: UBC Press.

Brandes, O., K. Ferguson, M. M'Gonigle, and C. Sandborn. 2005 *At a Watershed: Ecological Governance and Sustainable Water Management in Canada.* Victoria: POLIS Project on Ecological Governance.

Cantin, B., D. Shrubsole, and M. Aït-Ouyahia. 2005. Using Economic Instruments for Water Demand Management. Introduction, *Canadian Water Resources Journal* 30 (1): 1-10.

CGLG [Council of Great Lakes States Governors]. 2001. *The Great Lakes Charter Annex: A Supplementary Agreement to the Great Lakes Chapter Council of Great Lakes Governors.* Ottawa: Office of the Auditor General.

Commissioner of the Environment and Sustainable Development. 2001. *Report of the Commissioner of the Environment and Sustainable Development.* Ottawa: Office of the Auditor General.

—. 2005. *Report of the Commissioner of the Environment and Sustainable Development.* Ottawa: Office of the Auditor General.

Day, J.C., and Frank Quinn. 1992. *Water Diversion and Export: Learning from Canadian Experience.* Department of Geography Publication Series No. 36. Waterloo: University of Waterloo Press.

de Loë, R., R.D. Kreutzwiser, and D. Neufeld. 2005. Local Groundwater Source Protection in Ontario and the Provincial Water Protection Fund. *Canadian Water Resources Journal* 30 (2): 129-44.

DFAIT. 1999. *Bulk Water Removal and International Trade Considerations.* Ottawa: Department of Foreign Affairs and International Trade.

Environment Canada. 1987. *Federal Water Policy.* Ottawa: Environment Canada.

—. 2002. *Urban Water Indicators: Municipal Water Use and Wastewater Treatment.* http://www.ec.gc.ca/soer-ree/English/Indicators/Issues/Urb_H2O.

Gleick, P. 2000. The Changing Water Paradigm: A Look at Twenty-First Century Water Resources Development. *Water International* 25 (1): 127-38.

Horbulyk, T. 2005. Markets, Policy and the Allocation of Water Resources among Sectors: Constraints and Opportunities. *Canadian Water Resources Journal* 30 (1): 55-64.

IJC [International Joint Commission]. 2000. *Protection of the Waters of the Great Lakes.* Ottawa and Washington: International Joint Commission.

Kreutzwiser, R.D. 1998. Water Resources Management: The Changing Landscape in Ontario. In *Coping with the World around Us: Changing Approaches to Land Use, Resources and Environment,* ed. R.D. Needham, 135-48. Waterloo: Department of Geography, University of Waterloo.

Laing, R.D. 2002. *Report of the Commission of Inquiry into Matters Relating to the Safety of the Public Drinking Water of the City of North Battleford, Saskatchewan.* Regina: Office of the Queen's Printer.

Lilley, J. 2004. The CWRA Comments on Canada's Hydrologic and Meteorologic Networks. *Water News* 23 (4): 15-20.

Manitoba Water Stewardship. 2006. *Water Project Proposals: Garrison Diversion Devils Lake Outlet.* http://www.gov.mb.ca/waterstewardship/water_info/transboundary/north_dakota.html.

National Round Table on the Environment and the Economy. 1996. *Water and Wastewater Services in Canada.* Ottawa: NRTEE.

Nowlan, L. 2005. *Buried Treasure: Groundwater Permitting and Pricing in Canada.* Toronto: Walter and Duncan Gordon Foundation.

NWRI [National Water Resources Institute]. 2001. *Threats to Sources of Drinking Water and Aquatic Ecosystem Health in Canada.* Burlington: National Water Resources Institute.

—. 2005. *Threats to Water Availability in Canada.* Burlington: National Water Resources Institute.

O'Connor, D.R. 2002. *Report of the Walkerton Inquiry.* Toronto: Ontario Ministry of the Attorney General.

Parr, J. 2005. Local Water Diversely Known: Walkerton, Ontario, 2000 and After. *Environment and Planning D* 23 (2): 251-71.

Plummer, R., A. Spiers, J. FitzGibbon, and J. Imhof. 2005. The Expanding Institutional Context for Water Resources Management: The Case of the Grand River Watershed. *Canadian Water Resources Journal* 30 (3): 227-44.

Prudham, W.S. 2004. Poisoning the Well: Neo-Liberalism and the Contamination of Municipal Water in Walkerton, Ontario. *Geoforum* 35 (3): 343-59.

Renzetti, S., and D. Dupont. 1999. An Assessment of the Impact of Charging for Provincial Water Use Permits. *Canadian Public Policy* 25 (3): 361-78.

Schindler, D. 2001. The Cumulative Effects of Climate Warming and Other Human Stresses on Canadian Freshwaters in the New Millennium. *Canadian Journal of Fisheries and Aquatic Sciences* 58: 18-29.

Scott, D., T.R. Yuzyk, and C. Whitney. 1999. The Evolution of Canada's Hydrometric Network: A Century of Development. In *Partnerships in Water Resource Management: Proceedings of the CWRA 52nd Annual Conference.* Nova Scotia: CWRA.

Senate. 2005. *Water in the West, under Pressure.* Fourth Interim Report of the Standing Senate Committee on Energy, the Environment and Natural Resources. Ottawa: Senate of Canada.

SLDF [Sierra Legal Defence Fund]. 2001. *Waterproof: Canada's Drinking Water Report Card.* Vancouver: Sierra Legal Defence Fund.

—. 2004. *National Sewage Report Card 2004.* Vancouver: Sierra Legal Defence Fund.

Sproule-Jones, M. 2002. Institutional Experiments in the Restoration of the Great Lakes. *Canadian Journal of Political Science* 25: 835-85.

Water Strategy Expert Panel. 2005. *Watertight: The Case for Change in Ontario's Water and Wastewater Sector.* Toronto, ON: Ministry of Public Infrastructure Renewal.

Woo, Dennis, and Kim Vicente. 2003. Sociotechnical Systems, Risk Management, and Public Health: Comparing the North Battleford and Walkerton Outbreaks. *Reliability Engineering and Systems Safety* 80: 253-69.

PART 1

# Muddy Waters: How Well Are We Governing Canada's Waters?

Canada has just over 6 percent of global annual renewable water supply. Despite this, rumours persist that it is water-rich.

# 2
# Great Wet North? Canada's Myth of Water Abundance

*John B. Sprague*

The news media and most Canadians are convinced that our country is exceptionally rich in water. We often see statements such as "we have more than a fifth of the world's freshwater supply," or "one-quarter of world supply." Certainly, it is easy to go along with this idea when standing beside one of the Great Lakes or when flying over the north with thousands of little lakes down below. But sometimes the statements become excessively optimistic. West Coast Environmental Law, a group in British Columbia that is usually well informed, exemplifies this optimism in a publication that stated that "Canada has one of the most abundant supplies of fresh water in the world, exceeding the volume of US water resources by a factor of 10" (Shrybman 1999). This might be the greatest over-estimate on record, and there is no apparent set of data that would justify "a factor of 10."

## Volume vs. Renewable Supply

The common overestimation of Canada's water supply is probably derived from the volume of fresh water sitting in all Canadian lakes, which is about 20 percent of the water in all of the world's lakes (Environment Canada 2005). Of course, the water sitting in a lake is totally different from the *renewable supply*. The renewable supply is what falls from the sky and runs off in rivers, often passing through lakes as it moves to the sea. Some goes underground, replenishing aquifers that can be tapped by wells. These flows are renewed every year and count as the *water supply*.

This renewable fresh water can be defined as "salt-free water that is fully replaced in any given year through rain and snow that falls on continents and islands and flows through rivers and streams to the sea" (WRI 2003). This definition includes groundwater because an aquifer that gains water from seeping rainfall, beyond its equilibrium, will drain off its excess water by means of springs, which run into streams. The definition excludes water that is evaporated or transpired from vegetation. To use a financial analogy, the water sitting in lakes and aquifers is comparable to a capital resource of money that can be spent only once. The rivers running out of the lakes would represent interest and dividends that could be used every year for an indefinite time.

For a more homespun analogy, it doesn't matter if my back yard has a swimming pool full of water and my neighbour's yard has only a barrel. If the taps in my house dribble, but the neighbour's taps gush, then it is the neighbour who has the good supply of water. The size of the swimming pool is of no consequence; if I start using the water in my pool for laundry and watering the garden, I will have a dry pool in a few months, along with my dribbling taps. The giant lakes in North America are comparable: we can use them to boat and fish and swim, but if we start consuming their standing water, we will eventually dry them up. Recent complaints about low levels in the Great Lakes indicate why lowering of water levels is not an option.

Indeed, lake storage capacity is not as big as might be thought. The total volume of water resident in all the freshwater lakes of the world is only equal to about two years' worth of runoff in the world's rivers (Environment Canada 2005). Exactly the same principles and consequences apply to underground aquifers, if we deplete them faster than nature renews them. In other words, this would deplete the "capital" in our "bank account" of stored water.

## Canada: Third to Sixth in Renewable Supply

Canada appears to have lots of water, but this is because of a topography that creates a few large lakes and many shallow, small lakes, along with a cool climate and low evaporation of the water. In fact, our share of the world's renewable water supply is relatively modest. Recent data put us in third place (barely) among all nations. This estimate comes from the World Resources Institute, a quarter-century-old environmental research institute and think tank based in Washington, DC (WRI 2003, 2005).

The two countries with the largest renewable water supplies are Brazil, with 12.4 percent of the world's renewable supply, and Russia, with 10 percent. Below that, Canada is in a virtual four-way tie with Indonesia, the United States, and China, each with about 6.5 to 6.4 percent. Next down the

Table 2.1

**Internal renewable water resources for the ten countries with the largest water supplies**

| | Supply (km³ per year) | Percentage of world supply |
|---|---|---|
| Brazil | 5,418 | 12.4 |
| Russian Federation | 4,313 | 10.0 |
| Canada | 2,850 | 6.5 |
| Indonesia | 2,838 | 6.5 |
| United States | 2,818 | 6.4 |
| People's Republic of China | 2,812 | 6.4 |
| Colombia | 2,112 | 4.8 |
| Peru | 1,616 | 3.7 |
| India | 1,261 | 2.9 |
| Democratic Republic of Congo | 900 | 2.1 |
| World total | 43,773 | |

Source: WRI (2005).

list, from about 5 percent to 2 percent, are Columbia, Peru, India, and the Democratic Republic of Congo, respectively (Table 2.1).

The numerical estimates in Table 2.1 are intended to represent the surface water flows and groundwater recharge that originate within a given country. In other words, the numbers signify the amount of precipitation falling within a country's borders. Water flowing into the country from a neighbour is not included, nor is water that flows out, evaporates, or transpires.[1] Canada's annual renewable volume of fresh water seems large, at about 2,850 cubic kilometres, but that is only 6.5 percent of the world total of almost 43,800 cubic kilometres.

There is another factor to consider if one is assessing Canada's water supply, particularly with reference to the United States. The two countries are almost identical in total renewable water supply. However, about 60 percent of Canada's water flows north to arctic or subarctic regions (Environment Canada 2005). That water is largely unavailable for use in the southern part of the country, where most people live, work, and farm. Accordingly, the supply in southern Canada is only about 2.6 percent of the world supply. This is the number that should spring to the minds of Canadians when they contemplate the country's water resources. Notably, the figure is tenfold lower than the frequently used and mythical "one-quarter of the world supply."

This accurate assessment of Canada's available water supply might well influence thoughts about sharing Canadian water with neighbours (Chapter 8, this volume). The total US supply is more than twice the amount available in the populated areas of Canada. Even excluding the Alaskan part of US supply, the 48 contiguous mainland states receive 57 percent of the national supply, or about 3.7 percent of world supply, and, thus, more than southern Canada!

In other words, the United States and Canada have similar amounts of easily accessible fresh water for human uses. This contradicts the public statements often made about Canada's water. In media and policy reports, emphasis is usually placed upon the perceived abundance of Canadian water, and precise figures for the United States are rarely mentioned. It must be admitted that southern Canada's share of 2.6 percent of world water supply is a comfortable amount, particularly when taking into account our relatively small population; our per capita availability is well above the global average. Nonetheless, water availability in both absolute and relative terms, is less than is usually supposed.[2]

## Continuing Misconception

The difference between standing volume and water supply is obvious with a moment's thought, so it is rather surprising that, over the decades, inflated estimates of Canadian water supply have continued to be published in magazines and newsletters. Examples of this are:

- "Canada contains as much as 25 per cent of the global supply of fresh water" (*Maclean's* 1985).
- "Canada is blessed with at least one-fifth of the world's supply of fresh water. We're the envy of other countries" (Pollution Probe 1987).
- "one fifth of the world's freshwater supply is located in Canada" (CUPE 1999).
- "with 20 per cent of the world's fresh water, resource-rich Canada ... " (Kirshner 1999).

Even national newspapers are not immune. In editorials and feature articles, Toronto's *Globe and Mail* has reported that:

- "Canada has lots and lots of water" (*Globe and Mail* 1999).

- "Canada is home to roughly 40 per cent of the Earth's store" of fresh water (Mitchell 2000).
- "Canada, blessed with the world's largest freshwater supply ..." (Mitchell 2001).

It seems to be nearly impossible to put an end to this myth, even though the correct technical information has long been available. In the 1970s, for example, Dr. A.T. Prince, director of Inland Waters for the Canadian Department of the Environment, said that "the water supply in Canada is ... about average for the area of the country as compared to other parts of the world" (Bocking 1972). Dr. Frank Quinn of the same department pointed out the difference between water stored in lakes and the actual usable runoff, which he estimated "as just 6 percent of the world's runoff" (ibid.). That agrees closely with the most recent estimate given above. The Canada Water Year Book of 1975, based on measurements taken at that time, gave 9 percent as Canada's share of the world's river flow (Environment Canada 1975). A few years later, Foster and Sewell (1981) published an excellent book, the second chapter of which is entitled "The Myth of Superabundance." They gave Environment Canada's 9 percent estimate and followed it with a tabulation of all the regional river flows, along with a comparison showing that the United States had a total river flow that was only slightly lower than Canada's. In 1985, *Maclean's* published a letter that I wrote, pointing out the fallacy of their above-mentioned article. Today, the website of Environment Canada (2005) continues to distinguish very clearly between static volumes of lakes and renewable supply. The website assigns Canada 7 percent of the world's runoff, which is in keeping with the estimate used here.

Although this technical information has been available for decades, we see from the quotations above that the myth of abundance has a life of its own and continues to be repeated from one writer or broadcaster to another.[3] Recently, there has been some indication that the message is finally getting through to the popular press. An article in *Maclean's* (Maich 2005) was quite perceptive and used the correct numbers: "This country boasts more than 20 per cent of the world's fresh water, and the flow of rain, spring water and snowmelt that courses through our waterways represents seven per cent of the planet's renewable water supply." However, even though these are the right numbers, they are framed by the familiar – and misleading – theme of abundance in statements such as, "Canada is a country of unbelievable water wealth" and "Canada, the most water-rich nation on the planet." Hence, the

*Maclean's* article reaffirmed Canada's myth of water abundance, ignoring Brazil and Russia (who have more water than Canada) and overlooking the fact that the continental United States has a greater proportion of the global water supply than does southern Canada. The *Maclean's* message, in keeping with the myth, was that water could be exported to the United States for financial gain.

## Social and Political Implications

Inflated perceptions of Canada's water supply are problematic because many Canadians continue to think that they are in a surplus situation regarding water. This is evident in a public opinion poll published by *Maclean's* (2003), which was conducted at the time a new prime minister was taking office. People were asked how acceptable they considered each of eight options for improving relations with the United States. The option of allowing the export of fresh water was acceptable to 42 percent of respondents. ("Very acceptable" = 11 percent, "somewhat acceptable" = 31 percent, "not too acceptable" = 18 percent, "not at all acceptable" = 38 percent, and "don't know" = 3 percent.) The degree of acceptability for this option ranked between the 61 percent approval for common continental laws for immigration and border crossing, and the 33 percent approval for sending troops to Iraq. The opinions are open to many interpretations, but almost half of the people believed the country had water to spare.

Public perception and media misinformation are bound to influence political decisions, and some government leaders have evoked the myth of superabundance in their statements:

- David Anderson, then minister of the environment: "Although Canada has the world's largest supply of fresh water ..." (Anderson 1999)
- Raymond Chrétien, Canada's ambassador to the United States at the time, made the following statements to the annual meeting of New England governors and eastern Canadian premiers in Rockport, Maine: "Between Canada and the United States, water is bound to emerge in the new millennium as one of the four or five biggest issues we will have to deal with." Chrétien told the governors and premiers that "the demands of U.S. society are bound to increase. We have in Canada 20 per cent of the world reserves of fresh water. Together, our two countries manage 40 per cent of

the world's reserves in fresh water. We're bound to have difficulties in the years to come" (Chrétien 1999).

It is not clear what Mr. Chrétien meant by "world reserves." However, any implication by a Canadian government official that such numbers represent a usable supply of water can only make the country more vulnerable to proposals for transfers. Any credence given to such inflated estimates impinges upon all aspects of regulation and negotiation. A more realistic value – 2.6 percent of world supply – casts an entirely different light on thoughts about available flows, sharing among jurisdictions, and diversions.

There is no question that Canadian demands for water will increase with time. If nothing else, a presumed population increase will have that effect. About 30 percent of Canadians depend on groundwater for their household supplies (Rutherford 2004), and this groundwater needs to be recharged from the surface. Northcote and Hartman (2004) point out that an expanding human population results in major increases in logging and other forms of deforestation, which, in turn, reduces or profoundly changes the seasonal flows of rivers, with a concomitant decrease in fish and their habitat. Climate change is predicted to affect the distribution and/or regularity of precipitation, and this would likely increase demand for irrigation. In the face of all these potential needs for water, it is essential that we take a realistic view of Canadian water resources and their future role.

It is distressing to hear incorrect statements from Canadian leaders, especially since government technical experts know the real situation (as noted above). However, there is some hope. Recently, representatives of Ontario and Quebec have demonstrated that they clearly understand the difference between renewable supply and standing water. Their statements about the removal of water from the Great Lakes were made in terms of the time that would be needed for inflows to replenish these lakes – that is, they were made in terms of water *supply.* These statements came during discussions with US states about water diversions (the so-called Great Lakes Annex Agreement; see Chapters 7 and 8, this volume).

## Ecological Implications of Diversions

In addition to its sociopolitical influences, the myth of abundance has probably encouraged a cavalier attitude toward the use and manipulation of water within Canada and reduced the concern for environmental side-effects. There have been small and large diversions of water flow, often for generating power (Chapter 7, this volume; Quinn et al. 2004). Historically, human needs have received most attention, as is documented in much of this book, with ecological consequences being ignored or put aside as a secondary consideration. However, it is imperative that we take ecological effects into account as well as social effects.

It is all too easy to destabilize the balance of natural habitats. Diversion of a river's natural flow always carries some ecological penalty and can create major upsets, both downstream and upstream. Even water that "is just pouring into the sea," as it is sometimes put, is not wasted. It maintains the reduced salinities in estuaries, which are major areas of natural production on coastlines.

The effects of changed and regulated streamflow on species of Pacific salmon were examined for eighty-one case histories by Burt and Mundie (1986). Of the sixty-three histories that provided known outcomes, three-quarters showed decreased populations of salmon or complete disappearance. Most cases of damage (60 percent) were associated with decreases in flow. Other important side effects of flow manipulation were blockage of habitat, sedimentation, fluctuating flows, and change in water temperatures. In the remaining one-quarter of the case histories, there was little effect on salmon populations or an increase in numbers. Most (73 percent) of these "good" outcomes were associated with only slight changes in flows or else increased flows in the system – that is, diversions elsewhere brought more water into the system.

One of the more spectacular diversions in Canada occurred in tributaries of the upper Fraser River in British Columbia. Forty percent of the water from the Nechako River was turned around to flow into a sixteen-kilometre tunnel in order to generate electricity for smelting aluminum. The major changes in flow regime and the associated history of salmon populations have been documented in detail by Hartman (1996) and Day and Quinn (1992).

Even without diversion, a mere change in the regime of seasonal flow can lead to a spectrum of ecological impacts – a spectrum that is outlined by Prowse, Wrona, and Power (2004). A powerful example is the Peace-Athabasca delta in Alberta, the largest boreal delta in the world. It was once a paradise for 400,000 migrating or breeding waterfowl and a treasure house for local trappers and fishermen. When British Columbia constructed the Bennett Dam in 1967, more than 900 kilometres upstream on the Peace River, there were extreme changes in seasonal patterns of runoff; in particular, there was a big decrease in the normally heavy spring runoff. The result was ecological chaos in the delta, as it dried out due to lack of springtime flooding (Townsend 1975; Prowse and Conly 1996). As the dam filled, 38 percent of the water area in the delta disappeared, and 500 square kilometres of mudflats were exposed (Alberta Environment 1996). In 1971, a weir was built in the delta in an attempt to imitate seasonal flooding, and this showed some signs of success. Unfortunately, the weir washed out in 1973, hasty repairs raised it too high, and flooding continued for the following summer and winter. An attempt to breach the weir the following spring was not successful, and the result was severe flooding. Bison were floating around on ice-pans during spring break-up, and it is thought that about 1,000 of them drowned. Marooned trappers had to be rescued by aircraft. Since then, the weir has gone, the delta continues to be in a drying phase, and parts of it are becoming covered in trees. The cause is thought to be a combination of the changes in flow regime, normal climate oscillations, and global warming, particularly changes in ice cover because of warmer winters. All of these have probably acted on this community, which, like all deltas, is in a dynamic state of natural stress (Timoney 2002). It is not known what the status of the delta would be had a natural flow regime prevailed.

## Conclusion

Most Canadians seem to believe that Canada has approximately one-quarter of the world's supply of fresh water. This misconception apparently arises from confusion between standing water, which fills the country's many lakes, and renewable supply, which is represented by each year's rain- and snowfall. It is the amount of that precipitation that governs river flow and groundwater recharge.

In fact, Canada's renewable supply is about 6.5 percent of the world supply. There is a four-way tie among Canada, Indonesia, the United States, and China for the size of water supply, and all of these countries rank far behind Brazil and Russia.

The misconception surrounding water supply has deep implications for government decisions, as a number of political representatives have made statements indicating that they buy into the notion of mythical abundance. A misplaced belief that Canada has an excess of water will likely lead to decisions that will be detrimental to the country throughout future decades.

Although Canada has a relatively large supply of water per capita, it does not necessarily have a large supply per region. Most Canadians live in the south of the country, far from some of the larger sources. Certain regions already have water shortages, and climate change could well exacerbate this situation. As is outlined in Chapter 3, ongoing developments are likely to further constrain the water supply. No decision on water use or diversion should be based on a misapprehension of the national or local supply. In particular, it must be realized that Canada's nearest neighbour, the United States, has a supply of water that equals ours.

The myth of abundance has no doubt also contributed to a thoughtless disregard for the environmental consequences of manipulating rivers. There have been large and small diversions of water, some of them causing major destruction of natural habitat and species. Even dams that merely distort the seasonal pattern of flow have caused problems (e.g., ecological chaos in Alberta's Peace-Athabasca delta).

Thus, there must be a continued campaign to get Canadians to abandon the myth of water abundance and to adopt a realistic view of their water supply. Otherwise, we will make policy decisions on the basis of misinformation, and this cannot help but have serious ecological, economic, and political consequences.

## Acknowledgment

I thank Dr. Gordon F. Hartman for supplying information on ecological effects and making suggestions on the text.

## Notes

1 Values are taken from the data table "Freshwater Resources 2005" (WRI 2005). WRI took most of the data from AQUASTAT, an online database that is maintained and frequently updated by the Food and Agriculture Organization of the UN (FAO 2003). In turn, AQUASTAT shows a collection from diverse national and international yearbooks, surveys, and project documents. Critical evaluation and judgment were needed in the compilation, and WRI cautions that the values are approximate estimates. I asked the water experts of WRI about changes in successive publications, and they replied that adjustments of the database occur all the time with improved gathering of information. Water supplies and water flows are difficult to assess at the best of times, and information from some regions is far from perfect. FAO and WRI experts adopt the most recent sets of data

that they consider reliable, on the assumption that information is improving. If the tabulation were to include the quantities of water reserved for upstream and downstream countries, by treaty or other agreement, then the United States would edge slightly ahead of Canada in water supply (WRI 2005). However, the comparison in this chapter deals only with the "within-border" natural supply.

2 Moreover, discussions about water export often overlook how Canada's limited water resources might be used creatively within the country. General A.G.L. McNaughton addressed the topic as long ago as 1965. A wartime leader, he became chairman of the Canadian Section, International Joint Commission on Boundary Waters (as did Adèle Hurley, an author in this volume). On 4 October 1965, McNaughton said to the Montreal Canadian Club: "All of our water can be translated into growth somewhere. Let it take place in Canada" (Anon 1965).

3 The day after the appearance of the *Globe and Mail* item listed above (February 1999), the paper published my letter to the editor, which pointed out the misconception inherent in the myth of abundance. That did not stop the *Globe* from reprinting the myth in feature articles in 2000 and 2001. (It did not publish my letters, which were written in response to those two articles.) If the author of those feature articles had conducted a simple computer search of the *Globe and Mail* archives they would have found my original letter and, consequently, might have presented a realistic portrayal of water supply.

## REFERENCES

Alberta Environment. 1996. Section 1.4, The Peace-Athabasca Delta, section 3.5, Flow Regulation. In *Northern River Basins Study Final Report.* Edmonton: Alberta Environment for Governments of Alberta, Canada, and Northwest Territories. http://www3.gov.ab.ca/env/water/nrbs.

Anderson, D. 1999. Sound clip played several times on radio news. Canadian Broadcasting Corporation, 22 November.

Anon. 1965. General McNaughton speaks to the Canadian Club at Montreal. *News from the Field.* Toronto: Ontario Water Resources Commission, 15 October, no. 65-10: 7-9.

Bocking, R.C. 1972. *Canada's Water: For Sale?* Toronto: James Lewis and Samuel, Publishers.

Burt, D.W., and J.H. Mundie. 1986. Case Histories of Regulated Stream Flow and Its Effects on Salmonid Populations. *Canadian Technical Reports of Fisheries and Aquatic Sciences,* no. 1477. Vancouver: Canadian Department of Fisheries and Oceans, Habitat Management Division.

Chrétien, R. 1999. Speech to Annual Meeting of New England Governors and Eastern Canadian Premiers in Rockport, Maine, 4 October. In Dispute with U.S. Predicted over Water, *Vancouver Sun,* 5 October, A3.

CUPE [Canadian Union of Public Employees]. 1999. They Can't Buy the Air We Breathe ... So They Want to Buy the Water We Drink. Centrefold advertising insert in *This Magazine* 32, no. 6 (May-June).

Day, J.C., and F. Quinn. 1992. The Kemano Diversion. In *Water Diversion and Export: Learning from Canadian Experience*, ed. J.C. Day and F. Quinn. Waterloo, ON: University of Waterloo, Department of Geography, Publication Series 36, 85-106.

Environment Canada. 1975. *Canada Water Year Book 1975*. Ottawa: Information Canada.

—. 2005. Freshwater Website. http://www.ec.gc.ca/water.

FAO [Food and Agriculture Organization]. 2003. AQUASTAT Information System on Water and Agriculture: Review of World Water Resources by Country. Rome: FAO, Water Resources, Development and Management Service. http://www.fao.org/waicent/fao/agricult/aglw/aquastat/water_res/index.htm.

Foster, H.D., and W.R.D. Sewell. 1981. *Water: The Emerging Crisis in Canada*. Toronto: Canadian Institute for Economic Policy, James Lorimer and Co., 117.

*Globe and Mail*. 1999. Weirdness about Water. Editorial, 13 February.

Hartman, G.F. 1996. Impacts of Growth in Resource Use and Human Population on the Nechako River: A Major Tributary of the Fraser River, British Columbia, Canada. *GeoJournal* 40: 47-164.

Kirshner, M. 1999. Water Fight. *This Magazine* 33, no. 1 (July-August): 6.

*Maclean's*. 1985. The Crisis over Water. 26 August, 34.

—. 2003. The *Maclean's* Year-End Poll. 29 December, 32.

Maich, S. 2005. America Is Thirsty. *Maclean's*, 28 November, 26-30.

Mitchell, A. 2000. Water: Oil of the Future? *Globe and Mail*, 5 August, A1, A6.

—. 2001. The World's Single Biggest Threat, *Globe and Mail*, 4 June, A8.

Northcote, T.G., and G.F. Hartman. 2004. *Fishes and Forestry: Worldwide Watershed Interactions and Management*. Oxford, UK: Blackwell Publishing.

Pollution Probe. 1987. Newsletter (Autumn): 1. Toronto: Pollution Probe.

Prowse, T.D., and M. Conly. 1996. Impacts of Flow Regulation in the Aquatic Ecosystem of the Peace and Slave Rivers. *Northern River Basins Study Synthesis Report No. 1*. Edmonton: Alberta Environment.

Prowse, T.D., F.J. Wrona, and G. Power. 2004. Dams, Reservoirs and Flow Regulation. In *Threats to Water Availability in Canada: Environment Canada*. Burlington: National Water Research Institute, NWRI Scientific Assessment Report Series, no. 3, and ACSD Science Assessment Series no. 1. http://www.nwri.ca/threats2full/pubdoc-e.html.

Quinn, F., J.C. Day, M. Healey, R. Kellow, D. Rosenberg, and J.O. Saunders. 2004. Water Allocation, Diversion and Export. In *Threats to Water Availability in Canada. Environment Canada.* Burlington: National Water Research Institute, NWRI Scientific Assessment Report Series, no. 3, and ACSD Science Assessment Series no. 1. http://www.nwri.ca/threats2full/pubdoc-e.html.

Rutherford, S. 2004. *Groundwater Use in Canada.* Vancouver: West Coast Environmental Law. http:www.wcel.org/wcelpub/2004/14184.pdf.

Shrybman, S. 1999. *The Once and Future MAI: International Investor Rights, the Environment, and Your Community.* Vancouver: West Coast Environmental Law. http://www.wcel.org/wcelpub/1998/12601.html.

Timoney, K. 2002. A Dying Delta? A Case Study of a Wetland Paradigm. *Wetlands* 22: 282-300.

Townsend, G.H. 1975. Impact of the Bennett Dam on the Peace-Athabasca Delta. *Journal of the Fisheries Research Board of Canada* 32: 171-76.

WRI [World Resources Institute]. 2003. *World Resources 2002-2004: Decisions for the Earth – Balance, Voice, and Power.* Washington, DC: World Resources Institute in collaboration with United Nations Development Programme, United Nations Environmental Programme, and World Bank. http://governance.wri.org/pubs_pdf.cfm?PubID=3764.

—. 2005. *Freshwater Resources 2005.* Washington, DC: World Resources Institute. http://earthtrends.wri.org/pdf_library/data_tables/wat2_2005.pdf.

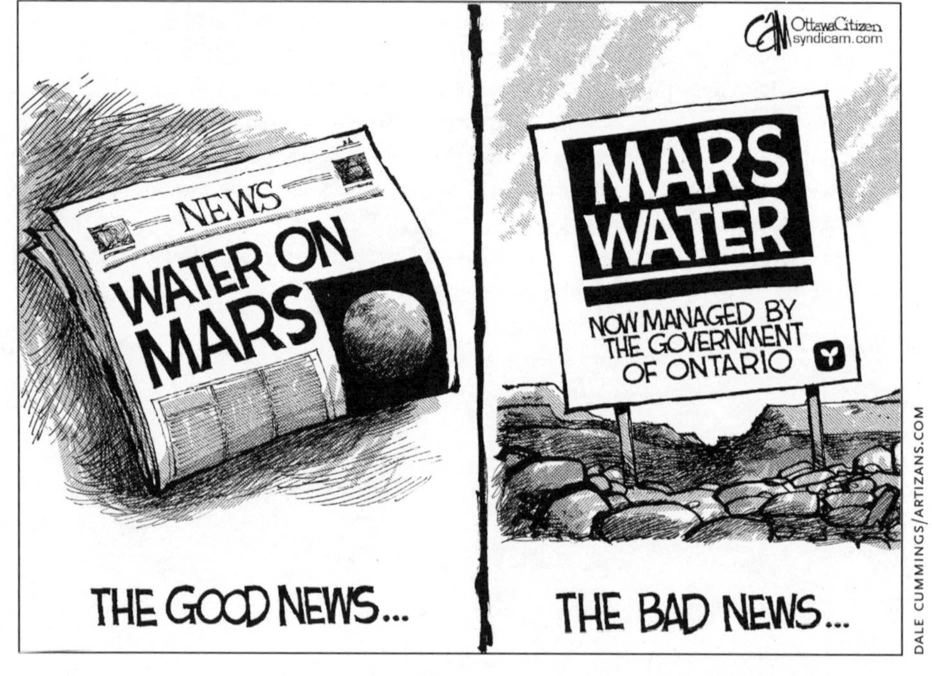


DALE CUMMINGS/ARTIZANS.COM

The contamination of the town of Walkerton's water supply in 2000 by *E. coli* from a local farm was the largest outbreak of its kind in North America. The Walkerton Inquiry identified negligent management by the town's water supplier, cutbacks in the Ontario Ministry of Environment, and outsourcing of water quality testing to the private sector as contributing factors.

# 3
# On Guard for Thee? Water (Ab)uses and Management in Canada

*Dan Shrubsole and Dianne Draper*

In a recent study of "water poverty" conducted by Sullivan (2002), Canada's water supply was ranked second best in the world, after Finland. Compared to those of most countries, Canada's water system is well financed and is operated by well-trained professionals. The majority of the population has access to clean water and treats its wastewater. However, chronic regional supply and quality problems persist. When surface or groundwater resources begin to show physical signs of abuse, ecological and financial consequences are usually not far behind. Doubts also exist regarding the ability of past strategies that are grounded in a supply management structure (with subsidized water prices), combined with water quality strategies based on government regulation, to continue to nurture the high quality of Canada's ecological, economic, and social life.

In order to provide a context for contemporary concerns over water management in Canada, this chapter presents a historical perspective on two interrelated topics: water use and water management. We begin by focusing on the characteristics of water use in Canada, describing some of the key terms related to the two main types of water use. We then examine patterns of water abstraction, use, and disposal in order to show how water uses have changed over time. As our analysis indicates, over the past few decades, Canada has not made significant improvements in its water use record.

This overview of water use is followed by a discussion of water management in Canada, which describes and briefly assesses the major government approaches to this issue. We discuss the shortcomings of our current approaches to water governance, focusing on the relative strengths and weaknesses of three management strategies that have been adopted throughout the country: command-and-control, citizen empowerment and information sharing, and economic instruments. We also consider the implications of the failure to acknowledge the interrelationships between water, humans, and the environment with reference to land use planning, zoning, and water quality management. In the final section, we argue that significant shortcomings in past approaches to water management are at the root of ongoing debate about new approaches to water governance in Canada.

## Sources and Uses of Water in Canada

Both withdrawal and instream water resources provide impressive economic benefits to Canada. In 2000 Canadian businesses earned $1.4 billion from water-related goods and services (Brandes et al. 2005), and, in 1992, Environment Canada (2002) estimated that water contributed between $7.5 and $23 billion annually to the economy. In addition to these economic values, aesthetic considerations must not be under-appreciated. Fresh water has played an important role in Canada's development. However, persistent pressures on water resources to provide an adequate supply to satisfy both withdrawal and instream uses, and chronic concerns about the quality of our water supply in the face of growing wastewater and stormwater treatment needs, highlights the importance of implementing effective water strategies.

Water, then, provides substantial economic benefits to Canadians; but this comes at a cost. Physical stresses on water resources are evident in Canada: between 1994 and 1999, one in four Canadian municipalities experienced water shortages due to increased consumption, drought, or infrastructure constraints (Environment Canada 2002). The federal Commissioner of the Environment and Sustainable Development (2001) found that fresh water in southern Canada was heavily used and overly stressed. Specific areas under stress included the Great Lakes, the Okanagan Valley, the South Saskatchewan River Basin, and the Assiniboine-Red River Basin (Brandes et al. 2005). Evidence of physical stress includes falling water tables and lower water levels as well as degraded water quality. In other situations, water is not returned to its original source but transferred to another water body. Cumulatively, both types of change produce direct and indirect consequences (Brookes et al. 2004): for instance, ecosystems may be altered and water that supports instream uses may be diminished. In turn, as water quantity is degraded, rising treatment costs generate increased financial costs.

### Water Use in Canada

Both groundwater and surface water supplies are distributed unevenly across Canada, as are the demands for water use from industry, agriculture, and municipal sectors. When we remove water from its ground and surface sources, we use it for a variety of purposes, including drinking, irrigation, manufacturing, mining, generating thermal electricity, and diluting waste. When we leave it in rivers and lakes for instream use, water acts as a support for transportation, recreation, tourism, and fish and wildlife.

Several measures reflect the state of water use in Canada, the simplest measuring total intake or withdrawal of water (Table 3.1). At least four important points emerge from considering the 1981-96 water withdrawal data. First, the relatively short period of record reflects Canada's recent interest in the systematic recording of water use. Second, thermal power production is the largest withdrawal use in Canada, followed by manufacturing, municipal, agricultural, and mining withdrawals. Third, total water use has varied. For the entire period, total water use increased from 36,717 million cubic metres (MCM) to 44,873 MCM. However, the downward trend from 1986 to 1996 suggests the situation may be improving. Our per capita use mirrors the same trends and was almost at the same point in 1996 as it was in 1981. Most other developed countries have reduced both overall and per capita water use levels during this time period (Brookes et al. 2004). Fourth, each Canadian uses about 1.50 MCM of water (about 4,400 litres per capita per day) to support our lifestyle. This far exceeds the amount of water used by

Figure 3.1

**Average daily domestic water use (per capita) for selected countries**

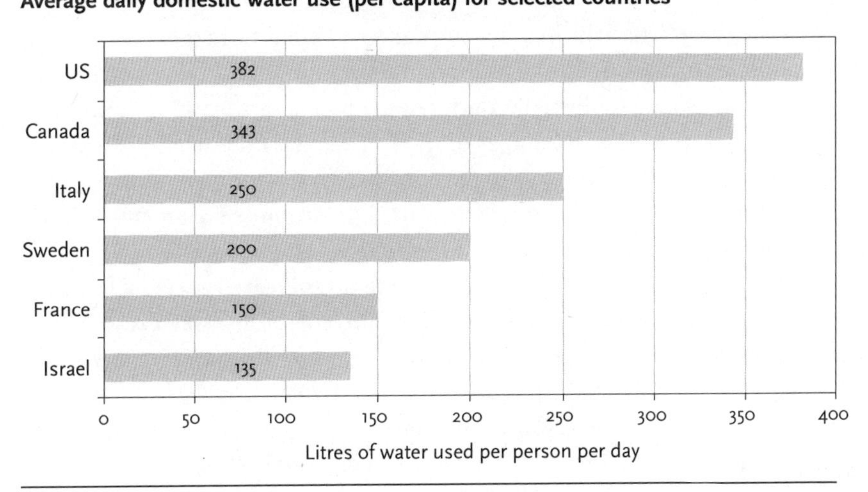


Source: Environment Canada's Freshwater Website, http://www.ec.gc.ca/water, 2006.

Europeans, who have a similar standard of living (ibid.); indeed, Canadians are some of the highest per capita water users in the world (Figure 3.1).

In 2001, average residential water use per person fell to 335 litres per day – the second lowest rate since 1991 – thus resuming a ten-year downward trend that has been interrupted only once (by a slight upswing in 1999; see Figure 3.1) (Environment Canada 2004). This positive trend notwithstanding, Canadians still rank among the most profligate consumers of water among OECD countries.

Additional insight into Canada's water use is evident in a review of withdrawals and measures of efficiency of water use in each sector (Table 3.2). The thermal power industry, including both fossil fuel and nuclear electrical generating stations, was responsible for 64 percent of Canada's total water intake in 1996 – more than four times greater than manufacturing, the next largest user. Producing just over 23,000 of the nation's 28,750 total kilowatt hours (KWh), Ontario dominates the country in use of water for thermal electric power generation.

A typical fossil fuel powered plant uses 140 litres of water to generate one kilowatt of electricity, while a typical nuclear generating station requires 205 litres per kilowatt (Statistics Canada 2002). Most of this water is used for cooling purposes and is returned to its source at an elevated temperature. Recycling of water is becoming a more widespread practice within the thermal sector, perhaps reflecting more stringent environmental regulations and/or a stronger environmental ethic on the part of energy producers (Schaefer et al. 2004).

Water supplied to the manufacturing sector differs from other withdrawal uses in at least two ways. First, much of the water supply comes from private, not public, water suppliers. Second, few members of the public appreciate fully the significance of the large volumes of water withdrawn by the manufacturing sector. For example, the production of a single car requires 250,000 litres of water, and 33,000 litres are required to assemble the average computer (Environment Canada 2004). Manufacturers also use water as a raw material, a coolant, and a solvent. The three main industrial users of water are paper and allied products, primary metals, and chemicals. Water is so important to most manufacturing industries that proximity to water is a determining factor in locating an industrial plant.

The manufacturing sector, like the thermal energy sector, is making steady progress toward the more efficient use of water: between 1981 and 1996, water intake declined from 11,042 to 7,508 MCM (Table 3.1) and water recycling

Table 3.1

**Major withdrawal uses of water and per capita water use in Canada**

| Major withdrawal uses (MCM) | Year 1981 | 1986 | 1991 | 1996 |
|---|---|---|---|---|
| Municipal | 3,760 | 3,719 | 3,802 | 3,922 |
| Agriculture | 3,125 | 3,559 | 3,991 | 4,098 |
| Thermal electric | 18,166 | 24,963 | 28,288 | 28,664 |
| Manufacturing | 11,042 | 9,298 | 8,410 | 7,508 |
| Mining | 624 | 544 | 489 | 681 |
| Total water use | 36,717 | 42,083 | 44,980 | 44,873 |
| Per capita water use (MCM/capita) | 1.48 | 1.61 | 1.60 | 1.50 |

SOURCE: Environment Canada (2002).

Table 3.2

**Major water withdrawals and water use efficiency in Canada, 1996**

| | Intake (MCM) | Recycling (MCM) | Gross water use (MCM) | Use rate | Discharge (MCM) | Consumption (consumption rate %) |
|---|---|---|---|---|---|---|
| Thermal power | 28,750 | 11,655 | 40,405 | 1.40 | 28,241 | 508 (1.8) |
| Manufacturing | 6,038 | 6,958 | 12,996 | 2.16 | 5,487 | 552 (9.1) |
| Municipal | 5,314 | – | 5,314 | – | 5,195 | 119 (2.2) |
| Agricultural | 4,098 | – | 4,098 | – | 1,062 | 3,036 (74.1) |
| Mining | 518 | 1,197 | 1,715 | 3.31 | 672 | 46 (8.9) |

INTAKE: amount of water taken (withdrawn) from a ground or surface water source by a water use.

RECYCLING (RECIRCULATION): water that is used more than once in a specific process or distribution system, or used only once and then recycled to another process.

GROSS WATER USE: total amount of water used (intake + recirculation) to carry out an activity.

USE RATE: the degree of recirculation in industrial, thermal, and mining operations. Gross water use divided by water intake.

DISCHARGE: amount of water returned to the source after use.

CONSUMPTION: water that has been removed from its source and is no longer available for use; the difference between water intake and water discharge.

increased. These trends also reflect changing environmental regulations as well as technological change and/or changes in other input prices (Renzetti 2003). Within the manufacturing sector, water consumption constituted just over 9 percent of withdrawals (Table 3.2). Beverage, wood products, and transportation equipment sectors had the highest rates of consumption

(Schaefer et al. 2004). Generally, the Atlantic provinces have among the lowest water use and consumption rates in Canada (a function of water availability and industrial make-up), while the Prairie provinces, particularly Alberta and Saskatchewan, show substantially higher use rates compared to the rest of Canada (a result of the semi-arid climate that necessitates enhanced water conservation efforts, including greater water recirculation, in plant operations) (Schaefer et al. 2004).

Municipal water use includes all water supplied by a municipal water system to residences and small commercial and industrial buildings. Municipal water use also includes water lost due to leaks in the distribution system (pipes) and water used for firefighting. In 1999, the average municipal per capita use was 638 litres/day (including a residential water use rate of 343 litres/day). This figure varied by municipality from a low of 156 litres/day in Charlottetown, Prince Edward Island, to a high of 659 in St. John's, Newfoundland (Environment Canada 2002). To place these data in context, the generally accepted minimum amount of fresh water required for human survival is approximately five litres per capita per day (lcd). To meet additional basic needs such as sanitation, food preparation, and bathing, Health Canada recommended 60-80 lcd, and Gleick (1996) suggested a minimum of 50 lcd. Boyd (2001) ranked Canada twenty-eighth out of twenty-nine on the basis of per capita water use for Organization for Economic Cooperation and Development (OECD) countries. Only the United States consumes more water per capita. "This poor showing relative to other OECD countries underscores how far Canada has to go before becoming a leader in *how* water is used, as opposed to *how much*" (Brandes et al. 2005, 25-26). Lack of a strong water conservation ethic helps explain the poor consumption habits of Canadians (McFarlane and Nilson 2003). Only a small number of small-scale operations using municipal systems recycle some or all of their water for multiple uses (Brandes with Ferguson 2003). A major management concern relates to the high (21 percent) rate of growth experienced within the residential water sector during the 1990s – a growth rate that far outstripped population increases (Brandes et al. 2005).

With 4,098 MCM of water withdrawn in 1996, Canadian agriculture accounted for only 9 percent of total withdrawals (Table 3.1). However, agriculture is the highest consumer of water because only about 25 percent of it returns to the source. About 85 percent of agricultural withdrawals are used for irrigation and 15 percent is used for watering livestock (Environment Canada 2002). About 75 percent of Canada's agricultural water withdrawals take place on the Prairies, mainly for irrigation. Alberta contains about 60 percent of the total irrigated cropland in Canada, and much of this land is located in the South Saskatchewan River Basin. Agriculture in the area consumes about 2,200 MCM of water per year from the river system, which is equivalent to 28 percent of the total annual river flow. As a result of increased water demands from all sectors and recent periods of low precipitation, some southern Alberta watersheds are near or at capacity (Corbett and Lalonde 2004).

At the national level, the mining sector withdraws relatively small amounts of water (Table 3.1). However, the scale of some withdrawals at the

BOX 3.1

**The cumulative effects of climate warming and other human stresses on Canadian fresh waters**

David Schindler, one of Canada's top water scientists, has warned that climate change will compound the stresses that humans are already placing on Canada's aquatic environments:

> Climate warming will adversely affect Canadian water quality and water quantity. The magnitude and timing of river flows and lake levels and water renewal times will change. In many regions, wetlands will disappear and water tables will decline. Habitats for cold stenothermic organisms will be reduced in small lakes. Warmer temperatures will affect fish migrations in some regions. Climate will interact with overexploitation, dams and diversions, habitat destruction, non-native species and pollution to destroy native freshwater fisheries. Acute water problems in the United States and other parts of the world will threaten Canadian water security. Aquatic communities will be restructured as the result of changes to competition, changing life cycles of many organisms, and the invasions of many non-native species. Decreased water renewal will increase eutrophication and enhance many biogeochemical processes. In poorly buffered lakes and streams, climate warming will exacerbate the effects of acid precipitation. Decreases in dissolved organic carbon caused by climate warming and acidification will cause increased penetration of ultraviolet radiation in fresh waters. Increasing industrial agriculture and human populations will require more sophistication and costly water and sewage treatment. Increased research and a national water strategy offer the only hope for preventing a freshwater crisis in Canada. (Schindler 2001, 18-29)

local level can be significant. Water is pumped from mineral, metal, and non-metal mines in order to gain access to the resources; at the mine site, water is used as a coolant and to wash and process the ore after it is extracted. This water is often recirculated, and, as a result, many mines are able to minimize water discharge during their operation (Table 3.2). Water that is used in mining and then discharged into freshwater bodies usually undergoes only primary treatment (settling of sediments) (Environment Canada 2002).

Although Canada's water resources provide a renewable supply that is the envy of many other nations, Canadians have the second highest per capita use of water in the world. Clearly, with increasing urbanization, industrial activity, and use of agricultural chemicals, the ability of freshwater resources to sustain our ecosystems, economy, and society is showing signs of stress – which Canadians would be foolish to ignore.

### Threats to Canada's Water Supply

Canadians have taken their access to safe drinking water for granted, viewing it as a right. However, the quality of this most precious of water resource uses and the utilities that provide for its treatment have received considerable attention following the events of spring 2000, when the small agricultural community of Walkerton, Ontario, found its water supply contaminated by *E. coli O157:H7*. This lethal strain of the usually harmless bacterium caused the deaths of seven people and sent over 2,500 individuals to hospital. In April 2001, *Cryptosporidium parvum* contaminated the water supply in North Battleford, Saskatchewan, inflicting over 5,800 residents with gastrointestinal illnesses and symptoms, including diarrhea, abdominal cramps, fever, nausea, and headaches. One lesson learned from these incidents is that we need to manage our drinking water supplies "from the source to the tap" rather than only at treatment plants. Effective disposal of our wastewater is also a primary concern.

Walkerton is an example of how public health may be affected when we ignore threats to water quality. Environment Canada (2001) has identified thirteen key threats to our drinking water sources: nutrients; acidification; endocrine disrupting substances (EDS); genetically modified organisms (GMOs); pathogens; algal toxins; pesticides; long-range atmospherically transported pollutants; municipal wastewater effluents; industrial wastewater discharges; urban runoff; solid waste management practices; and water quantity changes that result from climate change, diversions, and extreme events. Some of these threats, such as nutrients and acidification from agricultural and industrial activities, have been persistent problems during the past fifty years. Others, such as EDS and GMOs, are new technological risks emerging from actual food sources and lifestyle choices. Threats posed by climate change reinforce a long-standing and fundamental principle of water management: water quality and water quantity are interconnected and must be managed on an integrated basis.

## Water Governance: Who Controls Water Allocation and Management?

The following discussion focuses on how we allocate water within our legal structure and how management approaches have changed over time. As John Sprague argues in this volume (Chapter 2), Canadian governments historically did little to effectively respond to water use problems. In the 1950s, governments became more active in "controlling" water use and formulated some regulations and standards. The 1970s saw governments supplement these strategies with mechanisms whose purpose was to foster citizen empowerment and encourage their participation in decision making. In the 1980s, economic instruments were increasingly promoted as a means of influencing water use (Campbell 2005). These three approaches to water management – which are the core of current water use management strategies – are described below.

### Government Control

Governments control water uses and water discharges in two ways: by prior approval and through penal sanctions. This is often referred to as "command-and-control" regulation and is frequently achieved through having proponents complete permitting processes in order to undertake their activities. In granting a permit (also referred to as a licence or certificate of approval), a government body establishes conditions that applicants must meet. Prior approval requires that proponents seek permission of a government agency or body before undertaking any activity. Through penal sanctions, governments can charge water users with failure to comply with the requirements that are stipulated in a licence or certificate of approval. Problems with the implementation of command-and-control approaches include high costs, inconsistent cross-jurisdictional judgments, the long waiting periods required for approval, the relatively minor penalties applied to guilty parties, and inadequate monitoring. Some of these were evident in the unfortunate events in Walkerton and North Battleford.

Canada's wastewater management system is an example of a permitting system. In 1991, wastewater permits comprised the largest source of effluent authorized for discharge into Canadian waters (a total of about 4,300 MCM) (Servos et al. 2001). Substances within wastewater discharges may have a variety of impacts, including oxygen depletion and the disruption of endocrine system functions. While there is a need for decreased wastewater inputs, there is also a need for better knowledge of the nature of the substances entering the system and an increase in funding for infrastructure renewal and improved treatment technologies.

### Citizen Empowerment and Participation in Decision Making

In the 1970s, public involvement in decision making became more widespread in Canada. Currently, citizen participation is viewed as fundamental to the development of a "productive system" of ecological governance (Brandes et al. 2005, 20). Many federal, provincial, and territorial environmental statutes enshrine public participation in legislation. For instance, the Canadian Environmental Protection Act (CEPA) requires the minister of the environment to publish a list of proposed regulations in the *Canada Gazette*. Within sixty days of publication of a regulation, any person may file a notice of objection requesting a review. This is part of a "prior approval" approach. The CEPA also encourages the federal minister of the environment to consult with any interested person and enables joint work with representatives of different sectors interested in the protection of the environment. Some provinces, as well as the Yukon Territory (see the Yukon Environment Act), have adopted the prior approval approach to help enhance their effectiveness with respect to public involvement in environmental management. Frequently, the timing, nature, and mechanisms that support public participation are limited. While there is opportunity for citizen participation, legal venues are often confrontational and do little to build a sense of ongoing trust and cooperation. Given the typical thirty- to sixty-day periods to provide comments, it is also questionable whether there is sufficient time to promote an effective exchange of ideas. Sharing information on an ongoing basis is one way to promote an in-depth, thoughtful exchange of views.

Basic information about water quantity and quality, as well as patterns of instream and withdrawal uses, is a cornerstone in achieving effective water management (Rutherford 2004). Citizen access to balanced and accurate information is also fundamental to furthering effective, efficient, and equitable governance of water resources (Brandes et al. 2005). In terms of water quantity, Canada's hydrometric and meteorological network provides basic information to support informed decision making. Between 1990 and 1998, however, the number of stations across Canada dropped 21 percent (from 3,374 to 2,650). Some of these provided much needed information, and the loss of this component of the hydrometric information was viewed as "highly problematic" (Scott et al. 1999). This downsizing strategy reduced Environment Canada's capacity to monitor water resources and impeded its ability to formulate defensible policy (Bruce and Mitchell 1995).

In response to a recent letter from the Canadian Water Resources Association expressing concern about subsequent proposals that might further compromise the network's capacity to produce adequate information, all federal, provincial, and territorial levels of government acknowledged the network's importance and indicated that funding of their network budgets would be maintained or increased (Lilley 2004). Some governments expressed a desire to convert certain existing observer stations to automatic real-time stations. Despite this positive news, some areas of the country are inadequately served by existing arrangements. For instance, a recent BC assessment concluded that its "monitoring network fell short of international standards for station density and warned of a growing risk that the networks may not optimally support" resource management decision making (Lilley 2004, 17). Rutherford (2004) concluded that, relative to surface waters, we know much less about the basic conditions of Canada's groundwater resources. To make matters worse, since 1996 there has also been a lack of systematic data collection concerning withdrawal water uses. The need to continue to lobby governments to collect and share hydrometric and water use data is clear.

### Economic Instruments

Government control, even with the support of effective public participation, is achieved by establishing specific standards to be met and, often, the technologies to be used by a water user. This approach provides little incentive for innovation and does not ensure the implementation of effective, efficient, and equitable solutions. The use of economic instruments, which have been defined as using "market-based signals to motivate desired types of decision-making ... [that] either provide financial rewards for desired behaviour or impose costs for undesirable behaviour" (Stratos Inc. 2003), can address this shortcoming and promote greater water efficiencies. Four types of economic instruments may be identified: (1) property rights, (2) fee-based measures, (3) liability and assurance regimes, and (4) tradable permits (ibid.).

Economic instruments are created through legislation and should be viewed as an emerging approach to water management that complements rather than replaces traditional government control approaches. Some provinces and municipalities have implemented some economic instruments, but a nationwide economic water policy effort appears to be absent.

Part of the reason for our high use rate is a general lack of awareness, or poor understanding, of the pressures increasingly being placed on our existing water supply. This situation is exacerbated by the lack of a strong water conservation ethic, which is encouraged by the myth of water abundance. High use is also promoted by a very low price relative to those of other developed nations (Figure 3.2). This undervaluing and underpricing of water resources contributes to our water management challenges. Providing water services is costly. For example, in 2001, the City of Calgary spent almost $250 million to support its water supply and distribution systems, wastewater treatment and disposal, and solid waste management functions (McFarlane and Nilson 2003). However, most Canadian pricing systems do not adequately support routine water service costs (including pipes, pumps, water testing

Figure 3.2

**Municipal water prices in Canada and other countries**

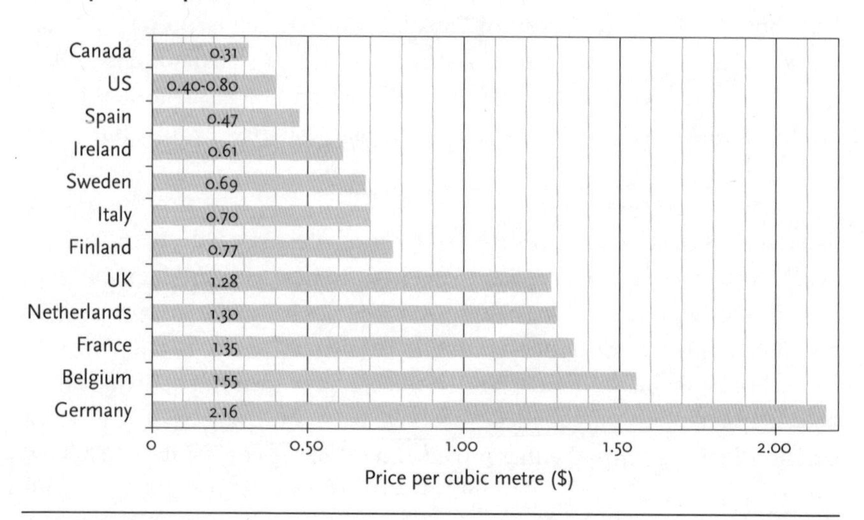


NOTE: All amounts are based on a 1998 survey of OECD countries and are calculated using a purchasing power parity (PPP) method.

SOURCE: Environment Canada's Freshwater Website, http://www.ec.gc.ca/water, 2006; World Water Council.

labs, and billing systems) and fail to acknowledge the true value of the resource. Municipal pricing policies in most municipalities have discouraged the achievement of efficient water use.

In 1996, the National Round Table on the Environment and Economy (NRTEE) estimated that Canada's water infrastructure deficit (the investment required to upgrade our aging water and wastewater facilities at that time) was between $38 and $49 billion. In looking at meeting future needs until 2016, the NRTEE (1996) suggested that $70 to $90 billion was required. The widespread use of a flat rate (or constant unit charge) pricing structure exacerbates the funding shortage and high use rate because it results in users' paying a fixed amount regardless of how much water they actually use (Brandes et al. 2005). Paying realistic charges for water and wastewater services is one potential solution to this problem.

While there is little debate regarding the need for increased levels of investment, the willingness of people to pay increased water rates is not clear. In western Canada, one study noted that over 84 percent of industrial users and over 63 percent of citizens would strongly or somewhat strongly support charging the full cost of water (Berdahl 2003). It is also not clear what kind of economic impacts increased water rates in the industrial and agricultural sectors would have on the Canadian economy or what costs consumers would have to bear for goods (ranging from milk and other beverages to cars and clothes) made in Canada.

Some critics of economic instruments suggest that pollution permits are, in essence, "licences to pollute." The same comment can be made with regard to certificates of approval that are based on standards that fail to protect the environment and human health. There is also some resistance to moving from a system that has treated many resources, particularly water and air, as free goods to a system based on "payments for use." According to Stratos Inc. (2003), other barriers to the implementation of economic instruments include lack of awareness, resistance to what are perceived to be new taxes, concerns over maintaining and improving the competitiveness of industry and agriculture, a lack of strong advocates, and a general lack of experience in dealing with them. Overcoming these barriers is critical if Canada is to apply economic instruments in an effective, efficient, and equitable manner. Privatization (partial or whole) is a controversial economic alternative that may be used to manage water demand (Sprague, Chapter 2). Issues pertaining to the commercialization and commodification of water resources are addressed in Part 3 of this volume.

Tradable permits have been the focus of efforts to increase water use efficiency in western Canada. As noted in Chapter 11 (Christensen and Lintner, this volume), water laws and institutions in western Canada are different from those in eastern Canada. The North-West Irrigation Act of 1894 established laws that promoted the agricultural development of the west through irrigated agriculture and allowed licensed water users to divert and use water from a surface water body and apply it to non-riparian lands (Percy 2004). Water rights were secured on a first-come, first-serve basis, and no charges were applied to the use of the water. When the responsibility for natural resources was transferred from the federal government to Alberta, Saskatchewan, and Manitoba in 1930, this approach continued. In 1925, British Columbia declared that it owned and had the right to use all of its water. In the Northwest Territories, the Yukon, and Nunavut, the Crown has similar powers (Percy 1988). In all cases, there is no charge for using the water (only an administrative fee for applying for use). This arrangement provides no incentive to promote efficient water use.

To compound the problem of a lack of incentive to conserve water, early licences in western Canada were granted for substantial volumes of water and were not transferable except as part of a land transfer. In contrast to the situation in Ontario, priority of use was initially established according to the seniority of licences, regardless of the type of use involved. In the 1990s, Alberta began to actively promote a water trading system in order to provide an incentive for existing licence holders to conserve water and lease or sell that savings to other users. While the theory is clear, de Loë (2005) maintains that the success of the water market in Alberta is likely to be modest because irrigation districts would prefer to expand current irrigation operations rather than to transfer water to other users, particularly when there is not a strong local demand for the water. Saskatchewan reorganized its approach in 2002, and Manitoba refurbished its approach in 1987 (Percy 2004). Provisions in Saskatchewan allow for the elimination and curtailment only of those water rights granted after 1984, and the issuing of licences is on a fixed-term basis. In 1987, Manitoba enacted fixed terms for licences issued after that date, and licences for lower priority uses may not be renewed if there is a need to serve higher water priorities. However, if licences are cancelled in Saskatchewan or Manitoba, then compensation must be paid. According to Percy (2004, 321), "this style of legislation tends to create only a cosmetic flexibility and at the same time tends to perpetuate the wasteful use of water."

## Conclusion

Canada has not made significant improvements in its water use record. This is so despite the clear focus on the environment that emerged in the 1970s, the investment of billions of dollars, and the incorporation of new mechanisms – citizen empowerment and economic instruments in particular – that supplemented the existing regulatory framework. There are at least two ways to proceed in terms of improving current water use management. The first is to refine current mechanisms. However, given the track record of "digging the same hole deeper," which is based on the assumption of a supply of water sufficient to meet all demands, this may not provide the level of use improvement the Canadian public both desires and requires.

The second way to improve water use management is offered by David Brookes (2005; Chapter 14, this volume), who suggests that greater efficiency lies in the "soft path" concept. The soft path relies on a multitude of relatively small-scale and renewable sources of supply, coupled with ultra-efficient ways of meeting end-use demands, in order to improve efficiency. Traditional approaches have been characterized as requiring a top-down administrative structure and have not considered using alternative resources (e.g., air) to provide the services (e.g., cooling, dilution) that are currently being provided by existing water sources. The soft path reflects a new way of thinking about the problem – essentially, it involves "digging a new hole." A mix of traditional and soft path approaches may offer a feasible route toward a better water future in Canada.

An ongoing challenge in water use management involves balancing withdrawal and instream uses. A number of initiatives, such as Ontario's 2002 Low Water Response, have recently emerged (Durley 2003). Temporary and voluntary restrictions, often in the form of municipal lawn watering bans, are now a regular feature in water conservation efforts in many provinces. Understanding how to advance stronger conservation ethics with regard to the management of human demands for water is essential if Canada is to establish an appropriate framework for the effective, efficient, and equitable management of its freshwater resources. Decision makers need to endorse the principle that conservation is a higher priority than is revenue generation. In addition, efforts to balance human and natural ecosystem needs can often benefit greatly from local citizen involvement and various stewardship actions that empower communities of water users. Given declines in government-supported water monitoring programs, for instance, the collaborative efforts of water users might enable them to devise and implement alternative monitoring systems in order to identify and measure threats to local freshwater supplies and ecosystems. Knowledge building – about a range of freshwater resource issues, including uncertainties such as climate change impacts on water supply and distribution as well as methods for protecting water quality and aquatic ecosystems – is important when it comes to identifying the principles according to which Canada's water resources need to be governed.

The sustainability of Canada's freshwater resources, and the social, economic, and environmental future of its population, depends on the management philosophies and approaches we employ in order to achieve effective, efficient, and equitable water management. Given the complexity of managing freshwater resources – and the considerable range of political, social, economic, and environmental obstacles that must be addressed – integrated planning and collaboration among those involved in and affected by water management will help ensure that land use planning, zoning, and water quality management efforts are undertaken with water sustainability in mind. Failure to do this is not an option: Canadians have a responsibility to themselves and to the international community to achieve improvements in freshwater resource allocation, use, and protection.

## REFERENCES

Berdahl, L. 2003. *Looking West 2003: A Survey of Western Canadians.* Calgary: Canada West Foundation.

Boyd, D. 2001. *Canada versus the OECD: An Environmental Comparison.* Victoria: University of Victoria Eco-Research Chair of Environmental Law and Policy. http://www.environmentalindicators.com/htdocs/PDF/CanadavsOECD.pdf.

Brandes, O., with K. Ferguson. 2003. *Examining Urban Water Use in Canada.* Victoria: POLIS Project on Ecological Governance. http://www.polisproject.org/polis2/PDFs/FlushingFuture.pdf.

Brandes, O., K. Ferguson, M. M'Gonigle, and C. Sandborn. 2005. In *At a Watershed: Ecological Governance and Sustainable Water Management in Canada.* Victoria: POLIS Project on Ecological Governance. http://www.waterdsm.org/PDF/atawatershed.pdf.

Brookes, D. 2005. Beyond Greater Efficiency: The Concept of Water Soft Paths. *Canadian Water Resources Journal* 30 (1): 83-92.

Brookes, D., L. Nowlan, O. Brandes, and A. Hurley. 2004. In *Controlling Our Thirst: Managing Water Demands and Allocations in Canada.* Edmonton: Walter and Duncan Gordon Foundation.

Bruce, J., and B. Mitchell. 1995. *Broadening Perspectives on Water Issues*. Ottawa: Royal Society of Canada.

Campbell, I., ed. 2005. *Canadian Water Resources Journal: Special Issue – Economic Instruments for Water Demand Management.* Cambridge: Canadian Water Resources Association.

Commissioner of the Environment and Sustainable Development. 2001. A Legacy Worth Protecting: Charting a Sustainable Course in the Great Lakes and St. Lawrence River Basin. *Report of the Commissioner of the Environment and Sustainable Development to the House of Commons.* Ottawa: Office of the Auditor General.

Corbett, B., and K. Lalonde., eds. 2004. *The State of Southern Alberta's Water Resources.* Lethbridge: Alberta Agriculture, Food and Rural Development.

de Loë, R. 2005. In the Kingdom of Alfafa: Water Management and Irrigation in Southern Alberta. In *Sustaining Our Futures: Reflections on Environment, Economy and Society,* ed. D. Shrubsole and N. Watson. Waterloo: University of Waterloo Department of Geography Publication Series.

Durley, J., R. de Loë, and R. Kreutzwiser. 2003. Drought Contingency Planning and Implementation at the Local Level in the Province of Ontario, Canada. *Canadian Water Resources Journal* 28 (1): 21-52.

Environment Canada. 2001. Threats to Sources of Drinking Water and Aquatic Ecosystem Health in Canada. *NWRI Scientific Assessment Report,* Series No. 1. Burlington, Ontario: National Water Research Institute. http://www.nwri.ca/threatsfull/toc-e.html.

—. 2002. *Urban Water Indicators: Municipal Water Use and Wastewater Treatment.* http://www.ec.gc.ca/soer-ree/English/Indicators/Issues/Urb_H2O.

—. 2004. *Municipal Water Use, 2001 Statistics.* http://www.ec.gc/water/en/info/pubs/sss/e_mun2001.htm.

Environmental Commissioner of Ontario. 2001. *Ontario's Permit to Take Water Program and the Protection of Ontario's Water Resources.* Brief to the Walkerton Inquiry. Toronto: Environmental Commissioner of Ontario. http://www.eco.on.ca/english/publicat/walker01.pdf.

Gleick, P.H. 1996. Basic Water Requirements for Human Activities: Meeting Basic Needs. Water International 21: 83-92.

Lilley, J. 2004. The CWRA Comments on Canada's Hydrologic and Meteorologic Networks. *Water News* 23 (4): 15-20.

McFarlane, S., and E. Nilson. 2003. *On Tap: Water Issues in Canada Discussion Paper.* Calgary: Canada West Foundation.

National Roundtable on the Environment and Economy. 1996. *Water and Wastewater Services in Canada.* Ottawa: NTREE.

Percy, D.R. 1988. *The Framework of Water Rights Legislation in Canada.* Edmonton: Canadian Institute of Resources Law.

–. 2004. The Limits of Western Canadian Water Allocation Law. *Journal of Environmental Law and Practice* 14: 315-29.

Renzetti, S. 2003. Municipal Water Supply and Sewage Treatment: Costs, Prices, and Distortions. *Land Economics* 32 (3): 688-704.

Rutherford, S. 2004. *Groundwater Use in Canada.* Vancouver: West Coast Environmental Law.

Schaefer, K., D. Tate, S. Renzetti, and C. Madramootoo. 2004. Manufacturing and Thermal Water Demands. In *Threats to Water Availability in Canada,* ed. L. Brannen and A.T. Bielek. Ottawa: Minister of Public Works and Government Services. http://www.nwri.ca/threats2full/perspective-e.html.

Schindler, D. 2001. The Cumulative Effects of Climate Warming and Other Human Stresses on Canadian Freshwaters in the New Millennium. *Canadian Journal of Fisheries and Aquatic Sciences* 58: 18-29.

Scott, D., T.R. Yuzyk, and C. Whitney. 1999. The Evolution of Canada's Hydrometric Network: A Century of Development. In *Partnerships in Water Resource Management: Proceedings of the CWRA 52nd Annual Conference.* Nova Scotia: CWRA

Servos, M., P. Chambers, R. Macdonald, and G. Van Der Kraak. 2001. Municipal Wastewater Effluents. In *Threats to Sources of Drinking Water and Aquatic Ecosystem Health in Canada.* Ottawa: Minister of Public Works and Government Services. http://www.nwri.ca/threatsfull/ch9-1-e.html.

Statistics Canada. 2002. *Human Activity in the Environment.* Ottawa: Minister of Public Works and Government Services.

Stratos Inc. 2003. *Economic Instruments for Environmental Protection and Conservation: Lessons for Canada.* Report prepared for External Environmental Committee on Smart Regulation. http://www.pco-bcp.gc.ca/smartreg-regint/en/06/01/su-11.html.

Sullivan, C. 2002. Calculating a Water Poverty Index. *World Development* 30 (7): 1195-10.

# 4
# Out of Sight, Out of Mind? Taking Canada's Groundwater for Granted

*Linda Nowlan*

> The oil and gas reservoirs in this country are better understood than the groundwater reservoirs, as you all know.
>
> – Karen Brown, Assistant Deputy Minister, Environment Canada[1]

Taken for granted by Canadians and the Canadian government, groundwater is a relatively unregulated and neglected resource, and this can have tragic consequences. Groundwater contamination in Walkerton, Ontario, caused seven deaths and thousands of illnesses in May 2000. Not only is the rise in groundwater contamination incidents a fear for millions of Canadians who drink this water but excessive groundwater withdrawals may also harm the environment and trigger disputes. Lester Brown (2005), founder of the Worldwatch Institute, believes that depletion of underground water resources poses a far greater threat to the future than does depletion of oil reserves.

Recently, the consequences of overpumping have become clear in the Great Lakes region. The US Geological Survey has issued a "wake-up" report illustrating that, on the US side of the Great Lakes Basin, overpumping can cause groundwater to flow away from major bodies of water rather than into them (USGS 2005). No similar research results have been reported on the Canadian side of the Great Lakes. The chief transboundary water cooperation agency, the International Joint Commission, highlighted groundwater as an important component of the Great Lakes Basin and pointed out that, while neither country fully understands the extent of groundwater resources, Canada lags behind the United States in data collection and monitoring (IJC 2000).

Groundwater is a topic Canadians know surprisingly little about, despite our reliance on it and despite its intimate and dynamic connections to wetlands, streams, and lakes. Incomplete knowledge and relatively minimal regulation can contribute to disputes between such competing water users as farmers and towns, developers and conservationists. The most prominent recent example of this type of dispute occurred in the Oak Ridges Moraine, a key groundwater recharge area in southern Ontario, where, after a lengthy controversy, the provincial government introduced new development controls.[2] Debate over the existence and scale of subdivision development on the moraine continues. Meanwhile, scientists warn that groundwater disputes are likely to increase in the near future (Senate of Canada 2005).

## Groundwater: "Buried Treasure" under Stress

Canadians rely on groundwater in both obvious and hidden ways. For about one-third of the total Canadian population, groundwater is drinking water, and this figure rises to about 80 percent for rural residents (Environment Canada 2001). Legally, bottled "spring" or "artesian" water must come from underground water sources, though the exact proportion of the skyrocketing sales of all bottled water originating from underground sources is unknown. Mining, oil, gas, manufacturing, and other industries extract groundwater too, often in surprisingly large quantities. Agriculture accounts for a varying share of groundwater use across the country, ranging from a high of 44 percent in Manitoba to a low of 3 percent in Saskatchewan (Rutherford 2005).

Canada's municipalities are also highly reliant on groundwater. Quebec has the greatest number of municipal systems reliant on groundwater (142), but Ontario has the highest population dependent on it (1.3 million) (Rutherford 2005). Prince Edward Island is entirely dependent on groundwater for its municipal supplies. Over one-quarter of Canadian municipalities have experienced water shortages in the last decade, with municipalities that rely on groundwater reporting more frequent shortages than those that use surface water (Environment Canada 2004).

Information on patterns of domestic use are difficult to determine. Many provinces do not collect data on actual use patterns, and reporting requirements often do not distinguish between groundwater and surface water. On a countrywide scale, it is impossible either to say whether overall groundwater use is increasing or decreasing or to pinpoint the trends being used by different sectors. Are farmers using more or less groundwater than they were five years ago? How much groundwater does the Canadian water bottling industry use? Is there any movement in groundwater use on the part of the energy production sector? There are no answers to these questions. The contrast between availability of information on groundwater in the United States and in Canada is striking. The United States is able to report that fresh groundwater withdrawals during 2000 were 14 percent higher than they were during 1985 (Hutson et al. 2004). The US Geological Survey first conducted water use compilations for 1950 and has published them every five years since. No comparable countrywide data compilations exist for Canada.

Though provinces have primary jurisdiction over water resources, the federal government, due to its constitutional powers, has a role over transboundary waters, navigation, fisheries, and international treaties. And it has been particularly criticized for its failure to conduct the studies necessary to properly guard groundwater. In its 1987 Federal Water Policy, the federal government pledged to develop a better understanding of groundwater; however, due to the elimination of the Inland Waters Directorate of Environment Canada and a consequent declining priority for water at Environment Canada, the policy remains largely unimplemented (Pearse and Quinn 1996). Canada's Commissioner of the Environment and Sustainable Development reported in 2001 that, despite the federal government's commitment to improving its understanding of groundwater aquifers, "it has gained little understanding of groundwater in the basin since [the 1987 federal policy]. Its knowledge has remained fragmented and incomplete" (Commissioner of Environment and Sustainable Development 2001). There have been many calls for the federal government to take a stronger role in freshwater management, given its jurisdictional powers and its central coordination role (Senate of Canada 2005; Smith 2002).

Not only is groundwater use in Canada undocumented but its environmental role is also poorly understood. No figures exist to show exactly how freshwater species depend on groundwater or how to calculate the amount of groundwater that can be pumped out of a spring before affecting the health of the river to which it is linked (Gartner Lee 2002; Rivera 2005). The environmental consequences of overwithdrawal can be profound and may lead to water shortages, harm to connected wetlands, and (in some cases) saltwater intrusion (Glennon 2002). Groundwater withdrawal can also affect groundwater quality as it may cause poorer quality surface water to be drawn into the aquifer.

This lack of knowledge poses management challenges as, across Canada, examples of groundwater stress emerge. When the local groundwater resources were no longer able to meet the needs of a growing population in the Lacombe/Ponoka area of Alberta, a pipeline was built to allow for an interbasin transfer of the treated water. This area of central Alberta is a groundwater "hot spot," as are areas in northeast Alberta where there is demand for groundwater to produce steam for the thermal recovery of bitumen (Nowlan 2005). Groundwater use in Alberta is likely to increase both in agriculture (due to increasingly intensive livestock operations) and the oil and gas industries (due to the rise in the exploitation of non-conventional energy resources). In British Columbia, water levels are declining in areas of intensive urban development, such as the Lower Mainland, the Okanagan, the southeast coast of Vancouver Island, and the Gulf Islands (BC Ministry of Water, Land and Air Protection, n.d.). Stress is also evident in Ontario, Manitoba, and Prince Edward Island, which have all issued moratoriums for a portion of their groundwater permits (Nowlan 2005).

Climate change threatens Canada's freshwater resources as glaciers are rapidly diminishing, receding in some places at rates not seen for as many as ten millennia. Since 1850, some 1,300 glaciers have lost between 25 percent and 75 percent of their mass, with most of this reduction occurring in the last fifty years. Along the eastern slope of the Rocky Mountains, glacier cover is receding rapidly, and total cover is now close to its lowest extent in 10,000 years. The full impact of climate change is unknown, but changes to water recharge patterns and the depletion of groundwater supplies in shallow unconfined aquifers are potential results (Statistics Canada 2003; Warren and Lemmen 2004).

Other stresses relate to diversions; agricultural, mining, and petroleum production and industrial practices; droughts; floods; urban development; manufacturing and thermal energy demands; forestry; and cumulative and integrated threats (Environment Canada 2004). Pressure on water supplies also comes from the profligate water use habits of Canadians, who are among the highest per capita consumers of water in the world (Statistics Canada 2003). Wasteful use affects groundwater supplies and groundwater-dependent ecosystems.

Knowledge about Canada's groundwater resources is incomplete. The Geological Survey of Canada, the chief federal authority on groundwater supplies, reports that Canada does not have more than localized problems with aquifer depletion and exploitation (Rivera 2005). However, in the same year, leading scientists testified to a senate committee that the government of Canada does not have the information required to manage water resources. They asserted that Canada's knowledge about the country's major aquifers was "pitiful" (Senate of Canada 2005). Groundwater regulation is in its infancy in most provinces, and groundwater has not been a priority for regulators, especially in an era of declining budgets for environmental protection (CIELAP 1996-2001; West Coast Environmental Law 2004, 2005).

In summary, Canadians don't fully know how much groundwater exists in the country; who is using and exploiting it and at what rates; how its extraction affects ecosystems; how stresses such as climate change, current practices, and overuse will influence this resource; or how well current provincial regulations protect it. Our knowledge lags behind that of other developed countries, most notably the United States.

### History of Groundwater Law: The "Unknowable Resource"

Water law in general has evolved along with society's priorities for resource use and the emergence of new hydrogeological knowledge. Yet groundwater law has evolved more slowly, perhaps due to its relative invisibility.

Initially, the law of groundwater allocation in Canada developed from court decisions based on English common law. Historically, it was impossible to see subterranean water resources and, consequently, impossible to detect whether they were moving, still, running out, or unimpaired. English judges applied different theories to water conflicts: riparian rights applied to surface water and to groundwater that flowed in "defined underground channels," while for all other sources of groundwater judges developed the Rule of Absolute Capture, allowing landowners to extract groundwater from under their land regardless of any injury caused to their neighbours. The rule was based on the proposition that everything that lies beneath the land belongs to the landowner. This was very different from the rules that applied to riparian landowners, who were forbidden to impair the rights of downstream water users either by fouling the water quality or by overly diminishing its quantity.

This legal distinction between surface water and groundwater was inappropriate because only rarely will groundwater flow in defined channels and act like streams. Why did the law treat these two classes of water differently? With the benefit of contemporary knowledge of hydrogeology, this distinction seems strange. But to judges who first heard disputes involving underground water, the matter was far from clear. Surface water that flowed above ground was visible, crossed land that was privately owned but was clearly not part of the land, and flowed in observable patterns. The activities that interfered with its flow were obvious. In contrast, groundwater could hardly be subject to the same rules as it "moved through the hidden veins of the earth," was "mysterious," and "unknowable."[3] The English common law rules that formed part of the laws of Canada and the United States treated groundwater as a mysterious substance.

Legal ignorance about groundwater reached a peak in a famous American case in 1904 in which the Texas Supreme Court ruled that a landowner, W.A. East, was not entitled to damages from a railroad whose drilling of wells for its steam locomotives had caused his well to go dry.[4] The Court said the railroad had a right to pump as much groundwater as it chose without liability as the origin, movement, and course of groundwater was "so secret, occult, and concealed" that any legal rules governing groundwater "would be involved in hopeless uncertainty, and would, therefore, be practically impossible." The other public policy reason for disallowing the claim was that attempting to apportion groundwater would discourage development.

This decision has been roundly criticized and has led to charges that Texas groundwater law is the "law of the biggest pump." Though there are recent signs of change, Texas is one of the few states to still maintain the Rule of Capture. The Texas Supreme Court has commented that "what was so secret [and] occult to us in 1904 – the movement of groundwater – was no longer so"[5] and held that unregulated groundwater pumping was unacceptable; however, it left it to the legislature to create the rules.

This characterization of groundwater as mysterious and unknowable, and the separate development of the law of groundwater and the law of surface water, continued well into the second half of the twentieth century. Provincial licensing provisions used for surface water were extended to groundwater in Ontario only in 1961 and to the Prairie provinces in the 1970s. The invisibility of the resource contributed to its neglect on the part of lawmakers and resource managers.

## Overview of Groundwater Permitting Systems

Permitting processes for groundwater resources vary significantly between the provinces, as Table 4.1 demonstrates. Although federal Aboriginal and local governments also play a role in water regulation, the provinces have primary jurisdiction. This is a result of their constitutional power to make laws concerning property, civil rights, "local works and undertakings," and natural resources (LaForest 1973; Nowlan 2005; Percy 1988). Local governments do not issue permits for water takings or allocations. In fact, local governments require a permit from the province for water takings to supply

Table 4.1

**Overview of groundwater permitting processes in provincial and territorial jurisdictions**

| Jurisdiction and primary regulator | No. of wells | Total no. of GW permits | No. of permits issued annually | GW licensing law | Date licensing applied | Regulation |
|---|---|---|---|---|---|---|
| BC: Ministry of Environment | 100,000+ estimated; submission of well records is not currently mandatory | n/a: in BC licensing does not currently apply to groundwater | n/a | Water Act could be extended for licensing groundwater; the necessary regulations do not currently exist. | no requirement permit | Ground Water Protection Regulation, 2004, focuses on well construction standards and GW quality protection. |
| AB: Alberta Environment | 500,000 total wells; approx. 5,000 new wells drilled each year. | n/a; numbers kept by regional offices. | n/a | Water Act | 1962 | Water (Ministerial) Regulation |
| SK: Saskatchewan Environment, Saskatchewan Water Corporation, and the Saskatchewan Watershed Authority share authority | n/a | approx. 3,600 | approx. 300 | Ground Water Conservation Act | 1959 | Ground Water Regulations |

▶

◄ TABLE 4.1

| Jurisdiction and primary regulator | No. of wells | Total no. of GW permits | No. of permits issued annually | GW licensing law | Date licensing applied | Regulation |
|---|---|---|---|---|---|---|
| MB: Department of Water Stewardship | n/a | 533 licences as of August 2004 | varies | Water Rights Act | 1972 | Water Rights Regulation |
| ON: Ministry of the Environment | approx. 500,000 | 2,800 | | Water Resources Act | 1961 | Water Transfer and Taking Regulation |
| QC: Ministère du Développement durable, de l'Environnement et des Parcs | n/a | more than 600 catchments > 75 m³ day | Ministry of Environment gets approx. 6,000 well reports/yr. | Environmental Quality Act | 2004 | Groundwater Catchment Regulation |
| NB: Ministry of Environment and Local Government | approx. 3,000 new wells each year | Approval (not permit) required to construct and operate wells above threshold. | n/a | Clean Water Act, Clean Environment Act | n/a | Environmental Impact Assessment Regulation, Water Quality Regulation |

| | | | | | | |
|---|---|---|---|---|---|---|
| NL: Department of Environment and Conservation | 17,000+ | n/a | approx. 10 | Water Resources Act | Water use authorization: May 1988 to May 2002. Water use licence: May 2002 to present | Water Resources Act |
| NS: Department of Environment and Labour | n/a | approx. 100 | less than 10 | Environment Act | 1973 | Activities Designation Regulations |
| PE: Ministry of Environment, Energy and Forestry | approx. 21,000 | 500-800 | approx. 30 | Environmental Protection Act | n/a | Water Well Regulations |
| YT: Environment Yukon | n/a | approx. 5 | variable | Waters Act | n/a | Waters Regulation |
| NT: Ministry of Environment and Natural Resources | n/a | n/a | n/a | Northwest Territories Water Act, Mackenzie Valley Resource Management Act | n/a | Northwest Territories Waters Regulations |
| NU: Department of Environment | n/a | n/a | n/a | Nunavut Waters and Nunavut Surface Rights Tribunal Act | n/a | Nunavut Water Board By-Laws |

their own systems. However, local governments need to be involved in decision making about water-taking permits or, at least, to be provided with the knowledge of other levels of government as water supply and land use decisions are closely tied.

A comparison of the provincial and territorial groundwater permitting requirements revealed striking variables: the existence of a groundwater permitting system, how regulators assess the environmental impacts of groundwater withdrawals, and public participation rights available under water laws (Nowlan 2005).

### Existence of a Groundwater Permitting System

Water use licences may specify the rate, quantity, duration, and time of use. A licence will also commonly state the purpose for which the water will be used. British Columbia remains the sole jurisdiction in Canada and one of the only jurisdictions in North America that has no general licensing requirement for groundwater extraction above a defined threshold level.[6] Many have decried the lack of groundwater controls in the province (Auditor General of BC 1998-99). The hydrogeologically connected provinces and states that share borders with British Columbia all license groundwater, but the provincial government still maintains that licensing is not needed.

### Environmental Impacts of Groundwater Withdrawals

Looking south across the border, excessive pumping has created "an environmental catastrophe known to only a few scientists, a handful of water management experts, and those unfortunate enough to have suffered the direct consequences" (Glennon 2002). It is true that Canada, due to a much smaller population and greater water resources, does not face the same supply problems as does the United States. However, because Canada's water laws are similar to those in the United States, and because only minimal attention has so far been paid to the environmental impacts of groundwater pumping, there is cause for concern. Also, the majority of Canada's population is highly concentrated at densities similar to those found in the United States, and Canadians rely mostly on confined aquifers. For these reasons, they are likely to face supply problems similar to those found south of the border.

Regulators consider the environmental impacts of groundwater withdrawals when they issue a groundwater licence. Applicants for a licence often need to submit a hydrogeologist's report, which details the environmental impact of the proposed withdrawal. Those provinces that do not routinely consider the environmental impacts of groundwater withdrawals commonly cite lack of staff and competing priorities (Nowlan 2005). The environmental consequences of groundwater withdrawals are dealt with through conservation requirements such as controls on wells, conservation plans, or statutory powers to establish conservation objectives. For example, under its new Water Taking and Transfer Regulation, Ontario requires that attention to water conservation be part of the process involved in reviewing permit applications. For all existing takings, permit holders will be encouraged to adopt water conservation best practices. Additional requirements may be invoked if the water taking is in a high- or medium-use watershed, is in a watershed with low water conditions, triggers the Great Lakes Charter, and/or is a large municipal residential supply (Maude 2004). To encourage municipalities to adopt more targeted conservation goals, environmental groups have suggested that municipalities should not be issued any new water-taking permits until they have a water conservation plan in place (CELA 2004).

Aquifers provide base flow for surface water sources; thus, another way regulators account for the environmental impacts of groundwater extraction is through instream or environmental flow protection requirements, which are needed to protect fish, wildlife, and recreation values. Environmental flows are a hot topic in environmental law (IUCN 2003). Reserving groundwater supplies can help maintain these flows. One possible consequence of setting instream or environmental flows is that it will prevent a licensed user from withdrawing water if a river flow drops below the threshold required to protect aquatic health. The role of groundwater in environmental flows is not well understood, and there is currently no standard approach to setting the stream flow limit (Gartner Lee 2002).

Groundwater protection may also be achieved through source protection legislation, the critical first barrier in a multi-barrier approach to protecting water against depletion and degradation. The judicial inquiry that probed the causes of Walkerton and examined how to prevent future tragedies gave regulators great guidance on source protection. Justice O'Connor's first recommendation in the Walkerton Inquiry report was that "drinking water sources should be protected by developing watershed-based source protection plans. Source protection plans should be required for all watersheds in Ontario" (Ministry of the Attorney General).[7] The shocking deaths from contaminated groundwater in Walkerton prompted regulators across the country to look more closely at their water laws and to revise them. The O'Connor Report was a thorough and detailed analysis of how drinking water laws should work. In light of this, many provinces felt pressured to measure their systems against its recommendations and, in doing so, found that they needed to put more emphasis on source protection. After all, protecting the water source may be the only type of protection available to consumers who rely on untreated well water. Source protection can involve many different actions, with the two most common being limiting the discharge of contaminants into a water supply area and leaving the area around a groundwater supply area untouched.

No jurisdiction in Canada has yet fully adopted source protection plans, though six provinces have laws that authorize or require the preparation of these plans: Ontario, British Columbia, New Brunswick, Quebec, Nova Scotia, and Prince Edward Island (Abouchar and Speigel 2004; Nowlan 2005). The provinces that do not have specific source protection planning requirements, such as Alberta and Saskatchewan, have other more general water management planning provisions. Justice O'Connor also made recommendations related to the content of source protection plans. In his view, they should address the management and protection of groundwater sources, identify all significant water withdrawals (including from municipal intakes), and be based on water budgets that compare the water flow into an area with the cumulative annual flow out of that area. As with other groundwater protection measures, source protection plans will require extensive data collection and interpretation, which is rare in most provinces. Knowing how many wells exist in a particular watershed and how much water they are pumping would help inform future allocation decisions. Some jurisdictions, like British Columbia, make no specific provision in their allocation decisions to maintain the health of groundwater-dependent ecosystems. Some, like Ontario, have a wide number of environmental criteria that regulators use in decision making. Considering the range of disparate approaches currently in use may help shed light on best practices that could be shared across provincial boundaries. The full environmental impacts of groundwater withdrawals are likely not fully accounted for in Canadian permitting decisions, in part due to a lack of commonly accepted methodology and in part due to a lack of knowledge about groundwater-dependent ecosystems. Until groundwater use is more closely monitored, regulators will be unable to determine the full range of environmental impacts of permitting decisions.

## Public Participation in Groundwater Decisions

Water is so central to people's daily lives that its management cannot be reduced to simple technical formulas. As with most environmental regulations, any type of water law will be more effective if it is developed and implemented with a high degree of user participation (Tuinhof et al. 2001). Public understanding, awareness, and participation in water use decisions are essential to ensuring that those decisions are sound. Yet, only seven of the thirteen jurisdictions surveyed make all groundwater records fully accessible to the public (Manitoba, Nova Scotia, Prince Edward Island, Newfoundland and Labrador, Yukon, Northwest Territories, and Nunavut), and even fewer have conflict resolution systems that give all affected people the right to be heard in water disputes (Nowlan 2005).

"Environmental democracy" can be a key to better environmental decisions.[8] When all those affected by the outcome of environmental decisions are involved in decision making, they tend to support the decisions and help implement them, with the result that joint solutions last longer than they otherwise might. The three components of environmental democracy are: access to information, public participation in decision making, and access to justice (i.e., the ability of an individual to seek a remedy for a violation of an environmental right).

Each of these components is important for all environmental regulations, including groundwater decisions. Some examples of how these principles come into play include the following: for access to information about groundwater, a resident should be able to find out both whether their water is safe to drink and whether there is enough available to water the garden, keep the fish in the backyard stream alive, give the household pets and farm animals a drink, wash the bikes, and fill up the swimming pool. While individual well monitoring is important, it alone cannot provide enough information on safety and supply. Only government agencies have the power and the resources to collect this type of data on a watershed basis. The same holds for public participation: if that same resident wants a say in whether groundwater extraction permits should be granted for intensive hog farms or coalbed methane extraction (both potentially very heavy degraders of groundwater), then there should be procedures that guarantee that she/he has a voice in such decisions. With regard to access to justice, if a resident cannot get the desired information, is denied a place at the decision-making table, is unable to compel the regulators to protect the water, and/or has no access to a judicial or administrative remedy, then the effectiveness of the overall regulatory system is reduced.

How do the Canadian provinces and territories live up to these environmental democracy principles in relation to groundwater decision making? As with other aspects of groundwater management, practice across the country varies.

### Access to Information about Groundwater Resources

Access to information about groundwater resources and laws are key to ensuring effective public participation. Yet many commentators have noted that the information about this resource is incomplete (Rivera et al. 2003; O'Connor 2002; Environmental Commissioner of Ontario 2001). Indeed, problems with access to information about a range of water governance issues are noted by other contributors to *Eau Canada* and pertains not only to groundwater but also to surface water (Box 4.1).

The report from the Walkerton Inquiry illustrated Ontario's problems with data on the extent of current draws on aquifers and watersheds. Dr. Ken Howard testified that, "in Ontario unfortunately, we don't manage water, the degree of management extends simply to issuing permits to take water, and to me issuing permits to take water is a little bit like writing cheques on my

Box 4.1

**Regulatory non-compliance: The Sierra Legal Defence Fund's quest for transparency in water quality reporting**

In the mid-1990s in Ontario, a newly elected Conservative government, bent on cutting back costs, slashed programs and staff at the Ontario Ministry of the Environment (MOE). One of the casualties was regular reporting of industrial and municipal non-compliance with air and wastewater pollution discharge limits. The MOE had published this information up until 1995. In the absence of government reporting, Sierra Legal Defence Fund (Sierra Legal) decided to make this information public on its own. For seven years, Sierra Legal has published annual reports on industrial and municipal facilities that do not comply with Ontario laws related to air and wastewater discharges.

Sierra Legal faced considerable obstacles, both procedural and financial, in its attempts to obtain detailed non-compliance information from the provincial MOE. It was necessary to submit repeated freedom of information requests under the Ontario Freedom of Information and Protection of Privacy Act. With each request, the MOE would issue fee estimates for the air and water pollution compliance information. One estimate requested payment of over $19,000. Faced with exorbitant fees, Sierra Legal applied for fee waivers under a provision in the act that allows fees to be waived where information, if disseminated, would benefit public health and safety. After MOE refused to issue a fee waiver, Sierra Legal appealed successfully to the Information and Privacy Commissioner of Ontario (IPC), and MOE was ordered to supply the information, charging only for photocopy costs. Sierra Legal successfully obtained repeated orders from the IPC before MOE final agreed to waive the fees. However, it still took almost a year before the information requested was delivered to Sierra Legal.

The annual non-compliance report was one of the most popular produced by the Sierra Legal Defence Fund and was picked up by media throughout Ontario. Making the information available increased the pressure on companies, particularly chronic violators, to clean up. Facilities that appeared regularly on the list became the subject of local concern and the target of local citizen groups and media.

In 1999, the MOE began producing summary environmental compliance reports on its website but the summaries failed to report important information like the number of pollution discharges that violated legal limits and the amount by which the regulatory limits were exceeded. The limit information published by the MOE made it impossible to assess overall compliance, compare facilities or analyze trends. At the time, the Ontario Commissioner of the Environment remarked that the Ministry's compliance reports "have deteriorated to the point where they are of little value to the interested public."

Over the years, Sierra Legal repeatedly requested that MOE return to providing full information regarding violations of both air and wastewater pollution laws through its website. Finally, in September 2005, MOE, after consulting with Sierra Legal regarding content, made more detailed non-compliance reports available on its Internet site. The information posted in September 2005, while dated (reporting on violations that occurred in 2003), included for each facility in violation: a list of the pollutants released above regulatory limits, the number of violations with respect to each pollutant, the month the violations occurred, the applicable discharge limits, the instrument type, actual measurements (or range of measurements) of the pollutant, the cause of the violation (if known), and MOE reaction. In effect, the ministry finally provided all the same information that Sierra Legal had reported over the previous seven years, the period of time through which MOE had failed to provide detailed air and wastewater reports.

While the Ontario government has (reluctantly) begun to report non-compliance information, the battle continues in other provinces. British Columbia regularly reported non-compliance information until 1999. Sierra Legal is currently seeking information to prepare a non-compliance report for British Columbia but has been stalled through the freedom of information process, including receiving a $24,000 bill for the information.

---

AUTHOR: Elaine MacDonald of the Sierra Legal Defence Fund, Toronto.

bank account when I don't know how much is going out to pay ... other bills" (O'Connor 2002).

Most jurisdictions map aquifers and wellhead protection areas and make the maps publicly available. Most provinces have freedom of information laws that apply to groundwater records, and some provinces, such as BC, also maintain a specific database or registry devoted to water records (BC Ministry of Land, Water and Air Protection n.d.). However, no province or territory in Canada has a central provincewide, user-friendly database, accessible to the general public, that consolidates all data on water extractions, water quality, and quantity indicators. At the very least, a complete database should include well-drilling records, notices of applications for permits, copies of licences or permits, and actual use records (Prairie Provinces Water Board 1995). Provinces, in contrast to many US states, score well on the first three criteria but not on the fourth.

### Access to Information: Reporting Actual versus Permitted Use

If the amount of water withdrawn from a watershed is to be fully understood, then not only should all licensed users be required to report their actual use data but unlicensed withdrawals should also be recorded, especially those related to agriculture and domestic use. One way to collect this comprehensive data would be to have all groundwater uses subject to reporting requirements, but not necessarily to licensing, in designated sensitive areas (CELA 2003). Staff resources would be required to process and manage this data. Actual use figures for groundwater permits are not widely collected or disclosed. At the end of 2004, five jurisdictions – Yukon, Manitoba, Prince Edward Island, Newfoundland and Labrador, and Saskatchewan – required reporting on actual use (Yukon Waters Regulation; Manitoba Water Rights Regulation; PEI Water Policy Requirement; Newfoundland Water Resources Act; Saskatchewan Policy Requirement).[9] In practice, however, only two provinces – Manitoba and Alberta – provided actual use data for a comparative report (and Alberta's data are incomplete) (Nowlan 2005; Rutherford 2005).

Some provinces require actual use reporting as a licensing condition. In Nova Scotia, the terms and conditions of Water Withdrawal Approvals require records of actual groundwater use to be maintained, and these must be submitted to the department upon request. In Manitoba, licences contain a condition requiring annual water use to be reported no later than 1 February of the following year, as well as a condition requiring meters to be installed to monitor discharge. Ontario now requires mandatory reporting of water use under its amended Water Taking and Transfer Regulation. Annual reporting of water takings by all permit holders will occur in three phases, starting with municipal drinking water systems, major industrial dischargers, and moratorium-type uses that remove water from the watershed. All of this must be reported on by 2006; all other uses must be reported on by 2008 at the latest. The new regulation requires every person to whom a permit has been issued to collect and record data on the volume of water taken daily. The data collected must be measured by a flow meter or calculated using a method acceptable to the ministry. On or before 31 March every year, permit holders must submit to the ministry the data collected and recorded in the previous year (Ontario Water Resources Act 2004).

## Public Participation in Decision Making

Participation procedures for licensing decisions apply to both surface and groundwater, except in British Columbia, where licensing does not apply to groundwater withdrawals. The federal government and many provinces (e.g., British Columbia, Alberta, New Brunswick, and Prince Edward Island) require environmental assessment of projects with significant groundwater impacts, and these procedures invariably allow public participation. Even if a formal environmental assessment process is not triggered, many provinces and territories require applicants for a licence to notify the public and to conduct public consultations.

Table 4.2 summarizes the various requirements and rights pertaining to groundwater licensing in the thirteen provincial and territorial jurisdictions.

### Access to Justice

Citizens whose information or participation rights have been breached should be able to review that decision before a court or another independent body (such as an environmental appeal board). Tribunals usually have discretion to decide whether an applicant has the legal standing necessary to pursue the appeal. This gives regulators great power to exclude potential appellants. For example, in Ontario, the test for granting leave is one's ability to demonstrate either significant environmental harm or an unreasonable decision by the director of the Ministry of the Environment. In Alberta, once a decision has been made authorizing or refusing a water licence, a person who is "directly affected" can appeal it. A narrow application of the definition of "directly affected" can limit the public right of appeal. In one recent case,

Table 4.2

**Summary of requirements and rights pertaining to groundwater licensing in provincial and territorial jurisdictions**

| Jurisdiction | Reporting requirements by location, source, and purpose of extraction | Notification requirements for permit applicants | Participation opportunities in permit decision-making process | Appeal rights in permit decisions | Public database of permit information |
|---|---|---|---|---|---|
| BC | n/a | n/a | No, except when environmental assessment procedures apply. | n/a | n/a. Well records submitted to government are public, but submission has been voluntary to date. Registry exists for surface water licences. |
| AB | Yes, Water (Ministerial) Regulation, ss. 63, 64. | Yes, 50(1)(d) of the Water Act, Water (Ministerial Regulation), s. 13(1) . | Yes, s. 109(1) of Water Act. | Yes, ss. 114-117 of Water Act. | No, but Part 4 of the Water (Ministerial) Regulation (ss. 15-17), "Access to Information," makes provision for full disclosure. |
| SK | Yes, ss. 31, 34, 35 of The Ground Water Regulations. | Yes, the Saskatchewan Watershed Authority Act requires the corporation (s. 61) to require the proponent to advertise the project by posting it. | Yes | Limited. Only a drainage approval may be appealed to the Water Appeal Board; appeal of approval or licence can be made to the courts. | No |

| | | | | | |
|---|---|---|---|---|---|
| MB | Yes, s. 8 of the Water Rights Regulation requires every licence holder to keep records of water use on a form approved by the minister. | Discretionary; s. 6(3) of the Water Rights Act requires a notice of an application be published if the minister directs, but this provision is rarely used. | No, except where environmental assessment procedures apply to the project. | Yes, s. 24 of the Water Rights Act allows any person affected by an order or decision to appeal. | No |
| ON | Yes, annual water use reports must be submitted each year to the Ministry of the Environment (phased in by 2008) (s. 9, Water Taking and Transfer Regulation).<br><br>Well contractor to complete a record for every new well constructed and submit it to the owner and the ministry (s. 11(5), Wells Regulation). | Mandatory notice to municipalities and conservation authorities; also can require applicants to consult with others who may have an interest and report back to the ministry (s. 7, Water Taking and Transfer Regulation). | Yes, Environmental Bill of Rights requires notice of application for certain types of permits to be posted. | Yes | Not currently available. |
| QC | Yes, s. 20 of the Groundwater Catchment Regulation. | No | No, but any person adversely affected by pumping may invoke s. 1 of the Groundwater Catchment Regulation and complain to the Ministry of the Environment. | Yes | Yes |

►

◀ TABLE 4.2

| Jurisdiction | Reporting requirements by location, source, and purpose of extraction | Notification requirements for permit applicants | Participation opportunities in permit decision-making process | Appeal rights in permit decisions | Public database of permit information |
|---|---|---|---|---|---|
| NB | Yes, s. 33, 34 of Water Well Regulation. | n/a | No participation in general groundwater approvals; environmental assessment requirements have participation. | Yes, s. 39 of the Clean Water Act establishes an appeal "for a person whose application has been refused or licence has been suspended or cancelled." | Yes, ss. 36 and 37 of the Clean Water Act. |
| NS | Yes, water use records must be kept as a condition of approval. | n/a | Yes, under the Approvals Procedure Regulations the minister or administrator may require public consultation. | Yes, Environment Act, ss. 137-40, part XIV. | Yes, on environmental registry, s 10, Environment Act. |

| | | | | | |
|---|---|---|---|---|---|
| PE | Yes, s. 4 of Water Well Regulations require a well contractor to complete a construction report and submit it to the owner and the department. | No, except when environmental assessment requirements apply. | No, except when environmental assessment requirements apply. | No | No |
| NL | Yes, s. 31(2) (3) of the Water Resources Act; annual water use reports must be submitted each year as a licence condition. | Yes, s. 14 of the Water Resources Act. | Yes, s. 14 (3) of the Water Resources Act. | Yes, s. 86(4) of the Water Resources Act. | Yes, ss. 13, 86 of the Water Resources Act. |
| YT | Yes, s. 14(1) of the Water Regulation. | Yes, s. 21(1) of the Waters Act. | Yes, s. 21(1) of the Waters Act | Yes | Yes, s. 23(1) of the Waters Act. |
| NT | Yes, s. 15 (1) of the Northwest Territories Waters Act. | Yes, s. 23 of the Northwest Territories Waters Act. | Discretionary, s. 21 of the Northwest Territories Waters Act. | Yes, s. 28 of the Northwest Territories Waters Act. | Yes, s. 14 of the Northwest Territories Waters Act. |
| NU | Discretionary | Yes, s. 55 of the Nunavut Waters and Nunavut Surface Rights Tribunal Act. | Discretionary, s. 52 of the Nunavut Waters and Nunavut Surface Rights Tribunal Act. | Yes, s. 81 of the Nunavut Waters and Nunavut Surface Rights Tribunal Act. | Yes, s. 78 of the Nunavut Waters and Nunavut Surface Rights Tribunal Act. |

the Southern Alberta Environmental Group was denied standing to appeal an amendment to a surface water licence on the grounds that its "policy concerns" were too generalized (Bradley 2004). Some provinces provide for appeals directly to courts. Judicial review of these administrative agency decisions is also usually possible.

When the first two elements of environmental democracy – access to information and public participation in decision making – are fulfilled, citizens will not need to go to the courts or administrative agencies to enforce environmental standards. An example illustrates this well. Artemesia Waters Ltd. (a large water bottling company) was blocked in its attempt to build a water storage and bottling plant on agricultural land in Artemesia Township (now the municipality of Grey Highlands), south of Owen Sound, Ontario. Residents opposed the plant due to their concerns that the proposed groundwater extraction would drain almost 500,000 litres of water per day from a wetland and a fish stream. The Environment Ministry did not allay their concerns, and the residents did not have any information to persuade them that the proposed groundwater extraction would not be harmful to the environment. They were forced to appeal to the Environmental Appeal Board and to fight challenges launched by the company at the Ontario Municipal Board. The citizens group also had to apply for a judicial review of the latter's decision. Ultimately, the municipality (which was heavily influenced by the advocacy of its citizens) changed the zoning of the disputed area (Abouchar 2004). This decision emphasizes the importance of both citizen participation and of local government regulatory controls over groundwater.

Pursuing reviews requires perseverance, dedication, time, and (often) substantial funds – which citizens may or may not have. By providing sufficient access to quality information about groundwater, allowing citizens to participate in the decision-making process, and then providing them with remedies to pursue if they are still concerned about environmental impacts, we can ensure the health of our groundwater and groundwater-dependent ecosystems.

## Bringing Groundwater Back into View

When groundwater is a priority, government funding decisions will reflect it. Ontario's recent investments are an example. The province has invested significant sums to support the mapping of groundwater conditions and wellhead protection areas, and it has supported the development of water budgeting tools. It assessed water availability in all 144 tertiary watersheds, comparing estimated water use by permit holders to estimated natural stream flows – information that was compiled on maps referenced in the new Water Taking and Transfer Regulation, which was used to justify specific requirements and restrictions in high use watersheds. The Ontario government is also funding source protection plans and watershed-based water budgets across the province. Many other good provincial examples of the sustainable management of groundwater exist, such as the precautionary approach to management as the PEI moratorium on new irrigation licences, the collection of actual use data in Manitoba, and the development and implementation of comprehensive water strategies in Quebec and elsewhere.

Promising examples of local government regulations to protect groundwater are also increasing. The local role in groundwater management is critical due to the interconnections between land use decisions and groundwater quantity and quality (Winfield 2002). For instance, land development may be restricted by the availability of groundwater, as in British Columbia, where at least one subdivision has been scaled down to meet these concerns (e.g., 200 houses taking the place of 350) (Nowlan 2005). Reducing municipal demands on groundwater sources could be achieved through greater water efficiency, improved conservation measures, and increased attention to two of the major factors that influence water consumption: the cost of water and residential density (Smart Growth on the Ground 2004).

The links between droughts, sprawl development, and groundwater depletion are also receiving more attention. Paved development can impair the landscape's ability to recharge aquifers and surface waters as paved-over land sends billions of gallons of water into streams and rivers as polluted runoff rather than allowing it to seep into the soil to replenish groundwater (Smart Growth America NRDC 2002). One way of limiting paving is to restrict the amount of impervious surface that can be built, thus allowing for greater groundwater recharge – a provision available under British Columbia's Local Government Act.[10]

One example of local government action on groundwater comes from Quebec's $H_2O$ Chelsea, a collaborative project involving three partners: the Municipality of Chelsea, Quebec; the University of Ottawa's Institute of Environment; and Action Chelsea for the Respect of the Environment (ACRE), a local NGO. The Municipality of Chelsea is a groundwater-dependent community consisting of low-density housing built directly on the Precambrian Shield bedrock of the Gatineau Hills in Quebec. In 2005, Chelsea's municipal council adopted a policy that requires developers to conduct pump tests to demonstrate that there is a sufficient quantity of water to support the number of homes planned for the development. $H_2O$ Chelsea's work was instrumental in the development of this policy as its yearly water questionnaire demonstrated that there were areas of Chelsea that had experienced higher than background levels of water shortages. This project illustrates how cost-effective citizen actions such as collating information through surveys and training residents on how to monitor their wells can influence local government initiatives to protect groundwater. Though the Municipality of Chelsea had no legal obligation to manage groundwater, it was in its best interest to adopt prudent groundwater policies because, if the community experienced water quantity shortages, then the municipality would have to provide an alternative source of water. And this would be extremely costly given that the community is located on the Canadian Shield.

## Conclusion

Groundwater remains a relatively invisible topic in Canada. As signs of stress from increased withdrawals and climate change in groundwater-dependent ecosystems surface, and as conflicts increase, we need to pay more attention to this "buried treasure." Unlimited or minimally regulated pumping is a recipe for disaster in the more arid parts of the country. If experience in the United States is any guide to what may happen in Canada, then conflicts over groundwater will increase (Glennon 2002). Scientists in Canada have been warning about the potential for increased conflict for the past decade, and they continue their cautions (Canadian Geoscience Council 1993; Senate of Canada 2005; Environment Canada and National Water Research Institute 2004). Hotspot areas in Canada currently include the rapidly urbanizing Okanagan, also British Columbia's chief productive agricultural area, where water resources are heavily allocated (Stephens et al. 2005); southern Ontario (Kreutzwiser et al. 2004); and southern Alberta (Senate of Canada 2005). Conflicts are also likely to arise over transboundary aquifers, as is demonstrated in the examples of the Great Lakes and the protracted negotiations over amendments to the Annex to the Great Lakes Charter and nitrate contamination in the Abbotsford-Sumas aquifer flowing south from Canada to the United States (Environment Canada 2004).

Since the early 1900s, the demand for water has increased and society's priorities for allocating water have changed, but water laws have not kept pace. Allocation and protection rules based on historical priorities – to encourage settlement and economic development as well as to maximize resource extraction – should be changed to reflect current thinking on sustainability, which incorporates environmental, social, and economic priorities. Historic laws favoured economic development. It is time to right the balance and to adopt laws that not only fully address the social dimension of water management through the use of environmental democracy principles but also fully address environmental protection through the use of source protection, watershed-based planning, and environmental flow requirements.

Integrated management will involve all levels of government, and there are some signs that groundwater is making its way up the policy agenda for the different regulators in the country, who produced the Canadian Framework for Collaboration on Groundwater in 2003 (Rivera et al. 2003). Progress in the vital areas of access to information and public access to decision making is more slow-going. The ability to obtain information and to participate in and challenge groundwater decisions varies markedly across Canadian jurisdictions. Though the Geological Survey of Canada is working with provinces to complete the mapping of select priority regional aquifers that have been identified by both levels of government, a publicly accessible, thorough database on aquifers is far from complete. Collecting and publicizing data on all water sources, including groundwater, is a valuable public service that should not be sacrificed to budget cuts (Senate of Canada 2005). Involving local residents in water decisions can help to improve management by building on local knowledge and bringing potentially adversarial groups to the table to work out solutions. And if citizens are frustrated by their regulators' decisions (or their complete failure to act), they should have ways of pursuing their complaints and, thus, of increasing the level of environmental protection. As water scarcity increases and groundwater stresses multiply, better use of environmental democracy procedures would help to bring this practically invisible resource back into sight and into mind.

## NOTES

1 Testimony before the Standing Committee on Environment responding to a question about the level of understanding of hydrogeology in Canada. See http://www.parl.gc.ca/infocomdoc/38/1/ENVI/Meetings/Evidence/ENVIEV03-E.HTM#Int-985504.

2 The Greenbelt Protection Act, SO 2004, c. 9, clarifies the Oak Ridges Moraine Conservation Act, SO 2001, and c. 31, to ensure that all planning approvals on the moraine are consistent with environmental protection.
3 *Acton v. Blundell* (1843) 152 Eng. Rep. 1223 at 1233.
4 *Houston and Texas Central Railway Co. v. East*, 81 SW 279 (Tex. 1904).
5 *Sipriano v. Great Spring Waters of America, Inc.*, 1 SW 3d 75, 78-83 (Tex. 1999).
6 The *BC Water Act* contains licensing provisions that could apply to all or designated areas of British Columbia, but they will apply only if and when cabinet makes such a designation. In 2004, a groundwater protection regulation was introduced, but it does not mandate licensing, instead focusing on standards for well construction and groundwater quality protection,
7 *The Report of the Walkerton Inquiry*, part 2, chap. 4, "The Protection of Drinking Water Sources," http://www.attorneygeneral.jus.gov.on.ca/english/about/pubs/walkerton, especially s 4.3.5.3 Groundwater Management.
8 There is a huge range of other public participation mechanisms and procedures, but this chapter focuses on the most legally oriented participation procedures, which are commonly considered together under the term "environmental democracy" as enshrined in different international treaties.
9 Yukon, Waters Regulation, s. 14(1); Manitoba, Water Rights Regulation, s. 8; Prince Edward Island, policy requirement; Newfoundland, Water Resources Act, s. 31(2); Saskatchewan, policy requirement.
10 RSBC 1996, s. 907(2), c. 323, as amended.

## REFERENCES

Abouchar, J. 2004. Water Taking: Municipal Role in Ontario. *Municipal World* (March): 23.

Abouchar, J., and B. Spiegel. 2004. Navigating Multi-Jurisdictional Compliance Obligations for Safe Drinking Water. Canadian Institute's 4th Annual Conference, 26 April, Toronto.

Bradley, Cheryl. 2004. *Government Board's Refusal to Hear Environmental Group's Appeal of Water Licence Decision Still Unexplained after Five Months: This Is Not Responsible Government*. Lethbridge: Southern Alberta Environmental Group (November). http://issues.albertawilderness.ca/WAT/archive.htm.

British Columbia. Auditor General of BC. 1998/1999. *The Absence of Groundwater Management Has Resulted in Increasing Problems: Protecting Drinking-Water Sources*. http://www.bcauditor.com/PUBS/1998-99/report-5/water.pdf.

—. Ministry of Land, Water and Air Protection. N.d. *Groundwater in British Columbia*. Victoria: BC Ministry of Land, Water and Air Protection. http://wlapwww.gov.bc.ca/soerpt/7groundwater/wellsglance.html.

—. Ministry of Land, Water and Air Protection. N.d. *Guide to Using the BC Aquifer Classification Maps for the Protection and Management of Ground Water*. http://wlapwww.gov.bc.ca/wat/aquifers/reports/aquifer_maps.pdf.

Brown, Karen. 2004. Assistant Deputy Minister, Environment Canada Testimony before the Standing Committee on Environment Responding to a Question about the Level of Understanding of Hydrogeology in Canada. http://www.parl.gc.ca/infocomdoc/38/1/ENVI/Meetings/Evidence/ENVIEV03-E.HTM#Int-985504.

Brown, L. 2005. *Outgrowing the Earth.* Washington, DC: Earth Policy Institute.

Canadian Geoscience Council. 1993. *Groundwater Issues and Research in Canada.* Waterloo: Canadian Geoscience Council.

CELA [Canadian Environmental Law Association]. 2003. *Submission on the Proposed Amendments to the Water Taking and Transfer Regulation.* Publication No. 444. Toronto: CELA.

—. 2004. *Protecting Ontario's Water Now and Forever: A Statement of Expectations for Watershed-Based Source Protection from Ontario Non-Governmental Organizations.* Publication No. 486. Toronto: CELA.

CIELAP [Canadian Institute for Environmental Law and Policy]. 1996-2001. *Ontario's Environment and the Common Sense Revolution: First, Second, Third, Fourth, Fifth and Sixth Year Reports.* Toronto: CIELAP.

Commissioner of the Environment and Sustainable Development. 2001. *Annual Report.* Ottawa: CESD. http://www.oag-bvg.gc.ca/domino/reports.nsf/html/c101sec3e.html/$file/c101sec3e.pdf.

Environment Canada. 2001. *Tracking Key Environmental Indicators, Freshwater Use.* Ottawa: Environment Canada. http://www.ec.gc.ca/TKEI/air_water/watr_use_e.cfm.

—. 2004. *Nitrate Levels in the Abbotsford Aquifer.* Ottawa: Environment Canada. http://www.ecoinfo.ec.gc.ca/env_ind/region/nitrate/nitrate_e.cfm.

Environment Canada and National Water Research Institute. 2004. *Threats to Water Availability in Canada.* National Water Research Institute, Burlington, Ontario: NWRI Scientific Assessment Report, Series No. 3 and ACSD Science Assessment, Series No. 1.

Gartner Lee. 2002. *Best Practices for Assessing Water Taking Proposals: Toronto.* Prepared for Ontario Ministry of the Environment, unpublished.

Glennon. R. 2002. *Water Follies: Groundwater Pumping and the Fate of America's Fresh Waters.* Washington, DC: Island Press.

Hutson, Susan S., Nancy L. Barber, Joan F. Kenny, Kristin S. Linsey, Deborah S. Lumia, and Molly A. Maupin. 2004. *Estimated Use of Water in the United States in 2000: Circular 1268.* Washington, DC: United States Geological Survey.

International Joint Commission. 2000. *Protection of the Waters of the Great Lakes.* Ottawa: International Joint Commission.

IUCN [World Conservation Union]. 2003. *The Essentials of Environmental Flow.* Gland, Switzerland: IUCN.

Kreutzwiser, R.D., R.C. de Loë, J. Durley, and C. Priddle. 2004. Water Allocation and the Permit to Take Water Program in Ontario: Challenges and Opportunities. *Canadian Water Resources Journal* 29 (2): 135-46.

LaForest, G.V., and Associates. 1973. *Water Law in Canada: The Atlantic Provinces*. Ottawa: Deptartment of Regional Economic Expansion.

Maude, Stephen. 2004. *Amendments to the Water Taking and Transfer. Ontario Regulation 285/99*. Toronto: Land Use Policy Branch. Regulation. http://www.ene.gov.on.ca/envregistry/023109er.htm.

Nowlan, Linda. 2005. Buried Treasure: Groundwater Permitting and Pricing in Canada. Toronto: Walter and Duncan Gordon Foundation.

O'Connor, Hon. Justice Dennis. 2002. *The Report of the Walkerton Inquiry*. Part 2, Chapter 4: The Protection of Drinking Water Sources (especially s 4.3.5.3 Groundwater Management). http://www.attorneygeneral.jus.gov.on.ca/english/about/pubs/walkerton.

Ontario. Environmental Commissioner of Ontario. 2001. *Ontario's Permit to Take Water Program and the Protection of Ontario's Water Resources.* Brief to the Walkerton Inquiry. Toronto: Environmental Commissioner of Ontario. http://www.eco.on.ca/english/publicat/walker01.pdf.

—. Ministry of the Attorney General. 2002. *The Report of the Walkerton Inquiry*. Part 2, Chapter 4: The Protection of Drinking Water Sources (especially s. 4.3.5.3 Groundwater Management). http://www.attorneygeneral.jus.gov.on.ca/english/about/pubs/walkerton.

Ontario Water Resources Act. 2004. Ontario Regulation 387/04. In *The Ontario Gazette* (December). http://www.e-laws.gov.on.ca/DBLaws/Source/Regs/English/2004/R04387_e.htm.

Pearse, P.H., and F. Quinn. 1996. Recent Developments in Federal Water Policy: One Step Forward, Two Steps Back. *Canadian Water Resources* 21 (4): 329-40.

Percy, David. 1988. *The Framework of Water Rights Legislation in Canada.* Calgary: Canadian Institute on Resource Law.

Prairie Provinces Water Board Committee on Groundwater. 1995. *A Review of Groundwater Legislation in the Prairie Provinces*. Regina: Prairie Provinces Water Board.

Rivera, A. 2005. *How Well Do We Understand Groundwater in Canada? A Science Case Study*. Quebec City: Geological Survey of Canada.

Rivera, A., A. Crowe, A. Kohut, D. Rudolph, C. Baker, D. Pupek, N. Shaheen, M. Lewis, and K. Parks. 2003. *Canadian Framework for Collaboration on Groundwater.* Ottawa: Government of Canada. http://ess.nrcan.gc.ca/2002_2006/gwp/pdf/cadre_canadien_collaboration_eau_souterraine_e.pdf.

Rutherford, S. 2005. *Groundwater Use in Canada.* Vancouver: West Coast Environmental Law.

Senate of Canada. 2005. *Water in the West, under Pressure*. Fourth Interim Report of the Standing Senate Committee on Energy, the Environment and Natural Resources. Ottawa: Senate of Canada.

Smart Growth on the Ground. 2004. *Technical Bulletin No. 2: Water Consumption in Maple Ridge*. Vancouver: SGOG. http://www.sgog.bc.ca/uplo/mr2wateruse.pdf.

Smith, Ross. 2002. Canada's Freshwater Resources: Toward a National Strategy for Freshwater Management. In *Water and the Future of Life on Earth.* Discussion Paper. Vancouver: SFU Department of Geography. http://www.sfu.ca/cstudies/science/water/pdf/Appendix_3.pdf.

Statistics Canada. 2003. Fresh Water Resources in Canada. In *Human Activity and the Environment: Annual Statistics.* Ottawa: Statistics Canada.

Stephens, Kim, Erik Karlsen, Ted van der Gulik, and Ron Smith. 2005. *Water Balance Management in the Okanagan: Now What Do We Do?* Kelowna, British Columbia, Water Sustainability Committee of the BC Water and Waste Association, at the Canadian Water Resources Association's Water: Our Limiting Resource conference, 23-25 February, Kelowna, BC.

Tuinhof, Albert, Charles Dumars, Stephen Foster, Karin Kemper, Hector Garduno, and Marcella Nanni. 2001. *Groundwater Resource Management: An Introduction to Its Scope and Practice.* Washington, DC: World Bank, GW Mate Briefing Note 1. http://lnweb18.worldbank.org.

US Geological Survey. 2005. *Groundwater in the Great Lakes Basin: The Case of Southeastern Wisconsin.* Washington, DC: USGS. http://wi.water.usgs.gov/glpf/index.htm.

Warren, Fiona, and Donald S. Lemmen. 2004. Water Resources. In *Climate Change Impacts and Adaptation: A Canadian Perspective.* Ottawa: Climate Change Impacts and Adaptation Program.

West Coast Environmental Law. 2004. *Please Hold. Someone Will Be with You: A Report on Diminished Monitoring and Enforcement Capacity in the Ministry of Water, Land and Air Protection.* Vancouver: WCEL.

—. 2005. *Cutting Up the Safety Net: Environmental Deregulation in BC.* Vancouver: WCEL.

Winfield, M. 2002. *Local Water Management in Canada.* Drayton Valley, AB: Pembina Institute. http://www.pembina.org/pdf/publications/local_h2o_mgmt_0202.pdf.

# 5 Challenging the Status Quo: The Evolution of Water Governance in Canada

*Rob de Loë and Reid Kreutzwiser*

Given our dependence on water for our lives and livelihoods, it is remarkable how blasé we can be about this essential resource. Inattention to water in Canada can be traced to a widely shared belief that we are a water wealthy country (Chapter 2, this volume). It is commonly observed that approximately 7 percent of the world's renewable freshwater supply flows through Canada's rivers and streams and that the Great Lakes (which Canada shares with the United States) hold 18 percent of the world's fresh surface water (Environment Canada 2005b). When these figures are compared to Canada's share of the world's population – 0.5 percent in 2004 (Population Reference Bureau 2005) – it is understandable that a perception of water superabundance exists.

Unfortunately, as farmers and urban residents coping with water shortages already know, we are not blessed with unlimited water supplies. In reality, Canada's water resources are not distributed evenly in space or over time. Simply put, much of Canada's water wealth is located far from the people, farms, and industries that demand water. For instance, approximately 60 percent of Canada's fresh water drains to the north, whereas 85 percent of the population lives within 300 kilometres of the Canada-US border. Additionally, precipitation varies from more than 5,000 millimetres per year along parts of the Pacific coast to less than 100 millimetres per year in parts of northern Canada and the Prairies (Environment Canada 2005a).

The traditional response in Canada to the mismatch between the location and timing of human needs and available water has been to increase supplies. We cannot make more water – but we can store, divert, and move water so that it is available where and when it is demanded. Canadians have been masters of this approach to managing water resources for well over a century. Across the country, we have built massive water storage and diversion projects to supply water for power generation, agricultural irrigation, and flood control (Chapter 7). Unfortunately, the economic prosperity that has resulted from these costly "supply management" projects has also produced considerable environmental and social disruption. As a result, in Canada today, there is increasing emphasis on water management approaches that adapt human needs to available resources rather than on those that modify the hydrological cycle to fit our needs. These kinds of approaches include demand management, which uses water conservation and pricing (Chapter 13); soft path strategies, which seek to change the ways in which we use water (Chapter 14); and watershed management, which recognizes the many ways in which water interacts with land and other resources.

## Water Governance

Shifts in the approach to water management in Canada have created new kinds of challenges. For example, where water management was once largely the responsibility of technical experts, usually in senior government agencies, the field has now been opened up to a diverse range of "stakeholders" inside and outside of government and, increasingly, at the local or community level. Reconciling the often conflicting needs, values, and interests of various stakeholders without further compromising environmental quality is a key challenge facing Canadian society – and, indeed, all of the world's citizens. Significantly, this is a challenge for *governance* rather than a challenge for science and technology, a point argued forcefully in the United Nations' recent call to arms, *Water for People, Water for Life*:

> At the beginning of the twenty-first century, the Earth, with its diverse and abundant life forms, including over six billion humans, is facing a serious water crisis. All the signs suggest that it is getting worse and will continue to do so, unless corrective action is taken. *This crisis is one of water governance, essentially caused by the ways in which we mismanage water.* (World Water Assessment Programme 2003, 4; emphasis added)

"Governance" is a term that came to prominence in the 1980s in connection with debates surrounding international trade (De Angelis 2003). In that context, "good" governance is seen to be a function of fiscal discipline, trade liberalization, and so on. In the environment field, the term often has a broader, less ideological meaning. For example, the World Resources Institute (2003,

viii) defines environmental governance simply as "the processes and institutions we use to make decisions about the environment" and suggests that governance is "good" when decision making is transparent, when all stakeholders participate, when full accountability exists, and when environmental decisions are integrated with economic and development decisions. These principles are also strongly evident in *Water for People, Water for Life*, which argues that, even though a widely accepted definition of governance does not exist, it is clear that good water governance depends on participation by all stakeholders, transparency, equity, accountability, coherence, responsiveness, integration of resource sectors and issues, and attention to ethical concerns (World Water Assessment Programme 2003). In this chapter, we adopt the perspective of the World Resources Institute and the World Water Assessment Programme and, thus, consider water governance (broadly) to consist of the decision-making processes through which water is managed.

We believe that growing interest in water governance is important in at least two respects. First, it emphasizes that the most important factors accounting for contemporary water problems relate to "people issues" rather than to a lack of scientific knowledge or adequate technology (World Water Assessment Programme 2003). For example, millions of people around the world have no choice but to drink poor quality water and are exposed to waterborne diseases. Inadequate public water supplies have vastly more to do with inadequate financial resources, weak standards or a failure to implement standards, lack of skilled staff, rivalries among agencies, and insufficient political will than with lack of effective treatment technologies (de Loë and Kreutzwiser 2005). Second, the concept of governance draws attention to the fact that it is not only *governments* – or the "state" – that can (and should) make decisions about the use and management of water. It recognizes that "non-state actors" – citizens, non-governmental organizations (NGOs), and businesses – are essential to effective water management. More important, it emphasizes that decisions about the use and management of water resources *should* involve non-state actors because the state – through its various agencies – simply cannot do everything and because some water management functions are best handled by other actors (Rogers and Hall 2003).

Water governance truly is a global challenge. Countries around the world are struggling to provide adequate water and sanitation, to share scarce water resources fairly, to balance environmental quality and economic development, and to cope with risks and uncertainty. Ensuring transparent decision making, effective and fair stakeholder participation, and integration of water management with other sectors are universal challenges that defy tidy distinctions between "developed" and "developing" countries. Thus, it should not be surprising that, in Canada – even with our supposed water wealth and our economic prosperity – we face these challenges too. This chapter briefly explores selected trends in water governance in Canada. We discuss examples of emerging approaches to water governance in order to highlight some of the specific ways in which major governance challenges are being addressed. Ontario's experiences with source protection planning provide a case study that enables us to highlight key challenges and opportunities.

## Water Governance in Canada

Several other contributors to *Eau Canada* explore in detail the legal foundations for water management in Canada (see, in particular, Chapters 6 and 12). Of relevance to this chapter is the fact that the state has been responsible for water in Canada since Confederation. Under Canada's Constitution Act, responsibility for water was divided between the federal and provincial governments, with the provinces having the most direct responsibilities. Local governments and organizations, being subject entirely to provincial jurisdiction, hold only those powers and responsibilities that their respective provinces have assigned to them.

The relative balance of responsibility and authority among the various levels of government in Canada has shifted back and forth during the past century – even though the constitutional basis for these responsibilities has remained largely unchanged (Paquet 1999). An example of this involves the case of the Prairies, where the federal government was directly responsible for water allocation between 1894 and 1930, at which point, the provinces of Alberta, Saskatchewan, and Manitoba took over this function (de Loë 1997).

From the perspective of water governance in Canada, an important, and relatively recent, development is the growing importance of *local* and *non-state* actors. Local actors include municipalities, local water management agencies (such as irrigation districts in Alberta and conservation authorities in Ontario), First Nations communities, and local NGOs. Local actors have been involved in various ways in water management in Canada for many decades. Ontario's conservation authorities have operated dams and reservoirs to control river flows since 1946 (Mitchell and Shrubsole 1992). Irrigation districts in Alberta, Saskatchewan, and British Columbia have owned and operated works for storing and distributing water to farmers since early in the twentieth century. Furthermore, municipalities and public utilities across Canada remain the primary suppliers of drinking water despite the recent involvement of private sector firms (Chapter 9). However, contemporary water management in Canada is distinctive relative to that of earlier years because aspects of water governance are being distributed beyond the traditional players. While key functions continue to be undertaken by the state, intriguing examples are emerging of "non-state" actors' undertaking functions that used to be the preserve of state agencies. Box 5.1 describes the involvement of several non-state actors. As is illustrated by these examples, partnerships and collaboration are becoming much more common, and the responsibilities of local state (or quasi-state) agencies are increasing.

What accounts for increasingly distributed water governance in Canada? Some analysts have suggested that, in concert with globalization, the central or national state has been "retreating" from its traditional functions and that a variety of non-state and local or subnational actors are filling the void created. For example, Strange (1996, 4) argues that:

> Where states were once the masters of markets, now it is the markets which, on many crucial issues, are the masters over the governments of states. And the declining authority of states is reflected in a growing diffusion of authority to other institutions and associations, and to local and regional bodies, and in a growing asymmetry between the larger states with structural power and weaker ones without it.

BOX 5.1

**SELECTED EXAMPLES OF LOCAL INVOLVEMENT IN WATER GOVERNANCE**

*Annapolis River Guardians, Nova Scotia:*

In Atlantic Canada, numerous community-based water quality monitoring programs have been developed to gather data about river water quality (Clean Annapolis River Project [CARP] 2005). For instance, the Annapolis River Guardians have been collecting water quality data in the Annapolis watershed since 1992, with the aim of establishing a long-term record for the river and providing an early warning system for environmental problems (CARP 2005). Community-based monitoring initiatives such as these are part of a national trend in Canada (Doyle and Lynch 2005). In some instances, these initiatives represent independent, grassroots efforts to fill gaps in government monitoring programs. In other cases, they are collaborative efforts with government agencies.

*Watershed Organizations, Quebec:*
Quebec's 2002 Water Policy identifies integrated management of water quality and quantity concerns as a priority (Rousseau et al. 2005). However, implementation of integrated water management is occurring primarily through the work of local watershed organizations (WOs). Thirty-three WOs are operating in various watersheds in Quebec, with boards of directors representing citizens and citizen groups, municipalities, agricultural and industrial water users, and government agencies (participating as non-voting members). WOs are expected to develop a water master plan, authorize and enforce "watershed contracts" (agreements made among stakeholders regarding watershed plans), communicate with the public and other stakeholders, and, in general, help to achieve integrated water management (Rousseau et al. 2005). These WOs have, in essence, been delegated responsibilities traditionally undertaken by state actors in other jurisdictions.

*Water Response Teams, Ontario:*
Despite being generally perceived as a water-rich province, Ontario has a lengthy history of droughts (Durley, de Loë, and Kreutzwiser 2003). In response to water shortages across the province in 1999, the provincial government developed a drought response strategy known as Ontario Low Water Response (OLWR) (Ontario Ministry of Natural Resources et al. 2003). Under this policy, locally organized, watershed-based water response teams (WRTs) are expected to devise and implement ways of sharing water and otherwise dealing with water shortages. The province plays an advisory role up to the point where severe drought conditions are declared and mandatory water restrictions have to be applied (e.g., through provincial water taking permits) (Ontario Ministry of Natural Resources et al. 2003; Durley, de Loë, and Kreutzwiser 2003). The OLWR is the first explicit and systematic drought contingency program in Ontario designed to deal with water shortages. It is instructive that it leaves considerable responsibility with local, primarily non-state, WRTs.

*Water Sharing in the Southern Tributaries of the Oldman River, Alberta:*
The majority of Canada's irrigated land is found in one of its driest regions – southern Alberta. The province's water allocation system was designed to resolve water shortages by allocating water to users on a first-in-time, first-in-right basis. During the summer of 2001, it appeared that the priority system would have to be enforced, and 336 licencees reliant on the limited surface water resources of the southern tributaries of the Oldman River Basin would be cut off entirely during the irrigation season (Stratton, de Loë, and Smithers 2004). Instead, water users worked with provincial agencies to devise a water sharing agreement, whereby each accepted less water than legally permitted in order to ensure that all had some. This agreement is an important example of local stakeholders effectively circumventing an established state water allocation system – with the full support of state agencies.

In a related argument, Paquet (1999) suggests that hierarchical or top-down governance has been less and less successful in solving complex problems since the 1960s. As a result, he claims, it has lost legitimacy and is being replaced by a "distributed" governance model. The role of the state in distributed governance is transformed as decision making shifts to the lowest, most local level. Reliance on the private sector and civic society increases, power is distributed more widely, and "the task of the higher order of government is to assist and support the individual and the local body in carrying out their tasks" (Paquet 1999, 76).

Whether or not the phenomena of state retreat and distributed governance are as significant as has been suggested is a matter for debate. Drache (2001, 38), for example, suggests that, in Canada, "the day of the sovereign state is far from over." In the context of contemporary Canadian water governance, it certainly is too much to say that the state has *retreated* and is now limited to assisting individuals and local bodies. Many more actors (including non-state actors) are involved than was the case in previous decades. Collaboration and partnership are commonplace, and the roles and approaches of the federal and provincial governments have changed considerably. The federal government, for example, began experimenting with collaborative partnerships and alternative modes of service delivery in the mid-1990s (Fyfe and Fitzpatrick 2002). Nevertheless, even with these important changes, the state remains a central player in Canadian water management.

While contemporary Canadian water governance is different in very important ways relative to that of past decades, these differences are subtle and (at this time) are evolutionary rather than revolutionary. Furthermore, experiences vary widely among Canada's provinces and territories – each of which is conducting its own, largely independent, experiment in water governance. Thus, the evolution of water governance in Canada is best understood in terms of specific events and circumstances. Three examples of factors accounting for changes in water governance in Canada are discussed below.

First, in line with an increase in the complexity of many other policy fields (Fyfe and Fitzpatrick 2002), the complexity of water management in Canada has increased dramatically since the end of the Second World War. In the 1950s, water management in Canada involved relatively discrete functions such as allocating water among competing users; providing water supplies for consumers in municipalities, for industries, and for agriculture; generating power; ensuring that waterways were navigable; and protecting communities from floods. In contrast, water management today encompasses a vast range of additional activities, including protection and restoration of aquatic habitat, provision of recreational experiences, and conservation of Heritage Rivers. At the same time, where once it was considered appropriate to manage water separately from land and other sectors, it is now widely understood that water must be treated as an essential element of the complex ecosystems in which it exists (Pollution Probe 2004). "Integration" in water management is now an almost universal byword, and increased complexity has, in many respects, overwhelmed the abilities of state water managers to "go it alone" in water management (Cohen et al. 2004).

Second, in the two decades leading up to 2000, even as the complexity of water management increased, attention to water as a policy field distinct from the federal and many provincial governments declined. The federal government's waning focus on water throughout this period has been documented (and bemoaned) by observers such as Bruce and Mitchell (1995) and Booth and Quinn (1995). Managing water in its own right gradually ceased to be a priority for the federal government during the 1990s. Ironically, this occurred in part because of a new commitment to sustainable development and the ecosystem approach, based on recognition of the increased complexity of environmental management.

> In the early 1990s, departmental management [in Environment Canada] favored the prospect of faster and more politically-attractive results from other initiatives, especially the Green Plan ... which became the guiding light and source of funds for environmental activities undertaken by several federal agencies. The thrust was directed toward the "big picture," to think and act in terms of ecosystems rather than their components, like water. (Booth and Quinn 1995, 71)

As a result of this shift in thinking, water eventually became subsumed in the notion of "ecosystems." Thus, as funds for water specifically declined, and as federal priorities shifted to other topics, the federal government lost considerable water expertise. In an appalling turn of events, fragmentation of water-related units in the federal government became so severe in the 1990s that a "Where's Water?" team had to be assembled to determine whether or not the government's water-related duties were still being performed (Booth and Quinn 1995).

Similar inattention to water occurred across Canada at the provincial level during this time period. This was demonstrated in Ontario, where successive provincial governments repeatedly promised a policy for groundwater management throughout the late 1990s – but did not deliver one (de Loë and Kreutzwiser 2005). At the same time, in an ideologically driven effort to reduce the size and role of government, the budgets of key water management agencies in Ontario were severely reduced in the mid-1990s by the Progressive Conservative government of Premier Mike Harris (Kreutzwiser 1998). Additionally, the number of staff with water-related duties in the provincial Ministry of the Environment was reduced from 168 in 1995 to 71 in 1996 (Winfield and Jenish 1999). In fairness, worthwhile initiatives pertinent to water were under way across the country during this period. Alberta stands out in this regard, having devoted considerable energy to developing a new water allocation law throughout the 1990s. Nevertheless, *sustained* commitment to water from political leaders seemed to be declining in most provinces with each passing year during the 1990s.

A third key consideration accounting for contemporary water governance in Canada is, of course, the contamination incidents that occurred in Walkerton, Ontario (2000), and North Battleford, Saskatchewan (2001). Complacency about water in Canada was shattered by these incidents, and the response across the country was dramatic. In Ontario, several new regulations and laws were passed (de Loë and Kreutzwiser 2005), and, as is discussed in the next section, additional major changes are under way. Similar legislative and organizational changes designed to enhance the safety of drinking water supplies were implemented in Saskatchewan in response to the contamination of the water supply of North Battleford (Saskatchewan 2003) and in several other provinces – for instance, Nova Scotia (Nova Scotia Environment and Labour 2002).

Significantly, it is not just drinking water safety that has received renewed attention from provincial governments after the Walkerton and North Battleford incidents. Water allocation laws have been overhauled in Saskatchewan and Ontario, with other provinces currently in the process of revising their systems. Furthermore, most provinces have developed or are developing new water policies and strategies. Examples include Alberta's Water for Life Strategy (Alberta Environment 2003); Nova Scotia's Drinking Water Strategy (Nova Scotia Environment and Labour 2002); Quebec's Water Policy (Ministère de l'Environnement du Québec 2002); Manitoba's Water Strategy (Manitoba Water Stewardship 2005); Saskatchewan's Long-Term Safe

Drinking Water Strategy (Saskatchewan 2003); and Ontario's proposed Clean Water Act (Ontario 2005).

In terms of governance, many of these initiatives are quite "traditional," in the sense that the state (through provincial governments) is firmly in control even though non-state actors are playing significant roles. Indeed, the sheer volume of recent activity at the provincial level in Canada seems to suggest a reinvigorated state rather than a state in retreat. Nevertheless, the flood of new laws, strategies, policies, and programs originating from provincial governments should not be taken as a sign that contemporary water management has returned to its former state-centric glory days. Many of the initiatives that are under way are distinctive – and signal a new approach to water governance – in that they reflect recognition that governments simply can no longer undertake necessary activities on their own. Or, more charitably, they reflect an understanding that some water management functions should best be undertaken by other stakeholders (local governments, NGOs, citizens themselves, or multiple stakeholders working together collaboratively). The case of source water protection in Ontario illustrates well this shift in water governance.

## Source Protection Planning in Ontario

Walkerton is a small southern Ontario town whose municipal water supply was contaminated in 2000 by runoff from a nearby livestock farm. The farm was being operated according to accepted farm practices, but the town's well was poorly located and maintained, and the operators of the system were not properly trained. When they detected contamination in the water supply, they initially concealed its extent from public health officials and the provincial government – an action made easier by earlier cutbacks to provincial monitoring and reporting systems. Thus, people in and around Walkerton continued to drink water that had been contaminated by *Escherichia coli* O157:H7 and *Campylobacter jejuni* long after the system should have been shut down. Seven people died because of the contamination, and at least 2,300 became seriously ill (O'Connor 2002a).

The significance of the Walkerton contamination incident as a catalyst for changes in water governance in Ontario and other parts of Canada cannot be overstated. Others have laid the blame for the tragedy squarely on the neoliberal agenda of the Progressive Conservative government of the day (Prudham 2004). However, water management was being neglected in

Ontario long before that government came to power. For example, in the mid-1980s, provincial governments in Ontario ceased mapping major aquifers, publishing maps of groundwater resources, and preparing maps of groundwater susceptibility to contamination (Neufeld and Mulamoottil 1991). Thus, while cuts to the budget and staff of the Ministry of the Environment and a weakened regulatory framework unquestionably contributed to the tragedy, the commission of inquiry into the incident, under the direction of Justice Dennis O'Connor, clearly signalled that numerous other factors were at play (O'Connor 2002a). Significantly, a lack of basic source water protection measures was identified by the commission as a serious gap in Ontario's water management framework (O'Connor 2002b).

In response to the Walkerton tragedy and the recommendations of Justice O'Connor's commission of inquiry, the provincial government instituted a host of changes relating to areas such as the regulation of large and small drinking water systems, the financing of water supply systems, and the construction of water wells. Additionally, studies were completed across the province to strengthen the basic database regarding groundwater resources and threats to groundwater (de Loë and Kreutzwiser 2005). These studies built on an earlier series of groundwater investigations partially funded by the provincial government under its 1997 Provincial Water Protection Fund (de Loë, Kreutzwiser, and Neufeld 2005). As suggested earlier, most of the measures instituted in response to the Walkerton tragedy – while extremely important – do not constitute a new approach to water governance; rather, the measure that has the most potential to reorient water governance in Ontario is the source protection planning system that is currently being developed in that province.

Watershed-based, locally organized source protection planning was identified by Justice O'Connor (2002b) as the first line of defence in a "multi-barrier" approach to drinking water safety. The Progressive Conservative government in power at the time of the Walkerton incident and its Liberal successor have been committed to developing a system to ensure that surface and groundwater sources providing water to drinking water systems are protected from contamination. The Advisory Committee on Watershed-Based Source Protection Planning released its report in 2003. A subsequent White Paper on watershed-based source protection and a draft drinking water source protection act were released for discussion in early 2004 (Ontario 2004a, 2004b), followed by two detailed technical reports in late 2004 (Ontario 2004c, Ontario 2004d). Also, in 2004, the provincial government provided

$12.5 million to enable conservation authorities across the province to begin technical studies relating to source protection planning (Ontario 2004e). Late in 2005, a reworked source protection act, the Clean Water Act, received first reading in the Legislative Assembly (Ontario 2005).

The final form of the source protection planning system that will be implemented in Ontario is not yet known. However, if it resembles the system outlined by the proposed act and supporting reports, then source protection planning in Ontario has the potential to be a groundbreaking example of a more distributive approach to governance for water management.

- Source protection plans would be developed locally by watershed-based source protection planning committees operating in source protection regions. The boundaries of two or more conservation authorities will normally define the source protection regions in which these committees would operate.
- Source protection planning committees would be composed of a broad range of stakeholders (municipalities, user groups, First Nations, NGOs). They would be supervised by source protection planning boards – comprised of the board of directors of the lead conservation authority in a source protection planning region. Thus, Ontario's watershed-based conservation authorities – already key agencies at the local level – would have important coordinating responsibilities.
- Plan development would occur through a collaborative process involving local stakeholders on the source protection committees. Additionally, responsibility for implementation of source protection measures would lie primarily with local organizations (municipalities, conservation authorities) and citizens (e.g., landowners).
- The province would approve plans and would retain the authority to develop plans in regions where local stakeholders fail to do so. The province could also require changes to a plan but only in relation to specific considerations (for instance, if plans fail to address water risks identified in assessment reports prepared for the source protection region).

The potential even exists for unprecedented (in Ontario) integration of water quality and water quantity concerns, and land use planning and water management. The province's White Paper indicates that water allocation and source protection planning will be coordinated (Ontario 2004a). More strikingly, the committee that considered implementation issues recommended that municipal land use planning decisions *and* provincial instruments such as certificates of approval for wastewater discharges and water taking permits should be required to be consistent with source protection plans (Ontario 2004d). If implemented, this measure could lead to an extraordinary integration of land use planning and water management in Ontario, and it would represent a new kind of relationship between provincial and local decision making.

However, the devil is in the details – and these will not be known until the Clean Water Act becomes law. At this point, what is clear is that implementation challenges will be significant, and the most severe ones will relate to stakeholder involvement, local leadership, reconciliation of conflicts, and other "people" problems. Technical problems – for example, development of water budgets, collection of baseline data, mapping of risks – are not insignificant but will be much easier to resolve. For instance, once the provincial government became committed to improving the groundwater database (initially in 1997, and then with considerable vigour following Walkerton), understanding of groundwater resources and threats to groundwater in Ontario improved rapidly and substantially (de Loë and Kreutzwiser 2005; de Loë, Kreutzwiser, and Neufeld 2005). Similar rapid progress should not be expected in the development and implementation of source protection plans. In rural areas, many farmers and farm organizations are deeply concerned about the impacts of new regulations on their activities (Ontario Federation of Agriculture 2004), while some rural residents see new regulations as an opportunity to constrain the growth of large livestock facilities (Citizens for the Environment and Future in Eastern Ontario 2004). Source protection plans that do not accommodate or reconcile these kinds of conflicts are unlikely to be implemented successfully. Enormous implementation challenges also exist in other contexts. For instance, with their direct responsibilities for land use planning, municipalities will be essential partners in any source protection planning processes. However, significant unresolved issues exist relating to the roles and responsibilities of municipal governments, the availability of implementation tools, ultimate responsibility for enforcement of plans, and the question of "primacy" (i.e., whether municipal official plans must be consistent with source protection plans or simply be guided by them) (Association of Municipalities of Ontario 2005). Similar primacy questions exist regarding the provincial government and its instruments. Such questions include whether source protection plans should have primacy over provincial instruments such as certificates of approval for sewage treatment facilities and water-taking permits. The Advisory Committee on Watershed-Based Source Protection Planning (2003) gave qualified support to this idea, while other stakeholders argue that source protection plans should always have primacy (Canadian Environmental Law Association 2005). While the proposed Clean Water Act may undergo revision in the Legislative Assembly, as introduced, it requires municipal official plans and zoning bylaws, as well as provincial instruments, to conform to approved source protection plans.

Resolving these kinds of issues will not be straightforward. Experiences with the comparatively straightforward technical studies completed under the 1997 Provincial Water Protection Fund Groundwater Management Studies demonstrated the importance of basic capacity considerations such as leadership, financial resources, skilled staff, data, interorganizational collaboration, public support, and appropriate institutional arrangements (de Loë, Kreutzwiser, and Neufeld 2005). These kinds of capacity issues were found to underlie the variable capacity of Ontario municipalities and conservation authorities to manage groundwater resources (de Loë, Di Giantomasso, and Kreutzwiser 2002; Ivey, de Loë, and Kreutzwiser 2002). Difficult questions of incompatible values, conflicts, fear, and uncertainty on the part of some stakeholders, and the relationship between the various legal and planning tools, will further challenge water governance.

Particularly important to successful governance regarding source water protection, we believe, will be the institutional, political, and social aspects of capacity. As noted, it is not certain at this time that municipalities and conservation authorities will be given the legal tools to effectively implement source protection plans. Moreover, they will also need strong local political leadership and aware and committed community members who are prepared to achieve consensus on key elements of source protection plans and to support plan implementation. While the province announced funding of $16.5 million over one year for conservation authority staff and resources and $51 million over five years for technical studies (Ontario Ministry of the Environment 2005), money alone is unlikely to enhance the political and social capacity for source water protection.

## Conclusion

Water governance in Canada has rarely been static for any length of time. During the past century, the role of the federal government has evolved to reflect the maturation of the provinces and territories and shifts in federal priorities. In turn, provincial and territorial governments have grown into their constitutional responsibilities and are clearly the primary water managers in Canada today. From this perspective of continual change, is it really appropriate to suggest (as we have in the title of this chapter) that contemporary water governance is *challenging* the status quo? After all, while there is evidence of a limited retreat of the state in the 1990s, the first years of the twenty-first century have seen a resurgence of the provincial role in water management in the form of new laws, regulations, and policies. Ultimately, we believe that water governance is evolving in new directions in Canada and that, in several important respects, the changes that are occurring do present a challenge to the traditional, state-centric approach.

Barring fundamental constitutional changes, both provincial and federal governments will continue to have key water-related responsibilities in Canada. We see this as appropriate. However, by establishing planning and decision-making processes based on partnerships, collaboration, and the principle of subsidiarity (which asserts that matters should be handled by the lowest competent authority), water governance is being distributed beyond the traditional state actors. This can lead to better governance – as long as all stakeholders have both the *opportunity* and *capacity* to participate; decision making is equitable, responsive, and transparent; and accountability exists.

Distributed governance represents a challenge to the status quo of state-centric governance because success will depend on a true letting go of authority by the state. Locally led processes such as source protection planning in Ontario will fail if participants come to believe that they do not actually have the power to design and implement solutions locally. At the same time, success in these processes will dilute state authority as governance becomes more widely distributed.

How much distribution of responsibility for water governance can and should occur is by no means clear. Local organizations and communities are not all equally capable of participating effectively in water management. For instance, in Ontario's emerging source protection planning system, enormous variation from region-to-region can be expected in the quality and comprehensiveness of plans as well as in the rate of success of implementation (de Loë and Kreutzwiser 2005). Similar variability can be expected in any governance process organized at the local level in Canada. Thus, great care should be taken to ensure that distributed governance does not lead to weaker water management and is not simply used by state agencies as a way to divest themselves of their responsibilities. The challenge will be to find the appropriate balance between state and non-state actors, and among the various scales at which water governance can occur – locally, regionally, provincially, and nationally.

## REFERENCES

Advisory Committee on Watershed-Based Source Protection Planning. 2003. *Protecting Ontario's Drinking Water: Toward a Watershed-Based Source Protection Planning Framework.* Toronto: Ontario Ministry of the Environment.

Alberta Environment. 2003. *Water for Life: Alberta's Strategy for Sustainability.* Edmonton: Alberta Environment. http://www.waterforlife.gov.ab.ca.

Association of Municipalities of Ontario. 2005. *AMO Responds to Source Water Implementation Committee.* News release, 17 February.

Booth, L., and F. Quinn. 1995. Twenty-Five Years of the Canada Water Act. *Canadian Water Resources Journal* 20 (2): 65-90.

Bruce, J., and B. Mitchell. 1995. *Broadening Perspectives on Water Issues.* Ottawa: Royal Society of Canada.

Canadian Environmental Law Association. 2005. Letter to the Ontario Ministry of the Environment Regarding the Final Reports of the Implementation and Technical Committees, 14 February.

Citizens for the Environment and Future in Eastern Ontario. 2004. *The Case against Intensive Hog Operations.* Unpublished brief.

Clean Annapolis River Project. 2005. Water Quality Monitoring Programs. http://www.annapolisriver.ca/monitoringprograms.htm.

Cohen, S., R.C. de Loë, H. Hamlet, R. Herrington, L. Mortsch, and D. Shrubsole. 2004. Integrated and Cumulative Threats to Water Availability. In *Threats to Freshwater Availability in Canada.* Environment Canada, ed. NWRI Scientific Assessment Report Series No. 3 and ACSD Science Assessment Series No. 1, 117-27. Burlington, ON: National Water Research Institute, Environment Canada.

De Angelis, M. 2003. Neoliberal Governance, Reproduction and Accumulation. *The Commoner* 7 (Spring/Summer): 1-28.

de Loë, R.C. 1997. Return of the Feds, Part I: The St. Mary Dam. *Canadian Water Resources Journal* 22 (1): 53-62.

de Loë, R., S. Di Giantomasso, and R.D. Kreutzwiser. 2002. Local Capacity for Groundwater Protection in Ontario. *Environmental Management* 29 (2): 217-33.

de Loë, R., and R.D. Kreutzwiser. 2005. Closing the Groundwater Protection Implementation Gap. *GeoForum* 36 (2): 241-56.

de Loë, R., R.D. Kreutzwiser, and D. Neufeld. 2005. Local Groundwater Source Protection in Ontario and the Provincial Water Protection Fund. *Canadian Water Resources Journal* 30 (2): 129-44.

Doyle, M., and M. Lynch. 2005. *Linking Ecological Monitoring to Decision-Making at Community and Landscape Scales*. Ottawa: Environment Canada.

Drache, D. 2001. The Return of the Public Domain after the Triumph of Markets: Revisiting the Most Basic Fundamentals. In *The Market or the Public Domain: Global Governance and the Asymmetry of Power*, ed. D. Drache, 37-71. London: Routledge.

Durley, J., R.C. de Loë, and R.D. Kreutzwiser. 2003. Drought Contingency Planning and Implementation at the Local Level in the Province of Ontario, Canada. *Canadian Water Resources Journal* 28 (1): 21-52.

Environment Canada. 2005a. Hydrology of Canada. http://www.wsc.ec.gc.ca/hydrology/main_e.cfm?cname=hydro_e.cfm.

—. 2005b. Quickfacts. http://www.ec.gc.ca/water/en/ e_quickfacts.htm.

Fyfe, T., and T. Fitzpatrick. 2002. Distributed Public Governance: Agencies, Authorities and Other Autonomous Bodies in Canada. *OECD Journal on Budgeting* 2 (1): 81-102.

Ivey, J., R. de Loë, and R.D. Kreutzwiser. 2002. Groundwater Management by Watershed Agencies: An Evaluation of the Capacity of Ontario's Conservation Authorities. *Journal of Environmental Management* 64: 311-31.

Kreutzwiser, R.D. 1998. Water Resources Management: The Changing Landscape in Ontario. In *Coping with the World around Us: Changing Approaches to Land Use, Resources and Environment*, ed. R.D. Needham, 135-48. Waterloo: Department of Geography, University of Waterloo.

Manitoba Water Stewardship. 2005. *The Manitoba Water Strategy*. http://www.gov.mb.ca/waterstewardship/waterstrategy/pdf/index.html.

Ministère de l'Environnement du Québec. 2002. *Quebec Water Policy: Highlights*. Québec, QC: Ministère de l'Environnment du Québec.

Mitchell, B., and D.A. Shrubsole. 1992. *Ontario Conservation Authorities: Myth and Reality*. Waterloo, ON: Department of Geography, University of Waterloo.

Neufeld, D., and G. Mulamoottil. 1991. Groundwater Protection in Canada: A Preliminary Inquiry. *Ontario Geography* 37: 15-22.

Nova Scotia Environment and Labour. 2002. *A Drinking Water Strategy for Nova Scotia*. Halifax: Nova Scotia Environment and Labour. http://www.gov.ns.ca/enla/water/docs/NSWaterStrategy.pdf.

O'Connor, D.R. 2002a. *Report of the Walkerton Inquiry*. Part 1: The Events of May 2000 and Related Issues. Toronto: Queen's Printer for Ontario.

—. 2002b. *Report of the Walkerton Inquiry*. Part 2: A Strategy for Safe Drinking Water. Toronto: Ontario Ministry of the Attorney General, Queen's Printer for Ontario.

Ontario. 2004a. *Drinking Water Source Protection Act*. Environmental Bill of Rights Registry Number AA04E0002. Posted 23 June 2004. Toronto: Queen's Printer for Ontario.

—. 2004b. *White Paper on Watershed-Based Source Protection Planning*. Toronto, ON: Integrated Environmental Planning Division, Strategic Policy Branch and Ministry of the Environment.

—. 2004c. *Watershed-Based Source Protection Planning: Science-Based Decision-Making for Protecting Ontario's Drinking Water Resources – A Threats Assessment Framework.* Technical Experts Committee Report to the Minister of the Environment. 4935e. Toronto: Queen's Printer for Ontario.

—. 2004d. *Watershed-Based Source Protection: Implementation Committee Report to the Minister of the Environment.* PIBs 4938e. Toronto: Queen's Printer for Ontario.

—. 2004e. *Investing in Source Protection Planning.* Backgrounder (17 November). Toronto: Queen's Printer for Ontario.

—. 2005. *Clean Water Act.* Environmental Bill of Rights Registry Number AA050001. Posted 5 December. Toronto: Queen's Printer for Ontario.

Ontario Ministry of Natural Resources, Ontario Ministry of the Environment, Ontario Ministry of Agriculture and Food, Ontario Ministry of Municipal Affairs and Housing, Ontario Ministry of Enterprise, Opportunity and Innovation, Association of Municipalities of Ontario, and Conservation Ontario. 2003. *Ontario Low Water Response.* Revised. Toronto: Province of Ontario. http://www.mnr.gov.on.ca/mnr/water/publications/OLWR_2003.pdf.

Ontario Ministry of the Environment. 2005. McGuinty Government Introduces Clean Water Act. News Release. http://www.ene.gov.on.ca/envision/news/2005/120501.htm.

Ontario Federation of Agriculture. 2004. Ontario Federation of Agriculture (OFA) Presentation to The Honourable Dalton McGuinty, Premier of Ontario, 20 April. Unpublished brief.

Paquet, G. 1999. Straws in the Wind: Innovations in Governance in Canada. *Optimum, the Journal of Public Sector Management* 29 (2/3): 71-81.

Pollution Probe. 2004. *The Source Water Protection Primer.* Toronto: Pollution Probe.

Population Reference Bureau. 2005. *Data by Country.* http://www.prb.org.

Prudham, S. 2004. Poisoning the Well: Neoliberalism and the Contamination of Municipal Water in Walkerton, Ontario. *GeoForum* 35: 343-59.

Rogers, P., and A.W. Hall. 2003. *Effective Water Governance.* TEC Background Papers, No. 7. Stockholm: Global Water Partnership.

Rousseau, A.N., V. Luyet, R. Schlaepfer, J.-P. Villeneuve, and A. Bédard. 2005. A Preliminary Assessment of the Implementation of Integrated Watershed Management in Quebec. In *Reflections on Our Future: A New Century of Water Stewardship.* Banff, AB/Cambridge, ON: Canadian Water Resources Association.

Saskatchewan. 2003. *Saskatchewan's Safe Drinking Water Strategy.* Regina, SK: Government of Saskatchewan. http://www.swa.ca/publications/documents/SafeDrinkingWaterStrategy.pdf.

Strange, S. 1996. *The Retreat of the State: The Diffusion of Power in the World Economy.* Cambridge: Cambridge University Press.

Stratton, E., R.C. de Loë, and J. Smithers. 2004. Adaptive Capacity to Climate Change in the Water Sector and the Role of Local Involvement in the Oldman River Basin,

Alberta. In *Knowledge for Better Adaptation.* Proceedings of the CWRA 57th Annual Conference, 16-18 June, Montreal, Quebec. Cambridge, ON: Canadian Water Resources Association.

Winfield, M., and G. Jenish. 1999. *Ontario's Environment and the Common Sense Revolution: A Fourth Year Report.* Toronto: Canadian Institute for Environmental Law and Policy.

World Resources Institute. 2003. *World Resources, 2002-2004.* Washington, DC: World Resources Institute.

World Water Assessment Programme. 2003. *Water for People, Water for Life: The United Nations World Water Development Report.* New York: UNESCO Publishing.

# Is Canada's Water Safe?

## A PHOTO ESSAY

COLIN PRICE/*PROVINCE*

KEVIN FRAYER/CP PHOTO

A sign warns people not to drink the water from a fountain at the hospital in Walkerton during the *E. coli* outbreak in 2000.

In recent years, there have been a number of water quality tragedies in communities across Canada that have raised fears about the safety of drinking water. In 2000, thousands of people became ill and seven people died after a deadly strain of *E. coli* bacteria contaminated the water supply in Walkerton, Ontario. Then, in 2005, residents were evacuated from the Kashechewan reserve in Northern Ontario due to grave health concerns related to poor water quality.

JONATHAN HAYWARD/CP PHOTO

Residents "welcome" the media to the Kashechewan reserve in Northern Ontario, as cases of bottled water pile up on the airport runway.

KIM STALLKNECHT/*PROVINCE*

Kathy O'Lane takes samples of water from an irrigation ditch in Chilliwack, BC. Concerned about the environmental causes of her deteriorating health, O'Lane says that the water contains trichothecene mycotoxin, a poisonous chemical that is produced by fungi and attacks the brain, lungs, and spinal cord.

Canadians have sought more information about the safety of their drinking water and have mobilized against threats. Water experts agree that the accessibility and quality of information on water in Canada is significantly lower than in other countries. Pollution from agricultural, industrial and residential sources is a growing source of concern. Few communities continue to dump raw sewage into the environment (Victoria, BC, is an exception). But many communities have storm sewers that bypass treatment plants during heavy rainfall.

Scientists have documented the effect of known pollutants on aquatic organisms, and are raising concerns about new categories of pollutants from unusual sources – such as pharmaceuticals and personal care products. These new categories of pollutants have only recently been recognized; evidence is growing that they may persist in the environment and cause health effects, such as the disruption to reproductive systems caused by "estrogen mimicking" compounds. In many cases, Canada's drinking water quality guidelines do not cover these new pollutants.

Abnormally high frequencies of deformities in amphibians (such as this frog found in Quebec by researchers from McGill University) are associated with pesticides, which weaken the organism's defenses against deformity-causing parasites. In some intensive agricultural areas, the McGill team has found abnormality rates approaching 100 percent in juvenile frogs.

(FROGS COLLECTED BY MARTIN OUELLET, PHOTO BY DAVID M. GREEN, MCGILL UNIVERSITY)

Pollutants in lake water and sediment on lake bottoms can cause tumours in fish, such as in this longnose sucker found by scientists from the University of Michigan's Great Lakes Fishery Laboratory

(*CATOSTOMUS CATOSTOMUS* SPECIMEN, COURTESY OF DAVID JUDE, UNIVERSITY OF MICHIGAN).

CAMERON CARDOW/SYNDICAM

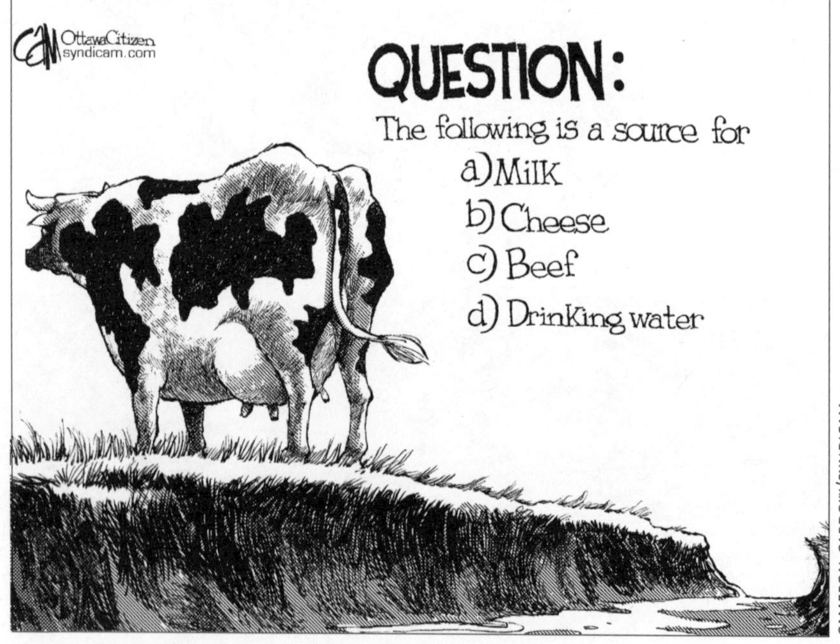


CAMERON CARDOW/SYNDICAM

As these satirical cartoons show, the Canadian media have commented cynically on the state of Canada's water management

INGRID RICE/NORTH VANCOUVER

DUKE: VINDICATED AT LAST

INGRID RICE/NORTH VANCOUVER

STEVE RUSSELL/*TORONTO STAR*

Bottles of water donated by elementary students from Hanover, Ontario, sit in the Walkerton Arena. At the height of the crisis, thousands of cases of bottled water filled the arena.

FRANK GUNN/CP PHOTO

Ontario Premier Mike Harris looks down before beginning his testimony at the Walkerton Water Inquiry in 2001.

Canadian tap water suppliers have run ads seeking to convince consumers that their drinking water is safe. Some consumers are choosing to use bottled water. Ironically, bottled water quality regulations are less comprehensive than those for tap water.

CITY OF TORONTO

Major municipal water suppliers, such as the City of Toronto's water utility, have run ads encouraging consumers to trust tap water.

MARK VAN MANEN/*VANCOUVER SUN*

For the love of water. In New Denver, BC, a nine-day anti-logging blockade of a local watershed by community residents concerned about effects on their water supply ended with the arrest of seven protestors, including a twelve-year-old girl.

COLIN PRICE/*PROVINCE*

Greenpeace members chain themselves to the sewage pumphouse on Iona Island, Vancouver, to draw attention to the city's pollution of the ocean with raw sewage.

Barry Lipton wears a bandanna covering his mouth at a vigil at Queen's Park in Toronto for the victims of the *E. coli* outbreak in Walkerton.

Unlike Europe and the United States, Canada does not have enforceable laws setting drinking water quality standards. Repeated calls have been made for all levels of government to strengthen water quality guidelines by implementing enforceable laws. Campaigners are also lobbying the federal government to amend the Charter of Rights and Freedoms to give Canadians a human right to water.

PART 2

# Whose Water? Jurisdictional Fragmentation and Transboundary Management

# 6
# Whose Water? Canadian Water Management and the Challenges of Jurisdictional Fragmentation

*J. Owen Saunders and Michael M. Wenig*

Wise management of water resources presents challenges for any government, but the challenges may be particularly difficult when they are aggravated by a fragmentation of constitutional responsibilities and/or where the water resources cross political boundaries. This is, of course, precisely the situation that faces Canadian policy makers. Federal and provincial governments both have constitutionally assigned powers touching on water management. Moreover, significant freshwater resources are shared across borders, both interprovincially (including, for these purposes, across provincial-territorial boundaries) and internationally with the United States. As a result, much of Canada's water resource endowment is subject to the demands of provincial, federal, and international management regimes.

In order to explore the implications of this jurisdictional fragmentation, this chapter takes an illustrative – and, necessarily, highly selective – approach. After providing a brief overview of how water is treated as a jurisdictional issue in the Canadian federal system, we turn to three types of jurisdictional fragmentation that Canadian policy makers have had to manage: federal-provincial, interprovincial, and international. In each case, rather than attempting to provide a comprehensive theory for managing fragmentation, we discuss the challenges within the context of specific examples – including issues related to both water quality and water quantity.

## Water and Canadian Federalism

As with other environmental and resource management areas, there is considerable debate about federal/provincial roles in water management. A major source of this debate is the complex and confusing constitutional allocation of water management powers. This section summarizes that constitutional context as well as the debate over whether federal or provincial governments should take the lead role in water management.

It should be noted at the outset that it is somewhat artificial to discuss policy in isolation from constitutional considerations. More typically, government articulation of policy is intertwined with an assertion of the appropriate limits of respective constitutional authority – so that one of the justifications of a particular policy is often (at least implicitly) the underlying division of constitutional authority as currently understood. Put differently, our understanding of what level of government is most appropriately placed to address a particular policy question is coloured to a significant degree by the particular Canadian consensus on what federal and provincial levels of government "should" be doing. In other federal states – for example, in the United States and Australia – there is a much greater, although hardly complete, acceptance of the appropriateness of federal initiatives in water management, particularly where transboundary waters are concerned.

### The Policy Context

One of the primary policy arguments for federal involvement in water management is based on the transjurisdictional character of surface water and groundwater bodies. According to this argument, individual provinces are functionally unable to adequately manage such water bodies because they lack the requisite authority to deal with the water bodies in their entirety. The counter-argument is that provinces can overcome their individual functional limits through interprovincial coordination, but even this approach leaves room for a significant federal role in encouraging and facilitating interprovincial processes, in resolving interprovincial disputes, and in addressing Canada-US water management issues.

A similar justification for federal involvement in water management is based on the national and international character of aquatic and terrestrial species. This extraprovincial character is most obvious for waterfowl and other migratory species that physically cross political borders, whether or not the waters they use do so. But, arguably, even locally occurring species have a national and international dimension in that they generally contribute to global biological diversity, which is considered essential for human existence (Wenig 2004). Indeed, virtually all water could be argued to possess an extraprovincial character because of the regional or global nature of hydrologic cycles. From this perspective, federal involvement in water management is needed to safeguard the national and international character of aquatic species and hydrologic systems. Provinces are simply presumed to lack sufficient interest in those extraprovincial aspects when making water management decisions.

Other policy arguments for federal involvement are based on the ubiquitous nature of water management issues and the fundamental importance of water for human existence – and, hence, on water's moral significance. Both of these factors arguably raise water management to a "national concern" that, in turn, calls for federal involvement. An example of such a federal role would be the establishment of national minimum standards through national processes that are typically less subject to local economic pressures than are provincial standard-setting processes. Of course, not all issues and concerns raised by water management are national and international in character; many are local. Hence, the competing claims for federal and provincial water management roles often depend on the type of interests that are of particular importance to the claimants.

There are also other arguments countering federal involvement in water management, one of which is that federal management would diminish the level of provincial autonomy that is deemed consistent with the political philosophy underlying Canadian federalism. Under this view, because of the ubiquitous nature of water – and of activities that affect water quantity and quality, and of human and environmental demands for water – the federal interests in water management would require federal management of virtually all waters and human activities. (Similar claims are made in the context of managing greenhouse gas emissions, but they are hardly unique to that issue.) Thus, taken to their apparent logical conclusion, the justifications for federal water management fall on their own sword because they would ultimately leave no room for provincial jurisdiction at all. The obvious response to this argument is that, precisely for political reasons, Ottawa does not and need not take such an intrusive water management role in order to promote extraprovincial interests in Canada's water resources.

The ubiquitous nature of water sources and threats also raises a practical, function-based counter-argument to federal involvement. Viewed from this perspective, Ottawa simply lacks the resources and expertise to adequately identify, monitor, and solve the myriad water problems that occur at local levels. In addition, top-down management may preclude the kind of local "buy in" that is often necessary for implementing management decisions. These functional arguments led a former Supreme Court of Canada justice and prominent federalism scholar to speculate that, "if the provinces didn't exist, the federal government would have to invent them because many of the [environmental] problems are best dealt with locally" (LaForest 2004). As with the political arguments, the functional arguments against federal involvement do not require black-and-white or all-or-nothing solutions. For example, in the regulatory arena, Ottawa should be able to set nationally binding parameters or targets that further extraprovincial interests in water resources while still leaving the provinces with considerable day-to-day water management functions. The interesting policy question, then, is not whether the federal or provincial level of government should exercise leadership in water management but, rather, what federal and provincial interests are implicated by particular issues and how those interests can be accommodated or (in the event they compete) resolved.

## The Constitutional Context

There is no dearth of literature on the constitutional aspects of Canadian water management. As is true with much of the Canadian experience with federalism, however, the Constitution hides as much as it reveals with respect to the management of Canadian water resources. This is particularly true given the relative sparseness of constitutional jurisprudence on water issues (in vivid contrast to the experience in the United States).

As with natural resources generally, the roots of provincial jurisdiction over water resources are the easiest to identify, lying in provincial ownership of resources and buttressed by corollary legislative rights in the Constitution Act of 1867.[1] Although federal proprietary rights also exist, these are much less significant in the provinces (and obviously extensive if one includes the territories), and the more difficult questions have arisen with respect to the extent of federal legislative authority. The bases of federal authority include such obvious, specifically assigned powers in the Constitution as navigation and shipping, and fisheries. However, federal jurisdiction over water may also be less obviously implicated through federal authority over agriculture (shared with the provinces), trade and commerce, taxation, and criminal law.[2] In general, in balancing federal and provincial interests in water management, the tendency of Canadian courts has been to read the federal interest narrowly and to restrict it to the particular power being invoked – for example to require that legislation passed pursuant to the fisheries power is indeed related to the management of fisheries rather than to wider goals of water management.[3]

In addition to the specific legislative powers assigned to it in the Constitution, the federal government may also, in some circumstances, rely on its general power to legislate with respect to "peace, order, and good government"

(POGG), which is often cited by those advocating a greater federal presence in water management. POGG has been the subject of extensive judicial and academic commentary over the years, both generally and with respect to water management in particular. While POGG itself comprises different doctrinal approaches, or "branches," the most likely candidate with respect to water management is that of "national concern," and, indeed, in some circumstances there is judicial support for a federal role based on POGG. The limits of how far the courts will extend POGG in water management are, however, anything but clear. For example, in a case striking down provincial water quality measures that had interprovincial implications (*Interprovincial Co-Operatives v. Manitoba*), three members of a four-three majority on the Court were prepared to conclude that pollution of interprovincial rivers falls within federal jurisdiction; however, the judgment refers to the fisheries power and is otherwise unclear as to whether a plenary federal power in this respect actually exists.[4] In a subsequent case (*R. v. Crown Zellerbach Canada Ltd.*), with potentially even more far-reaching implications for water management, the Supreme Court used POGG to support the constitutionality of a federal statute prohibiting marine dumping wholly within the province of British Columbia, even though there was no alleged effect on fisheries, because marine dumping was deemed a matter of national concern.[5] While the latter decision is important in its implications for potential federal environmental measures in other areas, it by no means gives carte blanche to the federal government to take sweeping measures with respect to water management. The majority stressed that, in order for the federal government to legislate on the basis of national concern, the subject matter had to have both "a singleness, distinctiveness and indivisibility that clearly distinguishes it from matters of provincial concern and a scale of impact on provincial jurisdiction that is reconcilable with the fundamental distribution of legislative power under the Constitution."[6] This is hardly suggestive of the sort of federal power that would support a wide-ranging role in water management in the provinces.

Unfortunately, the Supreme Court has recently given indications that it may be willing to rely more heavily on the criminal law authority of the federal government in validating federal legislation directed at environmental protection, including protection of water resources (*R. v. Hydro-Québec*).[7] While some may view favourably any additions to the federal tool chest to address environmental threats, the very nature of the criminal law power – prohibitions combined with penalties – makes it highly inflexible when compared to other bases of jurisdiction and, thus, often inappropriate as a regulatory instrument, particularly in the context of interjurisdictional cooperation.

As one would expect, the potential for federal jurisdiction with respect to water management is increased when the waters in question are interjurisdictional in nature or where the effects of water management are extrajurisdictional. This is true whether the borders are provincial or international, although the scope of federal authority varies somewhat in each case. In both cases, there is some limited potential for federal jurisdiction relating to works (including canals) extending beyond the limits of a province (Department of Justice 1867).[8] As suggested above, however, a more fruitful source of potential federal authority over interjurisdictional waters is POGG.

One other constitutional power, which has special relevance for federal management of waters that cross or form part of the Canada-US international boundary, is a provision that enables the federal government to implement treaties concluded on behalf of Canada by the British Empire.[9] There are a number of Empire treaties affecting the management of water basins shared with the United States, the most important of which – the Boundary Waters Treaty of 1909 – is discussed in more detail later in this chapter. As a result, for some very significant water bodies (most notably, the Great Lakes), the federal government has important powers that it may exercise, regardless of whether they would otherwise intrude on provincial jurisdiction. Unfortunately, this also leads to a somewhat asymmetrical scope of federal jurisdiction, given that there are other important treaties dealing with shared watercourses (e.g., the Columbia River Treaty) that do not qualify as Empire treaties. Because the federal government cannot rely on the Empire treaty section to assure their implementation, these post-Empire treaties may require a greater degree of provincial cooperation if they are to be implemented in Canada.

### Initiatives to Clarify the Federal Role in Water Management

Provincial governments have felt little need to clarify their role with respect to the management of natural resources, generally, and water, in particular.[10] As resource owners, they have naturally seen themselves as having the primary role in resource management. As is the case with respect to other natural resources, however, the federal government has been much more conflicted with regard to asserting a role in water management. Even where it does act under clearly established constitutional authority, the federal government typically feels the need to stress the lead role of the provinces as masters of their own resource endowments.

The federal government has, on a number of occasions and through different vehicles, attempted to clarify its appropriate role in water management, with varying degrees of success. One early attempt was the Water Powers Reference of 1928 (Commission of Conservation of Canada 1929), in which the federal government posed a series of questions to the Supreme Court as a means of clarifying the respective federal and provincial roles in water management. The questions were, however, very limited in scope (e.g., there is no reference to the possible implications of POGG for a federal role in water management), and, in any event, the response of the Court was so general that it did very little, if anything, to provide useful guidance for the resolution of specific disputes. The division of federal and provincial responsibilities with respect to water resources was also a matter taken up during the 1945 Dominion-Provincial Conference on Reconstruction, where the federal government took on leadership responsibilities for national water research but also asserted an interest in assuring the rational development of interprovincial water basins – an assertion that has been repeated on subsequent occasions but with little action to back it up.

A more significant initiative in the direction of asserting a federal role in water management was the Canada Water Act of 1970. While much of the act is devoted to providing a structure for cooperative work with the provinces on water-related issues, there are suggestions of a more assertive federal role in the event that cooperation fails. However, even these provisions largely reflect the traditional deference to provincial proprietary rights and, probably, to a degree beyond what is constitutionally necessary. For example, while the act permits (but does not direct) the federal government to conduct research and formulate management plans with respect to interjurisdictional waters of "significant national interest," no authority is provided for federal action to actually implement such plans directly. More ambitious is that part of the act dealing with water quality management, where unilateral federal action is permitted with respect to those waters where water quality management has become a matter of "urgent national concern" – albeit this first requires the failure of "all reasonable efforts" to resolve the problem with the respective province(s). In fact, however, no such unilateral action – nor even the antecedent step of designating a water quality management area – has ever taken place, and there is little prospect of this happening in the future, given the likely reaction of provincial governments.

A more recent initiative, which also provided the potential for a fuller articulation of the federal role in water management, was the Inquiry on Federal Water Policy of 1984-85 (Environment Canada 1985). The inquiry was given a broad mandate to "identify and substantiate the nature of emerging water issues, including the interjurisdictional dimensions thereof" (Environment Canada 1985, 189). In its final report, the inquiry recognized the diversity of federal interests in water and explicitly referred to the need to balance these with provincial interests. Accordingly, in its recommendations, the inquiry steered away from any concept of a federal master plan for water management and concentrated pragmatically on a series of particular issues where federal interests were clearest and where a federal role could most usefully be exercised. One of these areas of interest involved a potential federal role in dispute resolution where provinces could not agree on the management of interprovincially shared water resources.

The inquiry led to yet another initiative aimed at defining the federal role in water management – the federal government's 1987 issuance of the Federal Water Policy. The Policy stated that its overall objective was "to encourage the use of freshwater in an efficient and equitable manner consistent with the social, economic and environmental needs of present and future generations" (Environment Canada 1987, 5). To this end, it proposed five strategies (water pricing, science leadership, integrated planning, legislation, and public awareness) and twenty-four specific policy statements on issues of particular interest to the federal government. The statements tend to be worded very generally and to stress a cooperative approach with the provinces. Nevertheless, taken together, the undertakings promised in the Policy represent an ambitious agenda in many areas, and it was generally welcomed by the environmental community when it was issued. Even today, it represents a good starting point for defining a useful federal role in water management. Unfortunately, the undertakings in the Policy have largely been ignored by subsequent governments, and it remains little more than a statement of good intentions that have gone unfulfilled. For example, on the key recommendation in the 1985 inquiry final report – that the federal government exercise a role in interjurisdictional dispute resolution – the Federal Water Policy proposed "to develop appropriate procedures" to allow disputes to proceed to mediation or arbitration and "to negotiate with the provinces the development of [an appropriate dispute resolution] mechanism for the ultimate resolution of interjurisdictional disputes"(Environment Canada 1987, 33). Nothing has come of this. Another ambitious recommendation, to jointly draft with the provinces and territories "guidelines and criteria for assessing interbasin transfers within Canada"(Environment Canada 1987, 24), has met the same fate. With few exceptions, the Policy has had little practical impact on federal actions.

## Federal-Provincial Interaction: Water Allocation and Competing Uses

The above section describes a constitutional and political context for water management predicated on a balance of federal and provincial interests. In balancing these interests, the federal government has typically stressed deference to the interests of the province as resource owner, and it has preferred to work toward cooperative solutions (even if these are long in coming and not entirely satisfactory) over insistence on its constitutional authority. A current example of how federal and provincial levels of governments interact is provided by the issue of water allocation. With increased stress on water resources, particularly as the result of climate change, the question of how water will be allocated among different uses is likely to prove increasingly difficult, especially if federal and provincial priorities diverge. These pressures are already being felt in western Canada and are unlikely to decrease.

The management of human withdrawals of water from surface and subsurface water bodies has become increasingly difficult, even without the federalism challenges, owing to increased Canadian demand for water to sustain economic growth and the domestic needs of burgeoning regional populations. Meeting this growing demand is a special challenge in light of recent severe droughts in both western and eastern regions of the country as well as a heightened human concern for ensuring the health of watershed and aquatic ecosystems (which have their own water needs).

Because the Constitution gives the provinces both proprietary rights to water and considerable additional legislative water management authority, provinces have long been considered to be constitutionally entitled to allocate their waters for various private uses by granting water rights to private parties. However, provinces' water allocation function overlaps with, and may impinge upon, several of Ottawa's paramount constitutional authorities. In particular, federal roles in conserving fisheries and fish habitat, and in conserving biological diversity more generally, may be impinged upon by provincial allocations that diminish natural instream flows. This problem is exemplified in the Alberta portion of the South Saskatchewan River Basin (SSRB), where median flows in most of the SSRB sub-basins have been almost fully or, in several cases, more than fully allocated (Alberta Environment 2003a). Alberta granted the bulk of these allocations through water licences that are decades old, have no expiration dates, and were likely issued with little or no thought to their effect on watershed ecosystem health. Notwithstanding the antiquated management philosophies underlying these licences, to date, Alberta has generally refused to consider taking back those historic licences (or, at least, the portions of those allocations that are being used; it has taken back some of the unused portions).

To its credit, Alberta has generally embraced watershed management philosophies. In the specific context of managing allocations, the province has also conducted cutting-edge scientific studies to determine the levels of river flows necessary to meet the "instream flow needs" of aquatic ecosystems (Clipperton et al. 2003). Yet, as long as it insists on honouring its historic water licences, Alberta has arguably written off instream flow needs as meaningful targets for water management in most of the SSRB.[11] Given increasing human demands for water withdrawals and predicted declines in SSRB flows, Alberta's hands-off approach toward historic licences bodes poorly for the survival of fish and other aquatic species in the SSRB.

What, if anything, has Ottawa done about Alberta's water management strategy of essentially writing off instream flow needs in the SSRB? The answer is: almost nothing. To their credit, federal bureaucrats have contributed expertise and resources for Alberta's instream flow studies and watershed planning exercises. But the federal government has taken a largely hands-off approach toward the holders of the historic provincial water licences – particularly the irrigation districts – that are using the bulk of the flows in the SSRB. And Ottawa has not insisted that Alberta use instream flow needs as targets for managing water uses and allocations in general.

This allocation problem is hardly specific to Alberta or even to just the three Prairie provinces. The federal Department of Fisheries and Oceans' written policy arguably provides weak, if any, guidance on regulating water withdrawals, despite the broad prohibition in s. 35(1) of the Fisheries Act, providing that "[n]o person shall carry on any work or undertaking that results in the harmful alteration, disruption or destruction of fish habitat." It has, however, pursued an unwritten general policy that all water uses that originated before s. 35 was adopted in 1977 are grandfathered under the act, as long as each annual water withdrawal is roughly consistent with those pre-1977 uses. However, this policy appears to ignore the point that each annual withdrawal is arguably a new offence.[12] What the federal government *could* do is to reconsider its position on the inapplicability of s. 35 to historic water uses; consider the potential of applying Canada Water Act tools to address provincial water allocation strategies that fail to respect aquatic ecosystem health; and adopt new legislative tools to provide more leadership so that fisheries and biological diversity more generally are not sacrificed. However, there is little evidence that any of these actions will be seriously entertained.

## Interprovincial Cooperation on Water Management

While the challenges of interjurisdictional water management in Canada are usually cast in a federal-provincial context, there are important interprovincial dimensions to interjurisdictional issues. Indeed, the traditional deference by federal authorities to provincial interests in natural resources management generally means that interprovincial cooperation is more important than might be true in other federal states, where national governments (or constitutionally entrenched supreme courts) have been less timid about exercising a role in resolving interjurisdictional disputes than is the case in Canada.

In Canada, provinces have demonstrated a reluctance to resolve interjurisdictional disputes through litigation; this is at least as true for interprovincial disputes as it is for federal-provincial ones. The inclination of provinces to resolve disputes through negotiation is reflected in the practice with respect to interprovincial rivers. A number of agreements relating to the management of such rivers have been concluded. While the federal government is also a party to these agreements, it has tended to provide a largely facilitative role both in their conclusion and in their implementation (e.g., assisting with funding and technical expertise). The two most important sets of agreements are those relating to the eastward-flowing watercourses of the Prairie provinces (with Alberta, Saskatchewan, and Manitoba, in addition to the federal government, as partners) and those relating to the Mackenzie Basin (involving the federal government, Yukon, British Columbia, Alberta, Saskatchewan, and the Northwest Territories). In both cases, there is a blend of obligations and undertakings agreed to by all parties (spelled out in general agreements) and specific obligations owed bilaterally (which are set out in bilateral agreements). Both agreements also create boards (the Prairie Provinces Water Board and the Mackenzie River Basin Board, respectively) to supervise the implementation of the agreements. However, especially in light of the fact that the Prairie provinces' water agreement was generally considered to be the template for moving forward on the Mackenzie negotiations, there are important and discouraging differences between the two regimes.

The Prairie Provinces Water Board was created in 1948 as a body to advise on allocations of interprovincial waters. Owing to increasingly large requests for allocations in the 1960s, the parties agreed on a formula for sharing their waters, which was set out in a Master Agreement on Apportionment in 1969. Significantly, for a Canadian interjurisdictional agreement, the Master Agreement on Apportionment provides for binding dispute resolution by the Federal Court if the parties encounter irresolvable differences as to their respective rights and obligations (which has never, in fact, occurred). While the arrangement has generally worked well in managing the agreed apportionment, and has been expanded over time to include water quality issues, it is not clear how well it will perform in conditions of stress, when parties may choose to reconsider the somewhat ambiguous legal status of the agreements and the nature of their obligations under them.

Given the general comfort level that was achieved with the Prairie Provinces Water Board, one might have expected that, in the negotiation of an interjurisdictional agreement on the use of the Mackenzie Basin, the twin concepts of a specific formula for sharing of water resources and a procedure for binding dispute resolution would receive ready acceptance. The reality, though, has been disappointing. The negotiation of the Mackenzie agreement took a quarter century, and, in the end, agreement was possible only because of the highly general wording of the undertakings in the master agreement – which are largely "motherhood" in nature and leave the difficult issues to bilateral negotiations, which continue to drag on. Although there are provisions for dispute resolution, they have no legal teeth and depend upon the willingness of the affected jurisdictions to accept any recommendations. Although the Mackenzie Master Agreement came into effect in 1997, only one of the bilateral agreements (which are to provide the substantive obligations) has been signed (between Yukon and the Northwest Territories), with none of the others on the immediate horizon. This is twenty years after the Inquiry on Federal Water Policy noted the concerns expressed by downstream jurisdictions as to the future uses of the Mackenzie and recommended a more proactive role by the federal government to ensure that interjurisdictional disputes would not be left to drag on indefinitely – and nearly two decades after the federal government, in its Federal Water Policy, undertook to develop procedures to resolve impasses in negotiations. The experience with the Mackenzie Basin suggests that, in the absence of any serious indication on the part of the federal government that it will intervene in interjurisdictional disputes, upstream jurisdictions will either delay negotiations, or only agree to the most modest undertakings, so as not to constrain their own future uses of interprovincial waters.

## Managing International Transboundary Issues

Canada and the United States are inevitable partners in the management of significant water resources. The transboundary relationship with the United States is governed largely by a series of international agreements, with the Boundary Waters Treaty of 1909 standing as the centrepiece for bilateral cooperation. The discussion below examines that treaty in the context of recent concerns with respect to water quantity issues and, in particular, with respect to potential interbasin water transfers. We turn, finally, to an example of a transboundary water quality concern (on the Columbia River) that has been dealt with outside the structure of the Boundary Waters Treaty.

### Water Quantity: Water Exports and the Boundary Waters Treaty

The Boundary Waters Treaty (BWT) achieves two important goals: it establishes a core of legal principles to govern the management of internationally shared waters between Canada and the United States, and it establishes an institutional framework within which to supervise the implementation of these principles. The treaty is not without its problems, but, within the context of the times in which it was agreed to, it represents an important achievement in international cooperation on shared resources.

One of the primary limitations of the treaty is the asymmetrical treatment of boundary waters (i.e., waters that form part of the international boundary) and waters of rivers crossing the boundary or waters that are tributary to boundary waters. The principle that obtains for boundary waters is one of "equal and similar rights" on the part of each nation; for tributary or transboundary waters, each nation retains, in general, a sovereign right to use or divert the waters as it sees fit. Moreover, the treaty is silent on the status of groundwater. Another potential drawback of the treaty is the hierarchy of water uses it establishes, which very much reflects the priorities of 1909 but which would certainly appear dated if the treaty were being negotiated today.[13] Similarly, the heavy orientation of the treaty to water quantity issues would likely be cast much differently in a modern document.[14]

Apart from the deficiencies in water management principles, it is also possible to question whether the institutional framework for implementation of the treaty principles is adequate for modern needs. The key institution established in the treaty is the International Joint Commission (IJC), a binational body appointed by, but operating at arm's length from, the two national governments. In practice, the IJC has exercised two primary functions: (1) to act in a quasi-judicial capacity as the approving body for obstructions or diversions of boundary waters that affect their flow (and, in more limiting circumstances, for works in transboundary or tributary waters), and (2) to act as an investigating and reporting body on particular matters as requested by the two governments.[15] The commission does not, however, have an independent power of investigation analogous, for example, to that of the auditor general in Canada.

Despite the criticisms that might be levelled at the BWT, the possibility that it will be "modernized" through an amendment or renegotiation is highly remote, and neither government has shown any inclination to reopen it. There are good reasons for this. First, even if the parties were so inclined, there are technical obstacles that would militate against this course of action.[16] Second, in practice, the BWT has been applied in a much more flexible and robust manner than might be suspected from a mere reading of the treaty language. An example of how the BWT has been employed to deal with emerging concerns is provided by a public policy issue that has achieved some measure of public attention in Canada – the possibility of water exports, particularly from the Great Lakes Basin.

The potential of drawing on Canadian water resources as a means of supplementing shortages elsewhere is not new. Beginning at least in the 1960s, a number of megaprojects to divert northern Canadian rivers southwards into the United States were mooted as possible solutions to growing pressures on scarce water supplies in the arid southwest (Chapter 7, this volume). Ultimately, these schemes never went beyond the stage of dreaming on the part of the proponents, and, given today's understanding of the economic and environmental implications of such projects, there is no real possibility that they would attract significant support on either side of the border now or in the foreseeable future. There is, however, the real likelihood of pressures for more modest, but still significant, withdrawals from shared waters, especially in the areas bordering the Great Lakes Basin. The potential pressure for increased withdrawals from the lakes has been recognized for some years and had already begun to register as an issue of broad public concern during the free trade negotiations with the United States. What brought the issue from a simmer to a boil was a proposal that was neither linked to US demand nor, in any event, likely to have any realistic chance of proceeding. In 1998, a regional office of the Ontario government in Sault Ste. Marie granted a licence for the removal of water from Lake Superior for the purposes of export by tanker to an unspecified destination in Asia. Although the amount of water in question was not large, and although even a cursory analysis of the proposal would have made it plain that the costs involved were such as to make it virtually impossible for the scheme ever to go forward on a profitable basis, the very fact that such a licence could be granted served to act as a touchstone for lingering concerns as to the security of Canada's water resources.

In response to a public outcry, led especially by environmental groups, the federal government initiated a three-pronged strategy: first, to attempt to negotiate a moratorium on water exports with the provinces; second, to amend the legislation implementing the BWT so as to preclude the export of water from boundary waters; and, third, to initiate, in February 1999, a joint reference with the United States directing the IJC to investigate the broader issue of water uses in the Great Lakes, including not only the possibility of potential diversions and removals of water but also other issues relating to sustainability, such as the possible effects of climate change.[17] Following extensive studies and public hearings throughout the basin, the commission issued its final report in February 2000 (IJC 2000).

The particular findings of the commission – which, in brief, stressed that there was no "surplus" water available for removal from the basin and that the two governments could and should act to protect the basin from major new or increased consumptive uses – are of less interest here than are the lessons provided by the experience with respect to how the federal government approaches transboundary water management. For each of the three prongs of the strategy undertaken by the government, a great deal of care was directed at ensuring that the federal government did not stray from its historically restrained role in resource management. With respect to a moratorium on exports, for example, rather than relying on its clear authority to manage international trade and commerce to initiate such a moratorium, it preferred to simply urge provinces to act accordingly. Similarly, although it is true that the federal government passed legislation in 2002 to amend the legislation implementing the BWT so as to prohibit extrabasin removals, the legislation applies only to removals from boundary waters (and would not, for example, apply to tanker removals from coastal rivers or lakes not covered by the BWT) – again, despite the arguable ability of the federal government to act more broadly under either its trade and commerce power or POGG. Finally, of course, the mandate given to the IJC with respect to water uses was only to investigate and recommend and, as such, did not impinge on provincial authority.[18] Certainly, there are those who would have preferred a stronger Canadian response to the possibility of increased withdrawals from the Great Lakes (and, indeed, to the possibility of water exports more generally). Whatever view one takes of whether the federal government *should* have taken a stronger stance on constraining water exports, however, it is at least clear that the position it did adopt is consistent with historical preferences: first, internationally, for building a bilateral consensus rather than pursuing a unilateral solution; and second, domestically, for interpreting narrowly (at least in practice) its constitutional authority to legislate with respect to boundary and transboundary waters.

### Water Quality: Columbia River Pollution

While the history of the IJC illustrates one approach to dealing with the fragmentation of responsibility for internationally shared waters, a different and less cooperative template for addressing cross-border concerns is illustrated in the approach taken to a current transboundary dispute over pollution in the Columbia River. The dispute relates to heavy metals from an estimated ten million tons of slag material that was dumped over several decades into the Columbia River near Trail, BC, which is roughly sixteen kilometres north of the Canada-US border. The metals allegedly then migrated downriver across the border, causing substantial contamination of Lake Roosevelt, a 220-kilometre reservoir that stretches from the Grand Coulee Dam nearly up to the border. A large portion of the lake falls within a popular national recreation area and forms lengthy portions of the borders of the Colville and Spokane Indian Reservations in north-central Washington (USGS 2006).

The pollution emanated from a century-old metals smelter, which has international connections that go well beyond the transboundary nature of its water pollution. The smelter is owned and operated by Teck Cominco Metals Ltd., a subsidiary of Vancouver-based Teck Cominco Ltd., which has mining operations in both North America and South America. Even before the current dispute, the smelter enjoyed a certain amount of legal notoriety as the source of a landmark international law dispute arising from a previous

US environmental complaint involving transboundary air emissions. The complaint resulted in a 1941 international arbitration decision, which provided one of the most influential statements in international jurisprudence on the nature of states' international obligations and rights with respect to transboundary pollution.[19]

The genesis of the current dispute is a 2003 order issued by the US Environmental Protection Agency (EPA) under the US "Superfund" law – the Comprehensive Environmental Response, Compensation and Liability Act (CERCLA). The order requires Teck Cominco Metals Ltd. to investigate the nature and extent of the metals contamination and to report to the EPA. The EPA had issued this order after months of negotiations in which Teck Cominco had apparently committed to remedying the problem but not on the EPA's terms and not subject to US law.

Teck Cominco originally ignored the EPA's order, prompting the agency to commence its own investigations and to commit to recover its costs in a future court action against Cominco. However, Teck Cominco and the EPA subsequently reached an agreement committing the company to fund additional studies, so the EPA has withdrawn its order. In the meanwhile, the EPA's strategy has been overshadowed by a "citizen suit," filed in federal district court and seeking to require Cominco to comply with the EPA's order.[20] The suit was filed on behalf of two individual members of the Confederated Tribes of the Colville Reservation, and the State of Washington intervened as a plaintiff in the case. Cominco moved to dismiss the case, essentially on the ground that the Superfund law was not intended to cover non-US companies that release pollution from outside the United States. The US federal court judge rejected Cominco's arguments, focusing not on Cominco's Canadian nationality or on the extraterritorial location of the Trail smelter but, rather, on the fact – which the court assumed for purposes of its decision – that the smelter's pollution had contaminated a US site. Teck Cominco appealed this decision to the Ninth Circuit Court of Appeal, which affirmed the trial court's decision on the merits but remanded the case back to the trial court to decide whether the remedies requested by the plaintiffs had been mooted by the EPA's withdrawal of its order.

Canada has responded to the EPA order and citizen suit through diplomatic consultations that appear to have had little effect. Canada also filed an *amicus* brief in the Ninth Circuit in support of Cominco's appeal. The legal outcome of the current transboundary dispute may have considerable legal and policy reverberations for transboundary water management and environmental management more generally. In particular, the dispute indicates that, despite the presence of the BWT and the IJC, there are limits to Canada-US cooperation on transboundary water issues. There still remains the possibility of one country's acting unilaterally if bilateral solutions are either delayed or unpalatable. Canada and the United States have yet to fully demonstrate that they are willing to trust international institutions or processes with responsibility for transboundary watershed management.

In summary, jurisdictional fragmentation is an issue that touches international water management as much as it does domestic concerns. In the context of the US-Canada relationship, despite the apparent deficiencies of the BWT, there should be little appetite in Canada to consider its renegotiation. The principle of equal and similar rights to boundary waters enshrined in the treaty represents an important achievement for Canada, and the treaty's institutional architecture, as reflected in the IJC, has proved more robust than might have been anticipated in 1909. While much of the important work of the commission is investigative and recommendatory in nature, its stature is such that it plays a significant role in building bilateral consensus; it is hard to imagine that Canada would be able to achieve a similar impact on US water management practice in the absence of such an institution. As the current dispute with respect to Columbia River pollution suggests (and one might just as easily point to the dispute surrounding the draining of Devils Lake in North Dakota and its potential consequences in Manitoba), Canada is not likely to benefit from the adoption of unilateral approaches to transboundary water management.

## Conclusion

In dealing with the challenges of jurisdictional fragmentation with respect to water resource management, both federal and provincial governments have generally adopted a consensus that begins with the assumption that provinces, as the resource owners, should take the lead on domestic issues and should have a significant input into the resolution of international issues. Even where there is a clear constitutional basis to the federal interest – for example, in the protection of fish and fish habitat – there is a strong inclination to accommodate this interest to provincial management priorities. While in many respects this deference to provincial proprietary rights is consonant with a healthy cooperative federalism, there are grounds for believing that, in at least some situations, the federal stance has been overly timid and has not contributed to an optimal management regime. This is particularly true with respect to the passivity of the federal role in the management of interprovincial waters, where there is at least evidence that water management has suffered as the result of a disinclination on the part of the federal government to assert the role of referee in those cases where interprovincial negotiations stall or fail. Without the incentive of avoiding compulsory dispute resolution, it seems unrealistic to expect even well-intentioned upstream provinces to negotiate with a view to balancing the interests of the basin as a whole. This is likely to be even more the case as stresses on interprovincial waters increase in the future.

Fragmentation of responsibility for water management due to international boundaries is also an important challenge for Canada. Here, too, there is evidence that the federal government has been excessively cautious in construing its constitutional mandate with respect to water management. Certainly, there is much to be said for Canada's reliance on the BWT and its institutional provisions as a cornerstone of bilateral water management over most of the past century. It can be argued, however, that the federal government has become overly attached to the undoubted constitutional support provided by the Empire treaty clause of the Constitution, at the expense of exploring other water-related initiatives that might be justified on other bases – most notably under POGG. This reluctance to move forward on domestic measures that cannot be supported by reference to existing treaty commitments (even though such measures are not inconsistent with treaty commitments) was most obvious in the recent federal initiatives with respect to water exports. While the housekeeping amendments to the implementing legislation for the BWT were useful, it seems clear that the federal government wanted to constrain its legislative initiatives to those that could be characterized as treaty implementation measures rather than look to broader measures that might have been justified as matters of national concern pursuant to POGG. While such an approach may be conducive to federal-provincial harmony, it is less clear that it allows the federal government the necessary scope to fully address bilateral water issues. The ability of the federal government to act decisively is particularly crucial in the case of those issues that do not fall within the ambit of the IJC. As suggested by the Columbia River example, important transboundary issues that are not always resolved through existing bilateral institutions continue to arise.

In summary, Canada's current jurisdictional framework for water management is under strain. While the federal government has, at various times, indicated that it is willing to take a leadership role, especially in interprovincial waters or in cases where there were overriding national interests, its actual performance to date has been disappointing, even in utilizing legislation that is currently in place, let alone in pursuing new initiatives. As Canada's waters come under more stress in the next few decades, the federal government's stance of deferring to provincial interests in areas of legitimate national concern will become increasingly untenable, and the pressure for it to act decisively on a range of water quality and water quantity concerns will only grow.

## NOTES

1 The proprietary rights arise out of s. 109, conferring ownership of public lands on the provinces. Supporting legislative powers include primarily those relating to management and sale of public lands (s. 92[5]), property and civil rights in the province (s. 92[13]), and matters of a local or private nature (s. 92[16]). It might also be argued that the "resources amendment" (s. 92A) of 1982 has some limited impact on provincial powers in so far as it increases provincial legislative authority over electrical generation, including hydroelectricity, though it is difficult to see that this adds much in practice to the existing store of provincial powers for water management.

2 Another federal "power" that has been extremely important with respect to federal involvement in water management – although it does not appear specifically in the Constitution – is the power to spend. This has been of particular importance in the implementation of a number of shared-cost programs with the provinces.

3 So, the fisheries power by itself will not justify legislation regulating activities that may be environmentally undesirable but that are not in themselves harmful to fisheries: *Fowler v. R.*, [1980] 2 S.C.R. 213.

4 *Interprovincial Co-Operatives v. Manitoba,* [1976] 1 S.C.R. 477, per judgment of Pigeon J. One might infer from the judgment a very broad role for federal authority over interjurisdictional waters in his statement that "the basic rule is that general legislative authority in respect of all that is not within the provincial field is federal" (at 520). The judgment also analogizes interjurisdictional water quality issues to the management of a dam in one province that causes flooding in another province (at 511-512). In so doing, the judgment implies that this federal role extends beyond issues of water quality to those of water quantity.

5 *R. v. Crown Zellerbach Canada Ltd.*, [1988] 1 S.C.R. 401.

6 *R. v. Crown Zellerbach Canada Ltd.*, [1988] 1 S.C.R. 401, at 30 *per* Le Dain J.
7 *R. v. Hydro-Québec*, [1997] 3 S.C.R. 213.
8 Although the federal power in s. 92(10[c]) to exercise jurisdiction over works declared to be for the general advantage of Canada theoretically gives potentially much broader authority to the federal government, its practical implications are today highly limited in light of the disuse into which it has fallen. In the past, when the power had much more vitality, it was used explicitly on at least one occasion to justify important federal legislation with respect to regulation of the Lake of the Woods, a body that straddles both the Manitoba-Ontario and the Canada-US border: The Lake of the Woods Regulation Act, 1921, S.C. 1921, c. 38, s. 2(c).
9 The practice of Empire treaties ended formally in 1931 when Canada acquired full treaty-making capacity under the Statute of Westminster.
10 The western provinces are the exception because they did not achieve ownership of natural resources until the Natural Resources Transfer Agreements of 1930, which essentially placed the Prairie provinces in the same position as the others had attained under their terms of union, and which completed the process for British Columbia with the transfer to the province of the Railway Belt and the Peace River Block.
11 The province itself has stated bluntly that "managing the SSRB to meet instream flow needs (IFNs) for the aquatic environment is not possible because of existing allocations" (Alberta Environment 2003b, 22).
12 The Fisheries Department has taken a similarly hands-off approach toward historic dams throughout Canada by not requiring section 35 licences for most of these historic structures, notwithstanding that they continue to block fish passage. Once again, the department's (informal) position that these historic structures are grandfathered under the Fisheries Act is belied by the continuing harm caused by many of these structures.
13 The order of precedence is: domestic and sanitary uses, navigation, and power and irrigation. Glaringly absent from the perspective of modern water managers are instream uses, such as those incidental to aquatic and wildlife health.
14 There is one reference to water quality in the treaty. Article IV provides in part that "waters and waters flowing across the boundary shall not be polluted on either side to the injury of health or property on the other." There is, however, no elaboration on this obligation.
15 The treaty also provides for the possibility of an arbitral function if both governments consent; in practice, however, this has never been invoked.
16 In the case of the United States, there would be the challenges posed by Senate approval (and, in particular, reconciling possible difficulties as between border states and others, especially in the southwest); in Canada the status of the BWT as an Empire treaty would raise potentially serious constitutional difficulties for the

federal government if it were to attempt to amend it in a substantial way and still assert the application of s. 132 of the Constitution to justify its implementing legislation.

17 Although the reference was directed at the waters of shared basins and aquifers generally, the two governments asked the commission to address the issues in the Great Lakes Basin as a first urgent priority. And the commission's final report deals only with the protection of the waters of that basin.

18 Indeed, the commission's report has proved a useful complement to separate efforts by Great Lakes states and provinces to reach an agreement on limiting new or increased withdrawals from the lakes under the structure provided by the Great Lakes Charter.

19 *U.S. v. Canada,* 3 U.N. Rep. Intl. Arb. Awards 1911 (Ad Hoc Int l Arbitral Tribunal, 1941.

20 *Pakootas, et al. v. Teck Cominco Metals, Ltd.*, 59 E.R.C. (BNA) 1870 (E.D. Wa. 2004), *aff'd*, 452 F.3d 1066 (9th Cir. 2006).

## REFERENCES

Alberta Environment. 2003a. South Saskatchewan River Basin Water Allocation (May 2003, rev. Jan. 2005).

—. 2003b. *South Saskatchewan River Basin Water Management Plan Phase Two: Background Studies* (June).

Canada. 1867. *Constitution Act, 1867-1982*. s. 92(10a). Ottawa: Department of Justice.

Clipperton, G. Kasey, C. Wedell Koning, Allan G.H. Locke, John M. Mahoney, and Bob Quazi. 2003. *Instream Flow Needs Determinations for the South Saskatchewan River Basin, Alberta, Canada.* Edmonton: Alberta Environment and Sustainable Resource Development.

Commission of Conservation of Canada. 1929. *Waters and Water Powers, 1929, S.C.R. 200.* Ottawa: Commission of Conservation of Canada.

Environment Canada. 1985. *Currents of Change: Final Report.* Inquiry on Federal Water Policy. Ottawa: Environment Canada.

—. 1987. *Federal Water Policy.* Ottawa: Environment Canada. http://www.ec.gc.ca/water/en/info/pubs/fedpol/e_fedpol.htm#1.

IJC [International Joint Commission]. 2000. *Protection of the Waters of the Great Lakes.* Final Report to the Governments of Canada and the United States, 22 February. Ottawa: IJC.

LaForest, Hon. Gérard. 2004. Merv Leitch Q.C. Lecture at the University of Calgary Faculty of Law, 28 January, Calgary.

USGS [US Geological Survey]. 2006. Lake Roosevelt-Upper Columbia River. http://wa.water.usgs.gov/projects/roosevelt.

Wenig, Michael M. 2004. Federal Policy and Alberta's Oil and Gas: The Challenge of Biodiversity Conservation. In *How Ottawa Spends, 2004-2005: Mandate Change in the Paul Martin Era,* ed. G. Bruce Doern, 233-34. Montreal and Kingston: McGill-Queen's University Press.

DALE CUMMINGS/ARTIZANS.COM

Canadian experts say that North Dakota's Devils Lake Diversion should have been referred to the International Joint Commission under the terms of the 1909 Boundary Water Treaty, which governs shared watercourses between Canada and the United States. The failure by both Canadian and US governments to refer the diversion to the IJC raises concerns about unilateral action on the part of provinces and states and has led to debate about the future of the BWT.

INGRID RICE/NORTH VANCOUVER

The ability of Canada's federal government to respond adequately to water export proposals has been questioned by some Canadian water experts.

# 7
# Drawers of Water: Water Diversions in Canada and Beyond

*Frédéric Lasserre*

A truism of Canadian politics is that the Canadian public is wary about American water transfer schemes, and fears of water export have flared up in political debate over the better part of the past century. However, many Canadians do not realize that Canada is already one of the largest diverters of water in the world. Major diversion schemes – where water is transferred from one watershed to another – began in Canada at the beginning of the twentieth century. And more are still being built – notably in Quebec (Tables 7.1 and 7.2, Figure 7.1).

Most of the transfer schemes built in Canada were designed for hydroelectric development purposes, and very few were designed for urban consumption or irrigation. About 97 percent of all the volumes diverted in Canada are used for power generation. Therefore, water consumption is limited since little water is lost to evaporation or infiltration. However, bulk withdrawals do affect habitats from both the diverted and the augmented rivers. The scale of these impacts remains to be determined as the concept of minimum ecological flow remains controversial among biologists. The international academic literature reports that impacts begin when between 2 percent and 10 percent of the river flow is diverted (Lasserre 2005a). Although the environmental impacts of smaller diversions may be questionable, the sheer magnitude of some diversions leaves no room for doubt about harm to the environment. In northern Quebec, for example, the once mighty Eastmain River is now reduced to a shadow of its former self since it was diverted into the La Grande River for hydropower purposes (Table 7.3).

Canada is, in fact, a much larger diverter of water than is the United States, despite having a population only one-tenth the size of the latter. Actual volumes of water diverted from one river basin to another are far greater in Canada than they are in the United States, exceeding 3,500 cubic metres per second (in comparison, Niagara Falls has an average flow of 5,830 cubic

Figure 7.1

**Major bulk water diversions in North America**

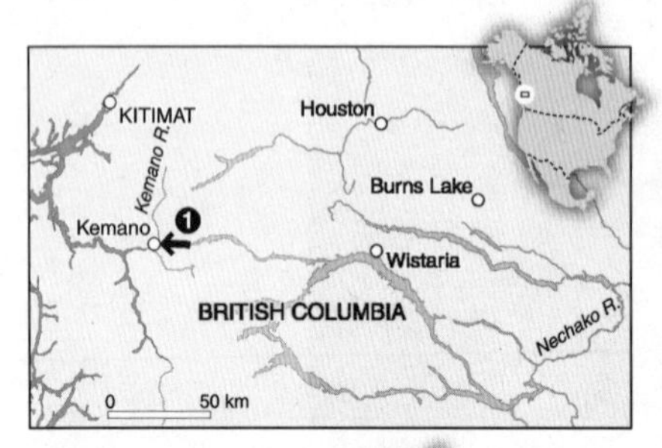

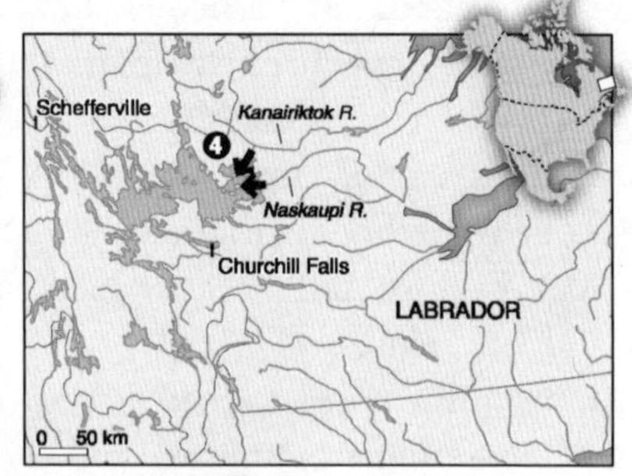


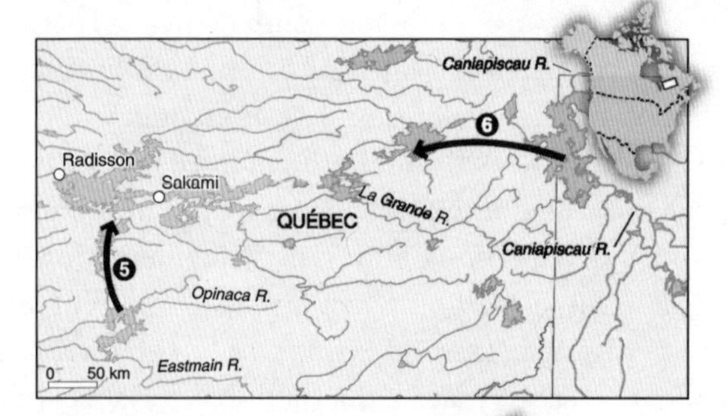


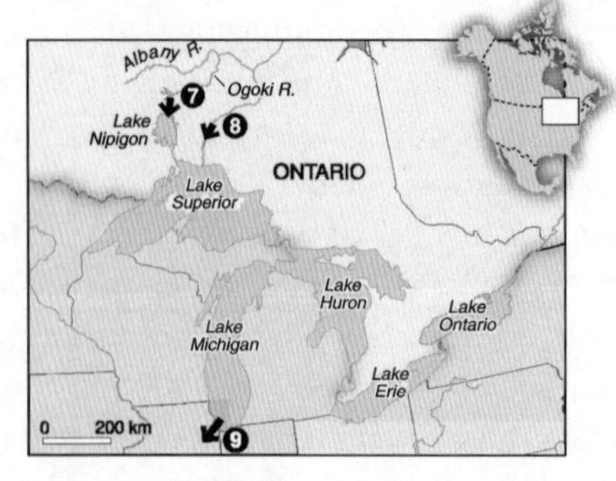


1 Kemano diversion
2 Churchill-Nelson diversion
3 Lake Saint-Joseph
4 Churchill Falls
5 EOL (Eastmain, Opinaca, and Petite Opinaca Rivers) diversion
6 Laforge diversion
7 Ogoki diversion
8 Lake Long diversion
9 Chicago diversion

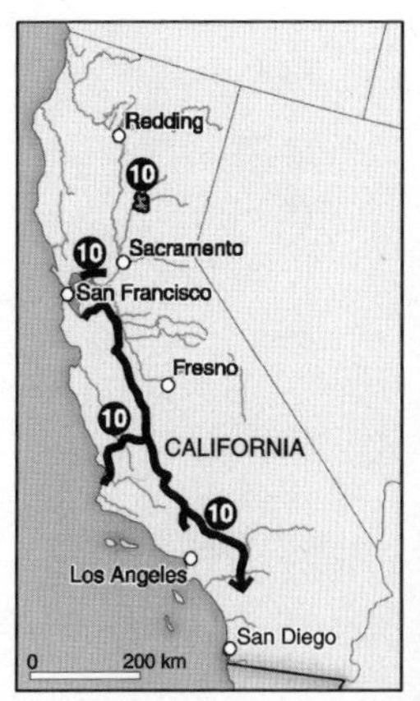


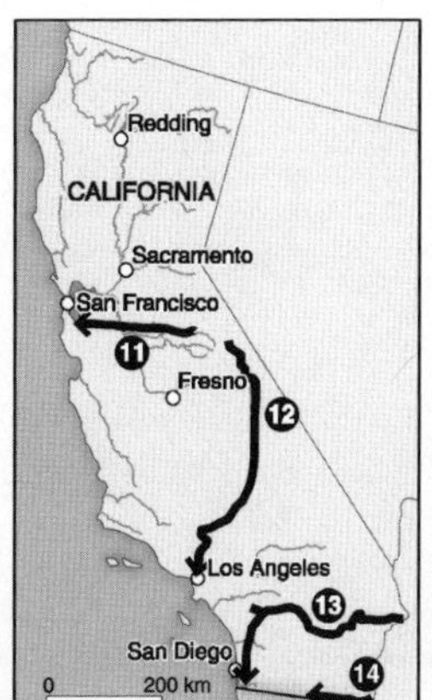


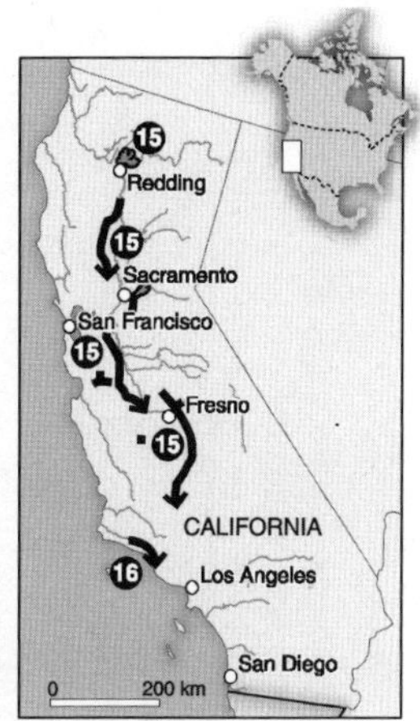

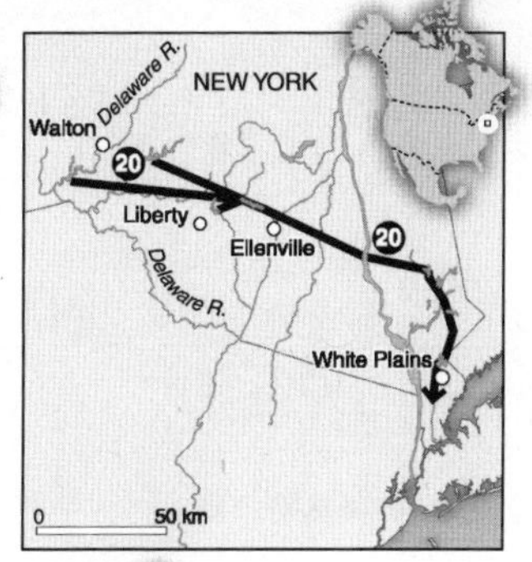


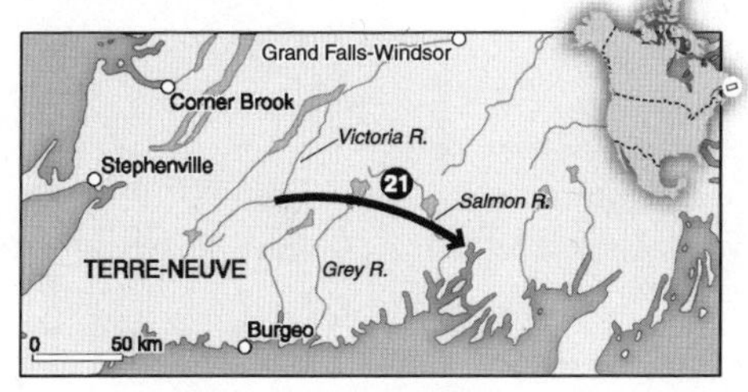


10 California State Water Project
11 Hetchy Hetchy Aqueduct
12 Los Angeles Aqueduct
13 Colorado River Aqueduct
14 All-American Canal
15 Central Valley Project
16 Santa Ynez
17 Central Utah Project
18 San Juan-Chama
19 Central Arizona Project
20 Delaware Aqueduct
21 Bay d'Espoir Development

TABLE 7.1

**Major existing large-scale water transfer schemes in Canada, 2004**

| Scheme | From (basin) | Destination | Beginning of operation | Transfer volume ($m^3$/second) | Length of diversion (km) | Objective |
|---|---|---|---|---|---|---|
| Kemano | Nechako River (Fraser) | Kemano River, BC | 1954 | 115 | 18 | Hydropower |
| Coquitlam-Buntzen Hydroproject | Coquitlam (Fraser) | Buntzen Lake and Indian Arm (Burrard Inlet), BC | 1912 | 28 | 4 | Hydropower |
| Vernon Irrigation District | Duteau Creek (Fraser) | Vernon Creek (Columbia Basin), BC | 1907 | 0.6 | ≅5 | Irrigation |
| Churchill | Churchill | Nelson, MB | 1976 | 775 | ≅40 | Hydropower |
| City of Winnipeg Aqueduct | Lake Shoal (Great Lakes) | Winnipeg (Red River), MB | 1919 | 3 | 137 | Urban |
| Great Lakes Basin | Long Lake | Lake Superior | 1939 | 42 | 0.4 | Hydropower |
| Great Lakes Basin | R. Okogi (Albany River Basin) | Lake Superior | 1943 | 113 | 8.5 | Hydropower |
| Root River | Lake St.-Joseph (Albany River Basin) | Root River (Nelson Basin), ON | 1957 | 86 | 7 | Hydropower |
| James Bay-La Grande | Caniapiscau River | La Grande River, QC | 1985 | 795 | 250 | Hydropower |
| James Bay-La Grande | Eastmain River | La Grande River, QC | 1985 | 835 | 150 | Hydropower |
| Bay d'Espoir Hydropower | Bear, Victoria, Salmon Rivers | Bay d'Espoir, NL | 1970 | 185 | 200 | Hydropower |
| Churchill Falls | Jultan | Churchill, NL | 1971 | 196 | n/a | Hydropower |
| Churchill Falls | Naskaupi | Churchill, NL | 1971 | 200 | 20 | Hydropower |
| Churchill Falls | Kanairktok | Churchill, NL | 1971 | 130 | 25 | Hydropower |

SOURCE: From several sources compiled by author, including Day Quinn (1992) and Hydro-Québec (2001).

TABLE 7.2

**Major existing large-scale water transfer schemes in the United States, 2004**

| Scheme | From (basin) | Destination | Beginning of operation | Transfer volume ($m^3$/second) | Length of diversion (km) | Objective |
|---|---|---|---|---|---|---|
| Croton Aqueduct | Croton (Hudson) | City of New York | 1842 | 449 | 66 | Urban |
| Catskill Aqueduct | Catskill (Hudson) | City of New York | 1915 | 768 | 147 | Urban |
| Delaware Aqueduct | Delaware | City of New York | 1952 | 1,271 | 169 | Urban |
| Chicago Diversion | Lake Michigan | Mississippi | 1848, 1900 | 2,870 | ≅40 | Navigation, urban, irrigation |
| Los Angeles Aqueduct | Owens River | City of Los Angeles | 1913 | 443 | 373 | Urban, hydropower |
| Second Los Angeles Aqueduct | Owens River and aquifer | City of Los Angeles | 1970 | 259 | 219 | Urban |
| Hetch Hetchy Water Supply | Tuolumne River | City of San Francisco | 1934 | 358 | 240 | Urban |
| "Continental Divide" diversions | Upper Colorado Basin | South Platte and Arkansas (Missouri) | as of 1892 | overall 592 | n/a | Irrigation |
| Colorado River Aqueduct | Colorado | Metropolitan Water District of Southern California | 1941 | 1,494 | 387 | Urban |
| All American Canal | Colorado | Imperial Irrigation District | 1942 | 3,827 | 325 | Irrigation |

▶

◀ TABLE 7.2

| Scheme | From (basin) | Destination | Beginning of operation | Transfer volume ($m^3$/second) | Length of diversion (km) | Objective |
|---|---|---|---|---|---|---|
| San Juan Chama Project | San Juan River (Colorado) | Chama River (Rio Grande) | 1970 | 136 | 14 | Urban, industrial |
| Southern Nevada Water Project | Colorado (Lake Mead) | City of Las Vegas | 1971 | 832 | 35 | Urban |
| Central Utah Project (CUP) | Colorado | Utah | 2004 | 333 | 242 | Irrigation, urban |
| Central Arizona Project (CAP) | Colorado | Arizona (Tucson) | 1993 | 1,852 | 528 | Irrigation, urban |
| Central Valley Project | Trinity, American, San Joaquin, Sacramento | Central California | 1951 | 8,638 | ≅600 | Irrigation |
| California Aqueduct, State Water Project (SWP) | Sacramento River | Central and Southern California | 2003 | 6,300 | 710 | Irrigation, urban |
| Rio Colorado-Tijuana Aqueduct | Colorado | City of Tijuana | 1985 | 102 | 150 | Urban |
| Santa Ynez Development Project | Santa Ynez River | City of Santa Barbara | 1956 | 32 | 42 | Irrigation, urban |

SOURCE: Form several sources compiled by author.

Table 7.3

**Share of the source river diverted by a specific development project**

| River | Project | % of natural flow diverted |
|---|---|---|
| Colorado (California) | California River Aqueduct | 8.1 |
| Colorado (California) | All American Canal | 20.7 |
| Colorado (Arizona) | Central Arizona Project | 10.0 |
| Owen (California) | Los Angeles Aqueduct | 90.0 |
| **North Thompson (BC)** | **North Thompson River Project** | **35.0** |
| Nechako (BC) | Kemano Project | 50.0 |
| Churchill (Manitoba) | Nelson Hydropower Project | 70.0 |
| Caniapiscau (Quebec) | La Grande | 40.0 |
| Eastmain (Quebec) | La Grande | 90.0 |
| **Rupert (Quebec)** | **La Grande, Eastmain 1A** | **82.4** |
| **Manouane (Quebec)** | **Complexe Bersimis** | **61.5** |

NOTE: **Bold type** indicates that project is under construction or study.
SOURCE: From several sources compiled by author.

metres per second), as opposed to about 990 $m^3/s$ in the United States. However, the length of American diversions is much greater: most Canadian diversions are smaller than 40 kilometres and only two exceed 150 kilometres, whereas American diversions easily exceed 250 kilometres, some conveying their water over more than 300 kilometres.

Water exports – where water crosses a geopolitical boundary – are another story. It is important to underline the fact that the diversions discussed above, whether in Canada or in the United States, take place within the limits of a single province or state. There are no interstate diversions, and those that are tapped from a river that mark state or provincial borders have proven to be the most complicated to solve politically. For example, in the case of the United States, there have been bitter quarrels between California and Arizona over the Colorado Aqueduct and the Central Arizona Project. As such, diversions were never thought of as water exports but, rather, as ways to mobilize a useful resource within a single political jurisdiction, whether to produce power, water fields, or supply cities.

So the water we divert does remain in Canada, and water diversions that cross political boundaries are, in fact, rare in both the United States and

Table 7.4

**Water export projects from Canada**

| Project | Year | Source | Annual transfer volume ($km^3$) | Cost of construction (billion current $) |
|---|---|---|---|---|
| North American Water & Power Alliance (NAWAPA) | 1952 | Transfer from Pacific and Arctic watersheds to the Great Lakes, the Mississippi, and California | 310 | 100 |
| Great Lakes Transfer Project | 1963 | Skeena, Nechako, and Fraser in British Columbia; Athabasca and Saskatchewan in the Prairies | 142 | n/a |
| Magnum Plan | 1965 | Peace River, Athabasca, and North Saskatchewan in Alberta | 31 | n/a |
| Kuiper Plan | 1967 | Peace River, Athabasca, North Saskatchewan, Nelson, and Churchill | 185 | 50 |
| Central North American Water Project (CENAWP) | 1967 | Mackenzie, Peace River, Athabasca, North Saskatchewan, Nelson, and Churchill | 185 | 30 to 50 |
| Western State Water Augmentation | 1968 | Liard and Mackenzie | 49 | 90 |
| NAWAPA-MUSĈHEC (Mexican US Commission for Hydroelectricity) | 1968 | Sources for NAWAPA, plus lower Mississippi and Sierra Madre Rivers | 354 | n/a |
| North American Waters | 1968 | Yukon and Mackenzie, Hudson Bay watershed | 1,850 | n/a |
| Grand Canal | 1983 | James Bay watershed; diversion toward the Great Lakes and the western United States | 347 | 100 |

SOURCE: Reisner (1993, 489); Day and Quinn (1992, 36-37).

Canada. But this does not preclude the possibility of future water exports; on the contrary, the scale of diversions in Canada makes our resistance to water exports less defensible. We collectively overlook the fact that our daily comfort and economic activity rests, to a certain extent, on existing water transfer schemes, which have significant negative environmental impacts. The federal government's policy that water exports should be prohibited on environmental grounds is, thus, relatively weak, given that large-scale water diversions are already a dominant feature of water management in Canada.

## Continental Water Transfer Projects

Water is a key ingredient in the fabric of western American society, as has been eloquently documented by environmental historian Donald Worster (Worster 1985). Twentieth-century American society, empowered by the industrial age, decided to harness rivers and aquifers on a large scale. In the 1960s, fears of exhaustion of water resources and ensuing economic catastrophe spurred politicians to consider large-scale water diversions from the Columbia and the Mississippi. The forceful rejection of the water exports proposal by the Columbia and Mississippi basin states triggered ideas about the much greater potential in Canada, and engineering firms started designing huge transfer projects from the north (Table 7.4).

The sheer size of these projects might lead us to dismiss them as unrealistic or as an illustration of engineering hubris. But Canadian politicians have not necessarily been opposed to large-scale water projects. Then federal prime minister Mulroney and Quebec premier Bourassa were favourable to the 1985[1] Great Recycling and Northern Development (GRAND) Canal proposal to dyke James Bay and transport fresh water south via the Great Lakes. This proposal illustrates that, to date, the very scale of continental water diversion has not deterred politicians. In fact, the GRAND Canal is still advocated by its designer, Canadian engineer Tom Kierans. In 2000, Kierans wrote letters in support of the canal to President Clinton and Prime Minister Chrétien; this, however, was to no avail thanks to the project's price tag (at least $100 billion), its huge suspected environmental impacts, and its political unpopularity.

## Pipe Dreams

The GRAND Canal proposal is illustrative of the fate of all large-scale North American water export proposals to date. Political opposition has played an important role in discouraging water exports. But an equally important deterrent has been the economic unfeasibility of these schemes. Water transported by aqueducts over long distances is costly not only because it is expensive to operate these infrastructures but also because the initial investment is huge (Table 7.5).

Moreover, alternative water sources are now becoming technologically feasible and economically affordable. Seawater desalinization technology, in particular, has improved quickly, and costs have been dramatically reduced. Desalinization is now a very affordable water-producing technology for urban and industrial consumers. However, it is rarely practical to use desalinated water for irrigation purposes, given the sheer volumes of water needs and the distance of most interior farmland from the coast; instead, water "swaps" are being considered for irrigation purposes. For instance, in the case of Arizona, the government has offered to pay for a desalinization plant near Los Angeles, in exchange for allowing Arizona to keep the equivalent amount of water from Colorado for its own use.

Most important, water demand has been stagnating in the United States for the past two decades (Gleick 2001); growth in US demand for Canadian water is not highly reliable (Chapters 7 and 8, this volume). Several factors explain why agricultural water use remained stable between 1990 and 2000 throughout the United States, especially in the west, though at unsustainable levels. First, water pricing and water costs, although still very low, are gradually increasing (Artiola and Dubois n.d.; CFWC 2005; Cline 2003). Second, water-starved cities are much richer than is the agricultural sector. Water transfers from the agriculture sector to the urban sector led to a much reduced water deficit, satisfying a large part of urban demand and rewarding water-efficient crops without requiring more water. An example is the agreement between the San Diego Water Authority and the Imperial Irrigation District (IID), through which irrigators agree to transfer up to 200,000 acre-feet (about 247 million cubic metres) per year for up to seventy-five years to the city of San Diego (San Diego County Water Authority 2002). Third, competition is fierce for American producers and is driving some out of business. In 2001, because of competition from other regions (mainly Asia and Mexico), sales of California cotton, prunes, and pistachios fell by one third; broccoli and plums by 24 percent; and tomatoes and lettuce by 22 per cent: "Californian farmers could soon realize there is more money to make selling their water rights than using them to grow crops" (*Economist* 2003, 11). Cost incentives also lure American producers to Mexico, further reducing water demand in

TABLE 7.5

**Cost estimates for water produced or transported by different means, 2002**

| Production or transportation method | Production cost, according to various estimates ($US/m$^3$) | Level of technology control | Advantages | Shortcomings |
|---|---|---|---|---|
| Transfer canal (500 km) | $.80 to $3.00 | High | Capacity to deliver large volumes. | Huge investment and operation costs. Environmental impact needs to be assessed. |
| Plastic bags | $.55 (Cyprus) to $1.35 (Greek Islands) | Average | Enables isolated islands or coastal communities to be supplied. | Technology needs to be improved. Small volumes. |
| Water-carrying ships | $1.25 to $1.50 | High | Simple technology. | Small volumes. Relatively high costs. |
| Iceberg transportation | $.55 to $.85 | Very low | Immense resource to be tapped. Acceptable cost for urban markets. | Technology needs to be much improved for a regular supply. |
| Desalination | $.75 for 40,000 m$^3$/day (Abu Dhabi)<br>$.85 for 40,000 m$^3$/day (Cyprus)<br>$.55 for 100,000 m$^3$/day (Tampa Bay)<br><br>*From brackish water* :<br>$.60 for 4,000 m$^3$/day<br>$.25 for 40,000 m$^3$/day | High | Immense resource to be tapped. Acceptable cost for urban markets. Fast-decreasing operating costs. | Large initial investment. Environmental impacts of salt residue. |
| Water recycling | $.07 to $1.80 | Average to high | Increases resource without developing new sources. | Required investments and operating costs are higher the more polluted the water is. Rarely socially acceptable as drinking water. |

SOURCE: Lasserre and Descroix (2003).

Table 7.6

**Evolution of water use in the United States, 1970-2000**

| | 1970 | 1975 | 1980 | 1985 | 1990 | 1995 | 2000 |
|---|---|---|---|---|---|---|---|
| Population (in millions) | 205.90 | 216.40 | 229.60 | 242.40 | 252.30 | 267.10 | 285.30 |
| Variation (%) | 6.20 | 5.10 | 6.10 | 5.60 | 4.10 | 5.90 | 7.00 |
| Total withdrawals (billion m³/day) | 1.40 | 1.60 | 1.67 | 1.52 | 1.55 | 1.53 | 1.55 |
| Variation (%) | 19.40 | 13.50 | 4.80 | -9.30 | 2.30 | -1.50 | 1.50 |
| Use (billion m³/day) | | | | | | | |
| Thermoelectric | 0.65 | 0.76 | 0.80 | 0.71 | 0.74 | 0.72 | 0.74 |
| Industrial | 0.18 | 0.17 | 0.17 | 0.12 | 0.11 | 0.11 | 0.08 |
| Irrigation | 0.49 | 0.53 | 0.57 | 0.52 | 0.52 | 0.51 | 0.52 |
| Public supply | 0.10 | 0.11 | 0.13 | 0.139 | 0.15 | 0.15 | 0.17 |

Source: Adapted from United States Geological Survey (1998, 2004).

the United States. On the whole, water withdrawals in the United States have been increasing slowly, much more slowly than has population, and could even begin a downward trend should competition from foreign fruits and vegetables producers increase (Table 7.6).

Therefore, although water is still used at an unsustainable rate in the western part of the United States, importing water from Canada is not as urgent as it once appeared to be to some American politicians, such as former senator Paul Simon. A stabilizing trend in water withdrawals and the availability of cheaper sources with desalinization plants and water savings has led public planners to forget about bulk water transfers from Canada to the Midwest and the west, and to the decline in popularity of such proposals amongst western politicians (Matheson 1984; Willardson 2005).

## The Great Lakes: The Focus of New Diversion Projects?

In the Great Lakes as well, bulk diversions already exist, and the balance of incoming and out-of-basin diversions is largely positive: about 66 cubic metres is diverted to the Great Lakes (Figure 7.2). After the Chicago diversion was built in 1900, through which water is diverted from the Great Lakes into the Mississippi River system by reversing the flow of the Chicago River, massive diversions from the Great Lakes Basin have been opposed by riparian states

Figure 7.2

**Major water diversions in the Great Lakes Basin**

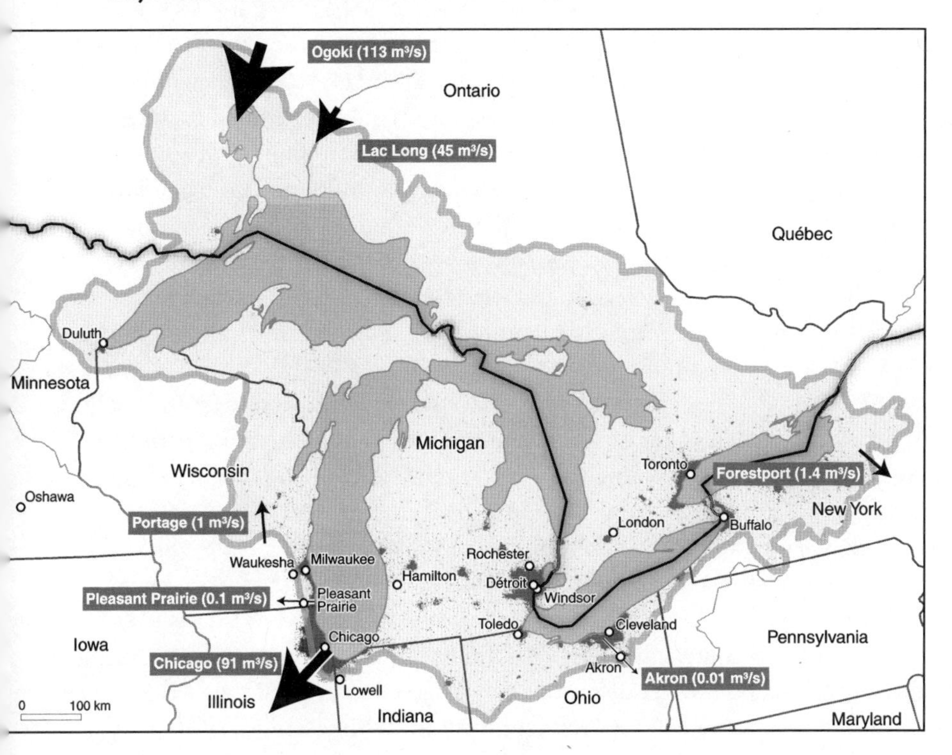


(Consulate General 1982; Lasserre 2001, 2005). However, such diversions have been proposed by both the Bureau of Reclamation and the US Army Corps of Engineers, agencies that dedicate most of their energies to promoting dam and hydraulic works projects. Nevertheless, it is worth emphasizing that the Corps concluded in 1982 that the project to divert water from Lake Superior or Lake Michigan to the Ogallala Aquifer was not to go ahead. This was due to the fact that it would result in major adverse environmental effects as well as the fact that there was simply no way it could make economic sense (Day 1992, 44; Keating 1986, 164; Micklin 1985, 54; Wright 2002, 8).

Great Lakes states wanted to resist these projects, both for environmental and political reasons. Why would the Great Lakes states give water to California at a time when many firms were leaving the area and relocating to the West Coast (Lasserre 2001)? Strongly influenced by these concerns, the Council of Great Lakes Governors (CGLG) has been central to the debate over exports from the Great Lakes region. The council, created in 1983, is a partnership of the governors of the eight Great Lakes states and the two Canadian provinces of Ontario and Quebec. The Great Lakes Charter stemmed from the growing concern that Great Lakes water could be diverted to water-scarce regions of the United States. The charter, a non-binding document signed in 1985 by the CGLG members, created a notice and consultation process for Great Lakes diversions. The signatories agreed that no Great Lakes state or province would proceed with any new or increased diversion or consumptive use of Great Lakes water that would exceed five million gallons per day without notifying, consulting, and seeking the consent of all affected Great Lakes states and provinces.

The riparian states and provinces believed that this charter was sufficient to prevent unilateral water export schemes by any signatories, but, in 1998, the Ontario government granted an export licence to the Nova Group to ship water from Sault Ste. Marie to Asia. Although the permit was soon to be revoked, the uproar in public opinion, both in Canada and the United States, and the fact that the charter had obviously not been enough to prevent export projects, underlined the need for clearer and more binding mechanisms to prevent water diversions. In addition, public concern in Canada about potential bulk water diversions was heightened by persistent – but so far unproven – rumours that President Bush told Prime Minister Chrétien, during their G8 July 2001 meeting in Genoa, that he wanted the two countries to consider a common pool of natural resources: energy, wood, minerals, and water.

At the same time, the issue of water supply to growing communities just outside the Great Lakes Basin became a pressing political problem for state governments. Their constituents wanted to tap the water that lay a few kilometres away, but governors also wanted to make good on their pledge not to divert water out of the basin – a pledge that could become politically difficult to fulfill. In 1992, Michigan governor John Engler vetoed Lowell (a town in Indiana) from getting Lake Michigan water, although its wells are now unsuitable for drinking water. And currently the town of Waukesha, a busy suburb of Milwaukee that lies outside of the Great Lakes watershed, also wants to tap into Lake Michigan (Chapter 8, this volume).

Therefore, the water diversions feared in Quebec, Ontario, and the Great Lakes Basin are no longer the kind of bulk diversions that were envisioned in the 1960s; rather, they are a multiplication of smaller, shorter diversions from the Great Lakes, all being perfectly justifiable, but the sum of which could amount to a large diversion (Lasserre 2005b). No new diversions have yet occurred, and the proposals remain controversial. Environmentalists have called for stringent rules that would prevent any diversion out of the lakes basin, while other pressure groups (urban, industrial, agricultural) have underlined that a vital resource was nearby and could be exploited without necessarily endangering the ecosystems.

In response, in 1999, the Canadian government moved on from considering laws banning water exports to promoting a three-point strategy to deal with the diversion issue. The first step was to ask for a legal opinion from the International Joint Commission, the management body for the Great Lakes instituted by the 1909 Treaty on Boundary Waters. The second was to convince the provinces to pass legislation banning water diversion since, constitutionally, water was a provincial responsibility. And the third and last step was to design a federal law that would provide a very tight frame of control for potential water diversion.

The provinces (except New Brunswick) eventually agreed on this project and passed laws that prevented water diversions or exports beyond their borders. However, this provincial legislation was limited in its potential to block bulk diversions. Should even one province allow water export schemes, all provinces would then be compelled to consider water as a tradable good under the North American Free Trade Agreement (NAFTA). Province-based legislation is, thus, only a very partial solution.

This trade-related dimension of water exports complicated federal attempts to deal with the water diversions issue. Since NAFTA prevents any member country from forbidding exports – once a good is traded it cannot be withdrawn from commerce by a political decision unless that decision applies to all parties (Article 315) – it was unwise for the federal government to tackle the issue of water diversions through a water export control mechanism. Besides, NAFTA's Chapter 11 precludes any regulation that would infringe on a party's investors rights. In a nutshell, Canada could not prevent American companies from diverting water if Canadian firms were allowed to do so.

However, Canadian federal officials noticed that an alternative strategy was possible. This was enabled by the geography of river basins along the Canada-US border as most large basins fall neatly onto one side of the border or another, apart from the Great Lakes Basin and the Columbia River

Basin (indeed, this was roughly the aim of negotiators when they settled the boundary at the forty-ninth parallel). Most major drainage basins – from the Fraser Basin to all the rivers draining into the Arctic and Hudson Bay – lie wholly in Canadian territory. Thus, at the provincial level, forbidding the transfer of water out of a province was roughly equivalent to preventing water export. At the federal level, forbidding inter-basin transfers on environmental grounds was tantamount to forbidding water exports to the United States without formulating the policy in terms that could be sued under NAFTA provisions. This seemed like an attractive strategy: laws forbidding interbasin transfers could be passed, with an ostensible justification of environmental protection, rather than being written as trade legislation explicitly banning water exports.

Accordingly, Canada's Commons voted the C-6 bill, An Act amending the International Boundary Waters Treaty Act, in December 2002. The new law effectively bans bulk water removals from the boundary water basins if they are greater than fifty cubic metres per day. Choosing to centre bill C-6 on watersheds also enabled an attempt to increase protection of the Great Lakes Basin, a body of water that was the focus of the 1909 Boundary Waters Treaty between Canada and the United States. The Boundary Waters Treaty made provision for the International Joint Commission, a binational body responsible for the arbitration of disputes between Canada and the United States regarding boundary water management, especially pertaining to the Great Lakes. When Canada and the United States jointly asked the IJC its legal opinion regarding water diversions from the boundary waters, the commission replied that these endeavours were to be allowed only with the explicit approval of their own governments (IJC 2000).[2] With the passing of the amendment to the Boundary Waters Treaty Act, protection for the Great Lakes seemed to have been strengthened.

Similarly, in the United States, efforts have been deployed to prevent large-scale water diversions from the Great Lakes. This underlines the fact that, contrary to public opinion and as I have already mentioned, bulk diversions are, *for now*, advocated only by a minority of lobbyists. In the year 2000, the government of the United States amended article 1962d-20 in the Water Resources Development Act (WRDA), which now prohibits:

> b.3. [...] any diversion of Great Lakes water by any State, Federal agency, or private entity for use outside the Great Lakes Basin

> unless such diversion is approved by the Governor of each of the Great Lakes States.

Because of this section, which effectively gives a veto to any of the Great Lakes governors, new diversions outside of the basin were also thought to be very difficult for the Americans to approve. Here again, diversions were considered at the basin level: the common unit that was highlighted by the act was the Great Lakes Basin.

At the Great Lakes scale, another approach to water diversion is leading to the use of similar tools. A parallel track of negotiations was undertaken by the Council of Great Lakes Governors (from eight American states and two Canadian provinces), triggered by the 1998 Nova Group export project and the Waukesha export proposal. The CGLG outlined a series of principles for reviewing water withdrawals from the Great Lakes Basin in the Great Lakes Charter Annex in June 2001 (Tonks 2004; Francoeur 2005). The annex showed a real political commitment to thwarting water export projects. In December 2005, the Annex Implementing Agreements were signed, banning water exports or diversions from the Great Lakes Basin and, thus, alleviating most concerns expressed by Canadians, specifically those living in Quebec and Ontario.

At the core of the Annex Implementing Agreements is a set of standards that would apply to withdrawals. Essentially, permits would be given for withdrawals only if the following criteria were met:

- There is no *reasonable alternative* to the proposed use, such as conserving existing water supplies.
- Withdrawals are limited to reasonable quantities for intended purposes.
- All water withdrawn is returned to the same Great Lakes watershed, less an allowance for consumptive use.
- There is a conservation plan (for major proposals) (CGLG 2005).

Just as for the Canadian and American federal governments, the CGLG decided to centre its response to potential excessive diversions on the Great Lakes watershed, underlining the understanding that use at one end of the basin would have effects downstream on other lakes and on the St. Lawrence River. In addition, all states and provinces will use a consistent standard to review proposed uses of Great Lakes water. Further to this, American states decided to sign a contract that, when ratified by Congress, will be binding upon them and not easy to alter without the consent of all signatories.

## Conclusion

Water diversions are already a fact, both in the United States and in Canada. Continental diversion projects came to the fore during the 1960s, when some western American states – California in particular – suddenly came to realize that locally available water was in finite quantity and of a smaller volume than was previously thought. The huge costs of these water diversions, coupled with a new emphasis on water efficiency in the United States, led to the decline of bulk diversion projects. It is unlikely, given the present trend in water use and management in the United States, as well as the strong opposition they meet in Canada, that bulk diversions will ever be implemented.

The Great Lakes, however, could witness the multiplication of small diversions, the cumulated volume of which could, potentially, be the equivalent of a large transfer. Urban growth on the southern limits of the watershed accounts for this trend, as do firms like Nova Group, who are waiting to get new licences to export water. Legislation has been passed in Canada and the United States that focuses on the Great Lakes Basin as a conservation unit. The Council of Great Lakes Governors finalized two sets of agreements, with the similar goal of designing a set of strict rules to control out-of-basin diversions. However, they underline the political debate about restricting even small diversions from the Great Lakes Basin. Unless a radically new vision of its management emerges among riparian states and provinces, it is likely that these diversions will be kept to a limited number and volume. However, all of these new regulations do not prevent water transfers from being developed within Canada. For example, new water diversions are under construction in Quebec, for hydropower purposes, in the Eastmain and Manouane River basins. New schemes are likely to be developed, in particular if climate change takes its toll on water availability for farmers in Alberta and Saskatchewan. In this case, it is possible that old diversion schemes such as PRIME (Prairie Rivers Improvement and Management Evaluation), which first saw the light of day in 1965, might be resurrected. Provinces supporting these projects would challenge any federal attempt to prevent them, given provincial jurisdiction over water resources and given the history of river diversions in Canada. In debating exports, then, we must not overlook the debate over diversions within Canada, which is likely to be as politically contentious as is the debate over exports to our neighbour to the south.

## ACKNOWLEDGMENTS

This research was made possible thanks to a grant from the Social Sciences Humanities and Research Council and the Fonds québécois de la recherche sur la société et la culture (FQRSC).

## NOTES

1 The project was also strongly advocated by powerful engineering, transportation, and financial firms such as SNC, UMA Engineering, Bechtel, Power Corporation, Rousseau, Sauvé and Warren, Louis Desmarais' Canadian Steamship Lines, and the nuclear energy sector.

2 The IJC confirmed it had a veto right on diversions on Great Lakes water that was divided by the boundary – that is, all the Great Lakes except Lake Michigan, which is wholly located in American territory.

## REFERENCES

Artiola, Janick, and Jim Dubois. N.d. *Arizona's Agricultural Ecosystems.* http://earthvision.asu.edu/acerp/section2/Chp_03ES.html.

CFWC. 2005. *California Farms at Risk: Myths and Facts about CVP Water Contracts.* Sacramento: California Farm Water Coalition. http://www.farmwater.org/should_know/cvp%20myths%20and%20facts.htm.

Cline, Harry. 2003. Flexible Management for Different Seasons. *Western Farm Press,* 27 March. http://westernfarmpress.com/news/farming_flexible_management_different/index.html.

Consulate General of Canada in Chicago. 1982. *The Great Lakes Water Diversion Issue.* Ottawa: Ministry of External Affairs, 23 April, SFAX 097/Restricted.

Council of Great Lakes Governors (CGLC). 2005. The Great Lakes-St. Lawrence River Basin Water Resources Compact and the Great Lakes-St. Lawrence River Basin Sustainable Water Resources. http://www.cglg.org/projects/water/annex2001 Implementing.asp.

Day, J.C., and F. Quinn. 1992. *Water Diversion and Export: Learning from Canadian Experience.* Canadian Association of Geographers. Department of Geography Publication Series, no. 36. Waterloo: University of Waterloo Geography Department.

*Economist.* 2003. Pipe Dreams, 11 January, 24.

Francoeur, Louis-Gilles. 2005. Les eaux des Grands Lacs seront protégées contre les États assoiffés: Les derivations massives seraient interdites, sauf exceptions. *Le Devoir,* 28 May.

Gleick, P. 2001. The Changing Water Paradigm. *Water International* 25 (1): 127-38.

Hydro-Québec. 2001. *Synthèse des connaissances environnementales acquises en milieu nordique.* Montréal: Hydro-Québec.

IJC. 2000. *Protection of the Waters of the Great Lakes: Final Report to the Governments of Canada and the United States* (February). Ottawa: International Joint Commission.

Keating, Michael. 1986. *To the Last Drop: Canada and the World's Water Crisis.* Toronto: Macmillan.

Lasserre, Frédéric. 2001. L'Amérique a soif: Les besoins en eau de l'Ouest des États-Unis conduiront-ils Ottawa à céder l'eau du Canada? *Revue internationale d'Études canadiennes/International Journal of Canadian Studies* 24: 196-214.

—. 2005a. Introduction. In *Les transferts d'eau massifs: Outils de développement ou instruments de pouvoir?* ed. F. Lasserre, 1-34. Quebec, QC: Presses de l'Université du Québec.

—. 2005b. La continentalisation des ressources en Amérique du Nord: L'ALENA oblige-t-elle le Canada à céder son eau aux États-Unis? In *Les transferts d'eau massifs: Outils de développement ou instruments de pouvoir?* ed. F. Lasserre, 463-87. Québec, QC: Presses de l'Université du Québec.

Lasserre, Frédéric, and Luc Descroix. 2004. *Eaux et territoires: Tensions, coopérations et géopolitique de l'eau.* Québec: Presses de l'Université du Québec.

Micklin, Philip. 1985. Inter-Basin Water Transfers in the United States. In *Large-Scale Water Transfers: Emerging Environmental and Social Experiences,* ed. Genady Golubev and Asit Biswas, 37-66. Tycooly: Oxford.

San Diego County Water Authority. 2002. Press Release, 31 December. http://www.sdcwa.org/news/123102Agreement.phtml.

Tonks, Alan. 2004. *The Great Lakes Charter Annex 2001.* Implementing Agreements Report of the Standing Committee on Environment and Sustainable Development. Ottawa (November).

Willardson, Tony. 2005. Deputy director, Western States Water Council. Correspondence with author, 22 February.

Worster, Donald. 1985. *Rivers of Empire: Water, Aridity, and the Growth of the American West.* New York: Oxford University Press.

Wright, Steven, and Jonathan Bulkley. 2002. *Diversion and Consumptive Uses of Great Lakes Waters: A Framework for Decision-Making.* Proceedings from the symposium Our Challenging Future: A Review of the State of Great Lakes Research, University of Michigan (November 5-6): 8. http://www.miseagrant.org/symposium.

# 8
# Thirsty Neighbours: A Century of Canada-US Transboundary Water Governance

*Ralph Pentland and Adèle Hurley*

In late 2004, at a workshop on Canada-US relations, a former US ambassador to Canada linked energy and water policy by suggesting that, in both cases, "you have it and we need it." In response, a former Canadian deputy prime minister made the astute political observation that, in Canada, "oil and water don't mix." He elaborated that Canadians are quite supportive of energy sales and fully understand that they make an important contribution to our national economy. On the other hand, he observed that any Canadian politician who openly supported the export of water would be out of a job very quickly, mainly because of Canadian sensitivity to the environmental implications of such a move. At the end of this brief exchange, the Canadian moderator quipped, "We have it, you need it: that's what worries us."

This good-natured exchange is probably a precursor to a much more serious dialogue that is likely to unfold between our two nations over the next several decades. In this chapter, we briefly explore some of the factors that may influence the outcome of that dialogue as well as the degree to which Canadian and bilateral institutions are equipped to lead those discussions to a mutually satisfactory outcome.

## Where's the Thirst?

Over the past several decades, the seventeen western US states have experienced the highest rate of economic growth of any region in North America (Quinn and Edstrom 2000). At the same time, there has been a significant migration of industries, jobs, and political power from the US Northeast and Midwest to the southwestern states. Western water resources have been almost fully exploited in support of this development. Along with the migration of jobs and political clout, there is a growing anxiety that water may follow and that waters from Canada and the northern tier US states will eventually be drawn into this regional tug-of-war.

This transformation in the southwest has been supported by numerous large-scale projects built to move water from its natural watercourses to areas of high demand. A good example is the Colorado River system, which has been stressed and transformed by seven decades of development (Commission for Environmental Cooperation 2001). Interbasin canals and pipelines carry water from the Colorado River to cities and irrigation districts in southern California and to sprawling metropolitan centres in Arizona. The twenty dams that turn the river on and off like a faucet, combined with the large-scale diversions, have drastically stressed the river's aquatic ecosystems. But the highest profile problem is that the long-term planned use of the river's water exceeds the available supply. Because the total legal entitlements to water are greater than the river's annual average flow, the river has been deemed "overappropriated" (Glenn et al. 1996).

Declining groundwater levels are widespread across the US southwest. Severe groundwater depletion in southern California has resulted in serious seawater intrusion near the coast and extensive land subsidence in the San Joaquin Valley. Groundwater levels in the huge Ogallala aquifer that underlies the seven High Plains states, and is arguably the most important source of groundwater in the United States, have been lowered by as much as 150 feet. Postwar technology made it possible to mine this resource but not to solve the costly yet predictable dilemma in which many irrigators now find themselves (Glennon 2002).

The apparent abundance of water in Canada combined with apparent shortages in the US southwest have, over the years, attracted the attention of engineers and planners to the possibilities of diverting Canadian waters southward. The two most ambitious proposals were the North American Water and Power Alliance (NAWAPA) scheme, which would divert water from the Mackenzie and Yukon river basins southward through the Rocky Mountain Trench, and the GRAND canal scheme, which would involve a dyke across James Bay and diversions via the Great Lakes southward and westward (Chapter 7, this volume).

However, attempts to import water to the US southwest have faced formidable obstacles, initially for political and economic reasons and, more recently, for environmental reasons. For political reasons, even within the United States, it is notable that interbasin diversion patterns have been severely constrained by the primacy of state jurisdictions. Diversions to date have crossed mountains, deserts, and other physical barriers but have rarely crossed state boundaries (Quinn 1968). The US federal government has always been sensitive to the primacy of states in this regard. For example, in 1968, in response to the concerns of the Pacific Northwest states, the US Congress imposed a ten-year moratorium on *studies* of major interbasin transfers. This moratorium was renewed for another ten years in 1978. Similarly, in response to concerns of the eight Great Lakes states, the federal Water Resources Development Act of 1986 placed a ban both on federal studies of interbasin transfers and diversions from the Great Lakes Basin without the approval of all eight governors.

A number of earlier studies have indicated that long-distance interbasin diversions are not economically feasible. These include: the US Army Corps of Engineers (1982) High Plains Study, which placed the cost of diverted Great Lakes water well above what even farmers would pay for irrigation water; and a cost-benefit study of the GRAND canal scheme (Muller 1988), which placed its benefit:cost ratio between 0.14 and 0.19. However, one should not automatically assume that lack of economic feasibility alone will always be an absolute deterrent to large-scale diversions. Huge federal subsidies have invariably been involved in developing irrigation projects and in supporting crop prices. In California's Central Valley, for example, the average price paid for water by farmers in 1981 was six dollars per thousand cubic metres, while the marginal cost of supplying it was $350 (Environment Canada 1985). However, subsidization of large, uneconomic water projects through "pork barrel" politics – previously commonplace in the US southwest and, to a lesser extent, Canada (Reisner 1993) – is now much less likely because decision making has become more transparent, particularly through environmental assessment processes.

The ultimate and perhaps most decisive impediment to very large-scale diversions is a keen and growing public appreciation of the adverse environmental consequences of such endeavours. They would all involve major changes to the natural environment by interrupting river flows, causing widespread flooding, transferring life forms, and even changing atmospheric and oceanic conditions. The ultimate consequences of many of these changes are uncertain. Not only are they vague, but, for all practical purposes, they are also irreversible. These concerns tend to be expressed much more frequently and forcefully in Canada than in the United States. For example, in the Federal Water Policy that was tabled in Parliament in 1987, the (then) minister's forwarding message noted that interbasin diversions, "would inflict enormous harm on both the environment and society, especially in the North, where the ecology is delicate and where the effects on Native cultures would be devastating" (House of Commons 1987).

The International Joint Commission's year 2000 report on Great Lakes diversions and consumptive use concluded that "there are no active proposals for major diversion projects either into or out of the Basin at the present time. There is little reason to believe that such projects will become economically, environmentally or socially feasible in the foreseeable future. Although the Commission has not identified any planning for or consideration of major diversions in areas outside the Basin, such diversions cannot be entirely discounted" (IJC 2000).

Most commentary on this subject tends to conclude with a note of caution similar to that offered by the IJC. Clearly, long-distance, large-scale diversions do not, and likely will never, make socioeconomic or environmental sense. Nevertheless, there are a few "wild cards" in the deck. Perhaps the most intriguing of these wild cards is global climate change and what this may mean for the food-producing "breadbasket" in the High Plains region. One could envisage a scenario under which severe climate change-induced water shortages in the Great Plains, combined with a continuing westward shift of political power, could tip the balance in favour of long-distance, large-scale diversions, despite their obvious drawbacks (see Chapter 7 for specific examples). Even though that prospect does not appear to be on the immediate horizon, it is one that Canadians may eventually have to face (Schindler 2001).

At present, we are encountering thirsts of a different kind. These involve attempts to shift the way in which we apportion water in the vicinity of the international boundary in favour of US interests. In the Great Lakes states, for example, where population is more heavily concentrated than it is in Canada, interbasin diversion possibilities are attracting the attention of several communities lying just outside or straddling the Great Lakes Basin divide. These pressures are being intensified by a trend toward the consolidation of water utilities over larger geographical areas, including both sides of the basin divide, and, in most cases, by urban sprawl and an over-reliance on limited groundwater resources.

Since 2001, the provinces bordering the Great Lakes and the eight Great Lakes states have been negotiating agreements. Representatives of these jurisdictions claim that these agreements would prevent both diversions to the thirsty US southwest and other forms of bulk water removals to thirsty nations around the globe. However, draft agreements (Council of Great Lakes Governors and Premiers 2004) released for public review in 2004 were widely criticized, especially in Canada, as being too permissive with respect to potential diversions (Chapter 7). If the agreements had been approved in their original form, they would likely have succeeded in promptly getting water to nearby communities such as Waukesha County in Wisconsin, Lowell in Indiana, and the urban sprawl around Chicago and Milwaukee. In and of itself, this may not have been a major problem as the amounts of water involved would have been very small. But the precedent of those small and formally sanctioned diversions, combined with what some reviewers concluded was a conceptually flawed regime, could conceivably have begun a very "slippery slope" to the long-term detriment of the economy and ecology of the entire Great Lakes region (Pentland 2004). Fortunately, agreements based on the principle of a prohibition on diversions with very minor exceptions were renegotiated and signed by the premiers of Ontario and Quebec and the eight Great Lakes governors in December 2005.

Another current example may be found in the St. Mary and Milk river basins in arid southern Alberta and northern Montana. Under Article VI of the 1909 Boundary Waters Treaty, the two rivers are to be treated as one stream and equally apportioned, "but more than half may be taken from one river and less than half from the other by either country so as to afford a more beneficial use to each" (Pentland 1996). In October of 1921, the International Joint Commission issued a subsequent order that was unanimously agreed to by all members of the commission after seven years of study and extensive public hearings. Even though the order is a fair and balanced interpretation of the treaty, perceived water shortages in Montana have prompted the state to lobby Canada and the IJC for more favourable treatment. It should be noted that most of Montana's problems relate to inefficiencies and inadequacies in their conveyance and irrigation systems rather than to the international apportionment arrangement. This matter is still under consideration, but, if any concession were to be made in that case, Canadians should expect similar requests in other shared basins with existing apportionment arrangements.

## How Real Is the Thirst?

Before getting into how the US thirst will be met, it would be instructive to first consider the extent to which the thirst is real. A 1999 report by the

Organization for Economic Co-operation and Development (OECD) noted that, per capita, water use by both the United States and Canada is two times greater than the OECD average. Although exact numbers are not available, it is likely that current water withdrawals in the United States represent only about 30 percent of renewable supplies, and the amount actually consumed is likely less than 10 percent of the amount available. Even though thirty-six states anticipate water shortages in localities, regions, or throughout their state within the next ten years, the US Geological Survey indicates that water use in the United States as a whole has stabilized since 1985. So, if there is a legitimate thirst in that country, it is on a regional or local scale and is definitely not national in scope. In most cases, it relates to the overpumping of groundwater rather than to an absolute water shortage. Although total water use has stabilized since 1985, groundwater withdrawals are up by 14 percent (Mehan 2005).

Even in the Colorado Basin, while media and other popularized reports tend to emphasize the "overappropriation" of water, they seldom mention either the fact that river flows were grossly overestimated at the time of the original apportionment agreement or the fact that human demands have not yet reached legal entitlements. It is also important to understand that very few of the massive projects that are now stressing the system would have been built had it not been for massive federal subsidies. Growing political resistance to that kind of heavy subsidization is already going a long way toward alleviating future water shortages.

In areas such as the Ogallala Aquifer and the St. Mary-Milk river system, where irrigation is the main consumer, many of the problems can be traced back to a legal regime that actually discriminates against water use efficiency. Most farming communities in the western United States are very well endowed with water rights. Their rights are senior and their entitlements are generally more than sufficient to meet their needs. However, under the prior appropriation doctrine, or "first in time, first in right" approach, there is absolutely no incentive for them to conserve because they can only keep the water right if they actually use the water.

In the case of communities just outside the Great Lakes Basin, residents receive the same three feet of precipitation and one foot of runoff every year as do those inside the basin. In fact, they do not generally have a water shortage problem. In most cases, residents are experiencing a shortage of water of sufficient quality for drinking, and this is related to urban sprawl and the overpumping of groundwater. The attempt to draw on the generally better quality Great Lakes water is a way to solve this drinking water problem. However, in most cases, there are more local solutions available. For example, a professor from the University of Wisconsin recently claimed that water withdrawals in Waukesha, Wisconsin, could be reduced by up to 80 percent by more careful management of groundwater withdrawals and reinjection of return flows (Cherkauer 2005). It would indeed be regrettable if the future of the Great Lakes were jeopardized in the interest of local, short-term political expediency.

There are, instead, other means for the United States to quench its apparent future thirst. A very strong technical argument can be made that, for the foreseeable future, US water needs can be met domestically. First, available water supplies can be stretched by various means, including rationing during drought periods; metering and pricing to cover the full cost of water services; system rehabilitation; upgraded technology, especially in the irrigation sector; and land use changes to reduce excessive mining of the resource. Second, it must be recognized that 90 percent of the water consumption in the arid west is in the agricultural sector and that most of that water is used to irrigate low-value crops like alfalfa and corn. A healthy process of reallocation is already under way, where legislation has been modified in some states to allow voluntary market transfers of water among different users locally, and water banks operate in Idaho and California to redirect water supplies to higher-valued uses.

Everywhere in North America there are local solutions available to deal effectively with local water shortages. But this does not mean that pressures to divert water will subside. Depending on public and political perceptions, interbasin diversion will often be viewed as the most obvious solution, and, in some cases, it may even be the lowest cost solution in the near term. This is the case, for example, for several communities lying outside but close to the Great Lakes Basin divide. In the immediate future, the most serious problems we are likely to face will involve dealing with a more aggressive, and perhaps more unilateralist, neighbour over the management of boundary and transboundary waters. Our first serious test in this regard began in earnest in July 2004 with the release of draft state-provincial agreements that appeared "to protect the Great Lakes, but which were in fact a thinly veiled effort for permitting the export of Great Lakes water" (Schindler and Hurley 2004).

## Will Canadian Exports Quench the Apparent Thirst?

Before discussing water export per se, it should be noted that Canada and the United States have always shared water on a local scale and developed water projects of mutual advantage in situations where this has been mutually beneficial. For example, the neighbouring communities of Coutts, Alberta, and Sweetgrass, Montana; the communities of Gretna, Manitoba, and Neche, North Dakota; and the communities of St. Stephen, New Brunswick, and Calais, Maine, have all shared common water supply systems. These transboundary water systems are all small, do not involve interbasin transfers, and do not invoke questions of national concern.

Canada's policies regarding water export have evolved, more generally, along with water management approaches. Canadian water management is sometimes described in terms of four distinct periods (Halliday and Pentland 2004): the early years (prior to 1945), the cooperative development period (1945-65), the comprehensive management period (1965-85), and the sustainable development period (1985-present).

During the cooperative development period (1945-65), international cooperation flourished with major projects such as the St. Lawrence Seaway and Columbia River hydroelectric development. A federal "pre-build" policy encouraged construction of major hydroelectric projects in Canada well before their output would be needed domestically, provided that markets could be found in the United States that would help to defray project costs. In the latter part of this period, several private interests expressed an interest in massive international diversion schemes such as NAWAPA and the GRAND canal. Interestingly, even though there was general support for projects of mutual advantage, public opposition to large-scale water export grew very quickly. It seems that Canadians have always attached a special heritage and sovereignty value to their water resources and that they have always instinctively reacted negatively to the notion of making water in its natural state a tradable commodity.

Even though Canada had no formal water export policy at that time, statements by opinion leaders in both countries very closely mirrored Canadian public opinion. For example, General A.G.L. McNaughton, who served as Canadian chairman to the IJC for more than a decade, insisted that "it is ... nonsense to talk about a surplus and it is dangerous folly even to contemplate selling water. All of our water can be translated into growth somewhere. Let this growth take place here in Canada."[1] William S. Foster, editor of American City, pondered the irreversibility of international exports: "I wonder what would happen to the sovereignty of Canada if a nation like the United States or a nation like Mexico would suddenly say that you no longer have the right to deny us the water even though you need it in your own industrial and agricultural development. I have often thought about that and wondered what the really long-range risks are to Canada, and I think they are considerable" (Foster 1996).

During the comprehensive management period (1965-85), water management was typified by comprehensive river basin planning and a more environmentally friendly framework. Several new research institutes supported these efforts, which included the consolidation of renewable resource and environmental agencies, and new international arrangements such as the Canada-US Great Lakes Water Quality Agreement. During this period, considerable water expertise was built up at the federal level, and this included government departments such as the internationally recognized Inland Waters Directorate of Environment Canada. These departments lent research expertise and credibility to the federal government's firm position in negotiations with the United States over controversial transboundary water-related issues such as acid rain.

Professional water managers of that time tended to express a water export position similar to that of their predecessors, but the more comprehensive water management approach led to more sophisticated arguments. For example, Bob Clark (2001), a senior Environment Canada official of that era, argues, "it is important that everyone understand that just as our land resources are a fixed quantity, so too is our water resource. Yes, water is a renewable resource in the sense that every year different water enters the system. However, the quantity is fixed in the long run – with some significant variations in the short run." He elaborated: "Our land and water resources are the birthright of all Canadians. We do not sell water – even to Canadians – we only give a right to its use. A water right can be cancelled. But, if Canada were to grant a right to the United States for the use of Canadian water we would never again regain its use unless a suitable alternative to water is found ... there is no substitute for water."

Toward the end of the comprehensive management period, media attention about possible water exports was high, and this topic was the focus of the public's attention during the Inquiry on National Water Policy. Interestingly, the inquiry members (Environment Canada 1985) themselves suggested that "the debate about water export and diversions, which has been gathering momentum, distracts us from the job of managing Canadian water in Canada for Canadians." They ended up advocating a cautious approach to export policy – one that does not reject the possibility entirely – but one focused on being better prepared when requests do come. When the Federal Water Policy itself was tabled in Parliament two years later, it was much less ambiguous. It stated that "the federal government will take all possible measures within the limits of its constitutional authority to prohibit the export of Canadian water by interbasin diversions and strengthen federal legislation to the extent necessary to fully implement this policy" (House of Commons 1987).

That political message was very much in tune with the instincts of Canadians. The government of the day did, in fact, table the Canada Water Preservation Bill in 1987, which would have banned water exports had it

Box 8.1

**Does NAFTA apply to bulk water exports from Canada?**

David Boyd is an environmental lawyer who has published widely on environmental issues in Canada. A former lead counsel with the Sierra Legal Defence Fund, Privy Council advisor to Paul Martin, and Trudeau Scholar, Boyd asserts in his recent book that NAFTA does apply to bulk water exports from Canada – despite assertions of politicians to the contrary (Boyd 2003). He argues that there are five areas where NAFTA applies to water:

1. *Does NAFTA prevent Canada from banning bulk water exports?*
   NAFTA limits (but does not entirely eliminate) Canada's ability to ban or restrict water exports. Canada could not ban or restrict water exports if water were defined as a tradable "good." Exemptions to NAFTA may be allowed where they relate directly to the protection of health or the environment. However, a total ban on bulk water exports would be too broad to meet the criteria for the necessary exemptions.

2. *National Treatment for Corporations*
   The requirement for equal treatment of Mexican, American, and Canadian corporations can be interpreted in two ways. On the one hand, one could infer that, if all corporations, both Canadian and foreign, were prohibited from exporting water, treatment would be "equal" and thus admissible under NAFTA. On the other hand, one could infer that American corporations in need of water for agricultural or industrial purposes would have the right to "equal treatment" along with their Canadian counterparts – and, thus, the right to take water from Canada regardless of whether their customers are in the United States. Until a dispute about water exports arises under NAFTA, neither interpretation will be decisively proven right or wrong.

3 *Can Canada Turn Off the Tap?*
"Once the tap is turned on, it cannot be turned off," is a phrase that is often used to describe NAFTA's proportionality provision, Article 315. The provision refers to the fact that, once Canada begins exporting water, it is not possible for it to reduce exports unless it also decreases domestic consumption to the same level. The threat of the proportionality provision's being invoked in the near future is low, given declining American demand and the currently low volume of Canadian water exports.

4 *Can Canada Enact Environmental Laws to Protect Water?*
The federal government maintains that NAFTA does not impede Canada's ability to legislate environmental protection. Critics such as the Council of Canadians assert that, as soon as water is exported, the United States has the right to take as much as they want without any limitations. In reality, the Canadian government is technically allowed to impose environmental laws; the catch is that these laws must follow the constraints imposed by NAFTA, which states that any law specific to environmental concerns is to have minimal impact on trade and cannot expropriate investments.

5 *NAFTA's Chapter 11: The Investor-State Dispute Settlement Mechanism*
Under Chapter 11 of the NAFTA agreement, foreign water investments are considered a trade provision regardless of whether water is considered a "good" or not. This is because Chapter 11 deals with investment and investor-state dispute settlement, which entitles investors to extensive protection with regard to government regulations. NAFTA's investor-state dispute mechanism permits foreign investors and corporations to challenge the Canadian government through binding international arbitration, without proceeding first through the Canadian legal system. Investments in water supply systems by an American company could conceivably be open to arbitration under Chapter 11. Currently, a Californian company (Sun Belt) is seeking more than $1.5 billion (US) in damages from the Canadian government under the provisions of Chapter 11. Sun Belt filed a NAFTA Chapter 11 claim in 1999 after British Columbia imposed a moratorium on the export of bulk water from that province after it had granted a licence to Sun Belt to export to the United States by marine water tanker. The company claims that British Columbia's law violates several NAFTA-based investor rights, including, in this case, its right to export BC water by tanker to California, claiming that it is entitled the same access to Canadian water as Canadians enjoy, thereby also invoking the principle of "national treatment" (point 2, above). In short, whether or not water is legally considered to be a "good," Chapter 11 applies to water investments.

been passed; however, with the calling of the "free trade" election in 1988, the bill died on the order paper. Observers have suggested that the Conservative government of the day had no intention of passing legislation banning water exports; rather, it was pressured into tabling the legislation as public opinion turned against it on NAFTA, partially because of fears over "opening the floodgates" on exports to the United States (Box 8.1). Indeed, the bill was not reintroduced following the re-election of the Conservatives. Ironically, as pointed out by environmental lawyer David Boyd (2003), the "no water export" policy that played at least some role in the re-election of the Conservative government – under whose auspices Canada signed the Free Trade Agreement – is now impossible to implement because of the provisions of NAFTA. Some suggest that a law like the Canada Water Preservation Bill could actually violate Canada's NAFTA obligations.

In the sustainable development era (1985-present), both legislative and program initiatives have become more integrative, anticipatory, and preventive. For example, toxic substances legislation now deals with the entire lifecycle of chemicals, pollution prevention has become a high priority, and environmental assessment legislation deals more comprehensively with matters such as cumulative impact. These changes are also reflected in the way we view Canada-US water management. In the Great Lakes Basin, for example, we are beginning to look not only at individual issues but also at the cumulative impacts of such issues as climate change, potential diversions, consumptive use, and modifications to the connecting channels. Unfortunately, we are not yet very good at translating cumulative impacts on water levels and flows into environmental quality and ecosystem impacts.

Three other very important changes took place during this period. The first was globalization of the world economy, along with trade agreements that may eventually restrict the ability of individual nations to preserve their water resources. For example, there is still some controversy about whether or not water in its natural state is covered under the provisions of NAFTA. The second change was the emergence of the United States as the world's only superpower, along with a trend toward a more unilateralist approach by that country in foreign affairs. It is only logical to consider just how far the United States is likely to go to get its way in future water disputes with Canada. Clearly, we should assume that the United States would act quite aggressively, but presumably within the rule of law, to assure its future water security. The third change was the decentralization of water governance within Canada, due, in part, to the significant decline of water expertise within the federal government (after years of underfunding), the decrease in priority of water issues within Environment Canada (the lead federal agency on water issues), and the growth of provincial capacity in water management. Decentralization is, in some cases, appropriate; the federal government has wisely stayed out of the way of the provincial management of small, local water management issues – even transboundary ones. However, this trend has, in some cases, been allowed to continue to the point of absurdity. For example, during recent high-impact, state-provincial negotiations regarding proposed Great Lakes diversions, the Canadian federal government played a remarkably passive role, even though the US Congress will ultimately have the final say on the matter.

In response to the new free trade realities, Canada has had to revisit its water export policy. We no longer speak about prohibiting the bulk export of water from the country; instead, we speak about prohibiting, "the bulk removal of water from all five major drainage basins in Canada." Four reasons are given for this new form of prohibition: "to sustain the natural supply within river basins; to prevent the introduction of non-indigenous biota, parasites and diseases, and pollutants into receiving water basins; to protect biological diversity and productivity within ecosystems; and to ensure the sustainable use of water to meet future needs of communities within watershed basins" (DFAIT 1999). Despite the reference to environmental issues, this policy was not developed for water management purposes; rather, it was a default position and was only taken once trade agreements were in place. But how ready is Canada to implement and defend this policy?

## Is Canada Ready?

What do we need to deal effectively with an apparently thirsty and aggressive, but, nevertheless, law-abiding, neighbour? First, we need the front line capability to analyze, and bring transparency to, all proposals to take water. Second, we need institutions and individuals who can bargain effectively with an aggressive but law-abiding neighbour. Third, we need to work with the American public and their political leaders to demonstrate that their own long-term interest would be best served by promoting water use efficiency and other local solutions. And, finally, we need domestic resolve to manage our own waters in an exemplary fashion in order to lead by example.

Canada-US water relations are governed primarily by the Boundary Waters Treaty of 1909. That treaty deals with a variety of issues involving boundary and transboundary surface waters; it provides for joint studies; and it establishes requirements for the approval of certain uses, obstructions, and diversions of waters that affect levels or flows in the other country. It also contains a provision against any pollution that would result in "injury of health and property" on the other side of the boundary. The treaty itself continues to be viewed by many as exemplary, and water managers all around the world consider it to be a useful model. It is the general opinion of most Canadian water professionals that Canada could not negotiate as favourable a treaty in today's political climate.

The Boundary Waters Treaty also established the International Joint Commission (IJC). The IJC's six commissioners, three from each country, are obliged to pursue the common interest of both nations rather than to adopt a nationalist perspective on each question. The IJC has two basic functions. First, it approves remedial or protective works, dams, or other obstructions in boundary waters and sets terms and conditions for the operation of projects. Second, it investigates and makes recommendations on questions or disputes referred to it by either or both governments.

While the IJC has had a long and distinguished record of achievement, there are growing concerns about its effectiveness. This is in part due to underfunding, as recognized by the Auditor General's Office in 2001 (Commissioner of the Environment and Sustainable Development 2001). Moreover, the process for appointing commissioners has been criticized for its lack of transparency and professionalism: appointments are not subject to oversight by Parliament. More fundamental is the perception that governments are not giving it sufficient support. For example, in 2001 the Commissioner of the Environment and Sustainable Development cited many examples of how the Canadian government's actions have limited the effectiveness of the IJC. In its November 2004 report on the Great Lakes Diversion issue, the Standing Committee on the Environment and Sustainable Development (SCEE 2004) commented as follows:

> The Committee believes that the power of the IJC could be increased by giving serious consideration to using all of the referral powers under the Treaty and by more fully supporting the IJC with timely information and resources. The Canadian government must make it absolutely clear that the IJC must remain the final arbiter for cases of withdrawals affecting levels and flows, but this can only be realized if it is provided with the proper resources and if it is backed up by the political will of government.

Canada's unwillingness to use its referral powers is, at times, paralleled south of the border. To a large degree, the effectiveness of the IJC is a question of politics. The IJC is only effective when both governments want it to be: where there is a clear mutual advantage (such as improving the quality of the Great Lakes); where both governments prefer to "study a problem away" (e.g., high or low water levels on the Great Lakes) rather than take action; or where projects are contentious and referral to an "independent" third party provides a political escape route. But the effectiveness of the IJC is often complicated by the asymmetrical power balance between the United States and Canada. In particular, when our southern neighbour is indifferent, or not inclined to cooperate with Canadian positions on shared waters (such as the recent controversy over Devils Lake [see Box 8.2]), then the IJC is likely to be sidelined. Nonetheless, most experts agree that Canada is infinitely better off with the IJC and the Boundary Waters Treaty than it would be without them, as they provide mechanisms for dialogue and negotiation that, given the right leadership, can be highly effective.

Canadians have always been vigilant and proactive concerning transboundary environmental and water issues. Examples include acid rain, eutrophication of the Great Lakes, and, more recently, climate change. Over the long run, that approach has turned out to be beneficial to both countries. However, serious questions now have to be raised about whether we have the national resolve to play that kind of progressive role in the inevitable regional tug-of-war that will take place in North America over regional water scarcity.

According to OECD figures, Canadian governmental expenditures on environmental issues declined from about 0.7 percent of GDP at the beginning of the 1990s to about 0.5 percent at the end of that decade. The same ratio for most other industrialized nations remained constant or increased. As a result, Canada probably moved from somewhere near the middle of the pack to somewhere near the bottom. A cursory examination of provincial budgets suggests a decline of capacity of about 25 percent over the decade, with the largest declines occurring in the industrial heartland of the nation, especially in Ontario (Pentland 2000). Although comparable statistics specific to water are not available, anecdotal evidence would suggest they are even less flattering.

Environment Canada is the lead federal agency on water matters. Water science has been severely cut in both the Department of the Environment and the Department of Fisheries and Oceans. This has clearly weakened Canada's ability to deal effectively with toxic substances and other forms of

box 8.2

### Devils Lake Diversion

author: Suzanne Moccia

The aptly named Devils Lake lies 160 kilometres south of the Manitoba-North Dakota border. With no natural outlet, the lake has become increasingly contaminated with human-made pollutants over the course of the twentieth century. Draining of wetlands for agricultural and residential development around the lake has resulted in rising water levels since the early 1990s, aggravated by higher-than-average precipitation levels in the past few years (Byers 2005).

Concerns about Devils Lake prompted the US government to request that the matter be referred to the International Joint Commission in 2002. The Canadian govenrment refused, citing a lack of environmental studies and taking a "wait and see" attitude that many now judge to be a mistake (Ibbitson 2005). Reacting to the rising water levels, and the flooding of 30,000 hectares of farmland, the State of North Dakota made the decision to build a diversion on Devils Lake in 2004. The diversion allows water from the lake to flow into the Sheyenne River, which crosses the border leading into the Red River, which, in turn, flows into Lake Winnipeg in Manitoba.

Diversion plans moved ahead, despite strong opposition from the provincial governments of Manitoba, Ontario, and Quebec; the eight Great Lakes states; and environmental and Aboriginal organizations on both sides of the border. Many believe the diversion project sets a precedent for unilateral water management decisions that will be made regardless of potential environmental problems downstream. Concern is not limited to the Canadian side of the border. US senator Lincoln Chaffe from Rhode Island appealed to Secretary of State Condoleezza Rice to encourage the US government to request a review of the diversion as many governors and congresspeople are concerned about its implications (Chaffe 2005).

North Dakota completed construction of the Devils Lake diversion in September 2005. The diversion operated for ten days but was shut down by the North Dakota Department of Public Health because of high levels of pollutants (Olivastri 2006). At the time of writing, the diversion has since reopened. Opponents to the Devils Lake diversion argue that the planners of the diversion should have respected the terms of the 1909 Boundary Water Treaty (BWT). They point to Article IV of the treaty, which states that cross-border water flows "shall not be polluted on either side to the injury of health or property on the other."[2] Opponents of the project also insist that the Devils Lake decision should be presented to the body that mediates conflict under the Boundary Waters Treaty – the International Joint Commission – for assessment (Olivastri 2006). However, although the Canadian government has requested that the case be referred to the IJC, the American government has not done so to date.[3]

This situation is unusual. Historically, Canada-US relations over transboundary water have honoured the BWT. The ability of the treaty to resolve future conflicts may therefore be at risk (Manitoba 2005b). This has resulted from a combination of factors. The governor of North Dakota, Republican John Hoeven, is a strong proponent of federal non-interference in "state's rights," which is supported by the current White House administration. Bilateral disputes over missile defence, softwood lumber, and bovine spongiform encephalopathy (BSE) make it more difficult for Canada to resolve cross-border issues (Byers 2005).

Aside from the political crisis that the Devils Lake diversion has caused, environmental risks are also pressing. Devils Lake has high concentrations of mercury, phosphorus, arsenic, and salt, making it too contaminated for local irrigation (Sawatzky 2005). The high pollution levels are made more severe because the lake has remained isolated from the Hudson Bay drainage basin for 1,000 years. These pollutants, along with non-native species and diseases, will flow down Manitoba's Red River into Lake Winnipeg, the tenth largest freshwater lake in the world and a popular recreation area. Estimates indicate that 40,000 pounds of phosphorus will reach Lake Winnipeg each year, resulting in a five-inch algae layer on approximately ten miles of beach (Manitoba 2005b).

Little official research has been conducted on the social and environmental effects of the diversion since no impact assessment was conducted before construction began. Those who oppose the diversion argue that, if an environmental impact assessment had been performed, then effective alternatives could have been developed to deal with rising water levels. Environmental groups, including Friends of the Earth, have called for a scientific review by the IJC (Olivastri 2006; Williamson 2005). Instead, the US government response has been to introduce an alternative to the IJC, the president's Council of Environmental Quality (CEQ), a national agency that was asked to study the environmental impacts of the diversion. Opponents have critiqued the inadequate terms of reference for the study, its limited testing sites, and its failure to conduct multi-seasonal testing.

In an attempt to address Canadian environmental concerns, North Dakotan officials built an interim eighteen-foot-deep pebble filter (Manitoba 2005a). Many argue that the filter is not sufficient to protect Canadian waters (CBC 2005; Olivastri 2006). The Manitoba government asserts that other possible alternatives that could effectively address flooding have not been adequately explored.

The Devils Lake case may signal a change in the binational approach to transboundary water governance between Canada and the United States. Other possible transboundary diversions are in the works, such as the Red River Valley Water Supply Project (also known as the resurrected Garrison Diversion Project). This project would allow water from the Missouri River to be transferred by pipeline to the Hudson Bay Basin via the Sheyenne and Red rivers – putting further pressure on the BWT governance framework (Canadian Embassy 2006).

pollution, its capability to deal effectively with interprovincial and Canada-US issues, and its ability to analyze international agreements before they are signed. The Inland Waters Directorate was disbanded in the early 1990s and has since been replaced with a much diminished freshwater organization. Funding in support of federal-provincial and Canada-US efforts under the Canada Water Act was cut to a few percent of expenditures in the 1970s and 1980s. A severely weakened water policy capability is already beginning to swamp Environment Canada as this strategic resource assumes increasing international and regional importance.

Fresh water is primarily a provincially managed resource, but it is at the same time an extremely important strategic national resource. At this critical juncture in the impending North American water debate, our water protection capabilities are unfortunately adrift, as was pointed out by the auditor-general of Canada in her year 2001 report. The federal government is under-resourcing a key department – a derogation of responsibility that should not continue. There is clearly an urgent need to reassess our national capacity to deal with the global, continental, national, and regional challenges now facing this strategic national resource. There are local solutions to water problems available everywhere in North America, but without an in-depth capacity to analyze, lobby, and negotiate, there is a distinct risk that these may be eclipsed by a short-sighted political expediency.

## NOTES

1 Cited in Clark (2001).

2 Article IV, Boundary Waters Treaty of 1909, quoted in *Protecting our Waters* (Winnipeg: Government of Manitoba, 2005). http://www.gov.mb.ca/waterstewardship/pdf/devilslakebro.pdf.

3 In order for a reference to actually occur, the United States and Canada need jointly to ask for it. The BWT does theoretically allow for a unilateral reference, but this has never occurred.

## REFERENCES

Boyd, David. 2003. *Unnatural Law: Rethinking Canadian Environmental Law and Policy.* Vancouver: UBC Press.

Byers, Michael. 2005. The Devil's Diversion. Editorial, *Globe and Mail.* 31 January.

Canadian Embassy. Washington. 2006. *Garrison Diversion and the Devils Lake Outlet: The Canadian Position.* http://www.dfait-maeci.gc.ca/can-am/washington/shared_env/garrison-en.asp.

CBC. 2005. *Filter Installed on Devils Lake Diversion*. Canadian Broadcasting Corporation, 3 August. http://www.cbc.ca/manitoba/story/mb_devils-lake-20050803.html.

Chaffe, Lincoln. Letter to Secretary of State Condoleezza Rice. 14 June 2005. http://www.lcnd.ca.

Cherkauer, Douglas. 2005. *The Hydrology of the Lake Michigan Region*. Straddling the Divide Conference, February, Chicago.

Clark, Bob. 2001. Concerns about Water Export. Unpublished paper (July).

Commissioner of the Environment and Sustainable Development. 2001. A Legacy Worth Protecting: Charting a Sustainable Course in the Great Lakes and St. Lawrence River Basin. *Report of the Commissioner of the Environment and Sustainable Development to the House of Commons*. Ottawa: Office of the Auditor General.

Council of Great Lakes Governors and Premiers. 2004. *Draft Great Lakes Basin Water Resources Compact and Draft Great Lakes Basin Sustainable Water Resources Agreement* (July). http://www.cglg.org/projects/water/Annex2001Implementing.asp#Background%20Documents.

Edward, P. Glenn, Christopher Lee, Richard Felger, and Scott Zengel. 1996. Effects of Water Management on the Wetlands of the Colorado River Delta, Mexico. *Conservation Biology* 10 (4): 1175-86.

Environment Canada. 1985. *Currents of Change: Final Report*. Inquiry on Federal Water Policy. Ottawa: Environment Canada.

DFAIT [Department of Foreign Affairs and International Trade]. 1999. *Bulk Water Removal and International Trade Considerations*. Ottawa: DFAIT.

Friends of the Earth Canada. N.d. Synopsis of Devils Lake Crisis. *Friends of the Earth Stop Devils Lake Campaign*. http://www.foecanada.org.

Foster, William. 1996. Statement on the CBC television program *The Politics of Water* (April).

Glennon, R. 2002. *Water Follies: Groundwater Pumping and the Fate of America's Fresh Waters*. Washington, DC: Island Press.

Halliday, Bob, and Ralph Pentland. 2004. *An Assessment of Modelling Needs and Opportunities*. Consultant report prepared for Environment Canada (March).

House of Commons. 1987. *Federal Water Policy*. Report tabled in the House of Commons by the Honourable Thomas McMillan, 1987.

Ibbitson, John. 2005. Canada Must Swallow Its Devils Lake Mistakes. *Globe and Mail*, 11 August: A15.

IJC. 2000. *Protection of the Waters of the Great Lakes*. Ottawa: International Joint Commission (February).

Manitoba. 2005a. *Advanced Filtration System to Be Built at Devils Lake, N.D*. News Release, 5 August. Winnipeg: Government of Manitoba. http://www.gov.mb.ca/chc/press/top/2005/08/2005-08-05-03.html.

—. 2005b. *Protecting our Waters*. Winnipeg: Government of Manitoba. http://www.gov.mb.ca/waterstewardship/pdf/devilslakebro.pdf.

Mehan, III, Tracy. 2005. A *Symphonic Approach to Watershed Management.* Straddling the Divide Conference, 15-16 February, Chicago.

Muller, Andrew. 1988. Some Economics of the Grand Canal. *Canadian Public Policy* 14 (2): 162-74.

North American Commission for Environmental Cooperation. 2001. *North American Boundary and Transboundary Inland Water Management Report, 2001*: 197-98.

Olivastri, Beatrice. Chief Executive Officer, Friends of the Earth Canada. Telephone conversation, 27 February 2006.

Organization for Economic and Cultural Development (OECD). 1999. *The Price of Water: Trends in OECD Countries*. Paris: OECD.

Pentland, Ralph. 1996. Unpublished Report on the Great Plains Region. Prepared for the Commission for Environmental Cooperation. Montreal, December.

—. 2000. Consultant Study on Canada's Environmental Capacity, carried out for Environment Canada. Unpublished. Ottawa: Environment Canada.

—. 2004. *Great Lakes Compact, Water for Sale*. Washington, DC: Woodrow Wilson International Centre for Scholars (September).

Quinn, Frank. 1968. Water Transfers: Must the American West Be Won Again? *Geographical Review* 58 (1):108-32.

Quinn, Frank J., and Jeff Edstrom. 2000. Great Lakes Diversions and Other Removals. *Canadian Water Resources Journal* 25 (2): 125-51.

Reisner, M. 1993. *Cadillac Desert: The American West and Its Disappearing Water.* London: Penguin Books.

Sawatzky, Wendy. 2005. *Devils Lake: Cross-Border Controversy.* Canadian Broadcasting Corporation. http://www.cbc.ca/manitoba/features/devilslake.

SCEE [Standing Committee on the Environment and Sustainable Development]. 2004. *The Great Lakes Charter Annex 2001, Implementing Agreements*. Ottawa: Government of Canada.

Schindler, D. 2001. The Cumulative Effects of Climate Warming and Other Human Stresses on Canadian Freshwaters in the New Millennium. *Canadian Journal of Fisheries and Aquatic Sciences* 58: 18-29.

Schindler, David, and Adèle Hurley. 2004. *Rising Tensions in Cross-Border Water Issues: The U.S. and Canada in the 21st Century*. Briefing Note. University of Victoria: Centre for Global Studies Conference on Canada/U.S. Relations. November.

US Army Corps of Engineers. 1982. *A Summary of Results of the U.S. Corps of Engineers Ogallala Regional Study, with Recommendations to the Secretary of Commerce and Congress*. High Plains Council. December.

Williamson, Dwight. 2005. *A Limited Survey of Biota in Devils Lake and Stump Lakes, North Dakota.* Manitoba's Contribution to a Multi-Jurisdictional Collaborative Assessment Coordinated by the United States Council on Environmental Quality. Winnipeg: Manitoba Water Stewardship, November. http://www.gov.mb.ca/waterstewardship/reports/transboundary/2005-10mb-devilslake_biota_rpt.pdf.

PART 3

# Blue Gold: Privatization, Water Rights, and Water Markets

JOSS MACLENNAN/FOR CUPE

The Canadian Union of Public Employees anti-privatization campaign has attracted national and international attention.

# 9
# Commons or Commodity? The Debate over Private Sector Involvement in Water Supply

*Karen Bakker*

Private corporations are playing an increasingly significant role as builders, owners, and operators of water supply systems around the world, and they have dramatically increased their market share in the water supply sector over the past two decades (Finger and Allouche 2002; Silva, Tynan, and Yilmaz 1998). Although most water supply systems in Canada remain publicly owned and operated, several Canadian municipalities have recently signed contracts with private companies for water supply and sewerage management. Typically, municipalities retain ownership of water supply networks but contract out their management to private, often international, companies.

This increase in private sector involvement has been highly controversial. Many Canadian cities (including Toronto, Vancouver, and Montreal) have experienced heated debate over the involvement of private companies in water supply management in the past few years. Canada's largest public services union – the Canadian Union of Public Employees (CUPE) – and the Council of Canadians have been running sustained campaigns against private sector involvement in the water sector (see, for example, Barlow and Clark 2002). In response, campaigns and publications in favour of privatization have been organized by groups such as the Canadian Council for Public Private Partnerships, Pollution Probe, and some government agencies and water supply utilities (see, for example, Brubaker 2002). Advocates of private sector involvement argue that the benefits include improved management and access to much-needed finance. Opponents of private sector involvement warn of decreased accountability, threats to public health, declining service levels, and degraded water quality.

This chapter analyzes the public-private debate in Canada, beginning with historical background on the three most common approaches to water supply: public, private, and cooperative.

## How to Provide Water? The Public, Private, and Cooperative Approaches

Debates over water supply governance revolve around three idealized models of resource management: the "public utility" (or municipal) model, the private sector "commercial" model, and the community "cooperative" model (Table 9.1). How did these different models emerge? The answer can be found, in part, through analyzing the history of urbanization and associated industrialization of water supply. As cities grow, some means of supplying large amounts of water and removing large quantities of sewage becomes increasingly necessary. In nineteenth-century cities, universal water and sewerage networks emerged as the preferred model (Goubert 1986; Hassan 1998; Melosi 2000). Water was no longer drawn from local wells and streams but, rather, was mass produced, abstracted in large quantities, and treated at plants before being distributed through networks in densely built up areas where economies of scale made supplying water feasible.

In many cities, private corporations built and operated the first water supply networks. Private companies operated in cities like Boston, New York, London, Paris, Buenos Aires, and Toronto, typically supplying water to wealthier neighbourhoods; the poor had to rely on public taps, wells, and rivers (or, in the most desperate cases, stolen water). Growing awareness of the role of water in the spread of disease, particularly in crowded urban environments, led to widespread support for the development of the provision of safe drinking water to all citizens through "universal" water supply networks (Luckin 1986; Hamlin 1990). The terrible cholera and typhoid epidemics of the nineteenth century, combined with an apparent inability or lack of interest on the part of the private sector to finance universal provision, led governments to take over the business of water supply infrastructure. In places where private companies continued to operate – as they did to a limited extent in France, England, and Spain – they were tightly regulated. Private water companies in the United Kingdom, for example, had dividends capped and were required to reinvest any remaining profits in the water supply business.

### "Public Utilities"

The "public utility" model of network water supply provision was thus, in many cases, a response to nineteenth-century experiences with the private provision of water supply. For much of the twentieth century, governments ran most water supply systems, particularly in industrialized countries and urban areas. With the aim of providing universal access and protecting public health, governments created public utilities that owned the infrastructure and, in most cases, provided services to consumers on a subsidized basis. Water was regarded as a public service, often run at the municipal level, and was frequently not metered. Where private companies continued to operate, their activities were strictly regulated. In Canada, few private companies continued to operate in urban areas (with exceptions in some smaller towns, such as White Rock, British Columbia). Many rural communities continued to run "private" systems independently of governments, but these were generally not-for-profit and on a small scale.

The justification for government control of water supply systems was made on economic and ethical grounds. In economic terms, water was thought to be subject to "market failures" – a characteristic that makes it difficult or impossible for markets to operate.[1] In particular, the fact that water flows through the hydrological cycle (circulating between oceans and rivers, lakes, streams, and groundwater through evaporation and precipitation) makes it difficult to establish private property rights. In the case of water supply networks, market failures also arise because networks are usually run as monopolies;[2] without competition, markets cannot operate effectively. Moreover, the public health risks that arise when even a few people do not have access to clean water (thus being subject to water-borne diseases like cholera and typhoid) mean that water can be considered a public good.[3]

In the nineteenth century, the idea that citizens had a right to a sufficient supply of clean water, and that this was essential to their being able to participate in civic life, became dominant in industrializing countries. Governments began to articulate and assume the duty of ensuring water supply, as a basic need, for all citizens. This, in turn, was understood to have economic benefits: healthier citizens who felt included in society would be more productive workers. Thus, ethical arguments were intertwined with economic justifications that favoured universal water supply, controlled by governments.

As a result, in most industrialized countries and urban areas, governments dominated the business of water supply throughout much of the twentieth century. Where governments set up corporations to run water supply systems (as in the Netherlands), these tended to be non-profit and publicly owned. Private companies played a minor role. In Canada, they tended to be small and owned by users or local investors rather than publicly traded corporations. The situation is similar in the United States: the National Research

Council found in 2002 that only 15 percent of the population was served by privately run systems, the majority of which were small and local. This figure has remained stable since 1945 (NRC 2002).

### Private Sector Management

The private sector model in water supply is, in contrast to the public utility model, characterized by the management (and sometimes ownership) of infrastructure by private, for-profit corporations. There are many different types of private sector models (for a review, see Bakker and Cameron 2003). Privatization involves the sale of water supply networks to the private sector; this approach to water management is rare and has only been attempted in a few countries, such as England and Chile (Bakker 2004; Bauer 1998). The most common approach in Canada is "public private partnerships" (P3s), in which the government retains ownership of supply networks and private companies are contracted for a defined period of time (usually less than thirty years) to design, build, operate, or manage components of a public water supply system. Cities such as London, Ontario, and Moncton, New Brunswick, have experimented with P3 contracts for water supply system management and sewerage system management. Sometimes municipalities may delegate management to another public operator, as is the case with many Ontario municipalities, which delegate management of their water supply systems to the Ontario Clean Water Agency (a provincial Crown corporation).

In some countries, this approach to water management is widespread (Johnstone and Wood 2001). For example, about 70 percent of the French population is served through different types of public-private partnerships. Over the past decade, there has been a rapid increase in the number of P3s (Bakker 2003). However, the majority of formal water supply systems around the world (and in Canada) remain publicly owned.

Private sector participation usually entails commercialization, through which private sector norms (such as efficiency and profit incentives) are applied to water supply management. Commercialization frequently involves the introduction of metering and associated changes in water rates. The principle of full-cost pricing (prices should reflect the full cost of the service) and economic equity (consumers should pay for what they use) are usually applied, in contrast to subsidized pricing and social equity (pricing according to ability to pay), which frequently characterize public utility systems.

### The Cooperative Approach

In areas where a strong tradition of community-run services exists, or in areas too sparsely populated to interest governments or private companies, communities often build and run their own water supply systems. Community-run water supply systems are frequently managed as cooperatives. There are many types of cooperative: a simple definition is "an enterprise owned and democratically controlled by the users of the goods and services provided" (Co-operatives Secretariat 1998, 1). Users can be consumers, employees, or producers of products and services. In most cooperatives, users are actively involved in various aspects of management and decision making. The goal of most water supply cooperatives is effective (not necessarily efficient) management, in line with community norms.

In Canada, this model is most widely used in rural areas (there are approximately 200 water supply cooperatives in Canada, mainly in Alberta, Manitoba, and Quebec) (Co-operatives Secretariat 2001). Water cooperatives tend to be rare in urban areas in developed countries, but they are widespread in Denmark and Finland. These Scandinavian countries have a long-standing tradition of private participation in water services through not-for-profit and self-sufficient "water associations" and cooperatives that are owned and managed by consumers, especially in sparsely populated areas. Water cooperatives are usually small-scale but may also be large-scale. In Wales in 2001, the regional water and waste water company, which had been privatized in 1989, was restructured into a non-profit corporation that is owned by its members and serves more than three million customers (Bakker 2004).

### Diversity in Practice: Mixing and Matching Models

There are significant differences between the three models outlined above. For example, the role of the consumer is different under public, private, and cooperative management, with consumers being treated, respectively, as citizens, customers, or as community members. Each role implies different rights, responsibilities, and accountability mechanisms.

Despite these differences, these three models often overlap in practice. For example, municipally run water supply systems can have a commercial approach to water pricing. Publicly owned utilities may even choose to mimic a private company through "corporatization"; that is, through creating a for-profit private corporation that remains publicly owned. Edmonton and Kingston, for example, have corporatized their water supply departments, creating private corporations with the city as sole shareholder. Or municipal governments may choose to contract the services of another publicly owned water supply utility (such as the Ontario Clean Water Agency) rather than a private company.

Water supply is a locally managed resource, which allows different communities to evolve different ways of implementing specific models. The result is a great diversity of approaches. Another reason for diversity is the fact that provinces have responsibility for water supply under the Canadian Constitution, with each province adopting a slightly different approach to regulations governing water supply. In addition, since provinces usually delegate water supply to municipalities, approaches to water supply differ strongly even within provinces. Table 9.1 gives examples of the range of business models applied across Canada.

Despite the range of possible approaches, it is important to note that, during the twentieth century in Canada, municipal ownership and management of water supply systems was dominant. Some private water supply systems were built in large cities (such as Montreal and Toronto), but these were relatively quickly incorporated into growing public systems (Benedickson 2002; Jones and McCalla 1979). Outside of large cities, some private systems continued to operate, but these are typically small; most large Canadian municipalities run their water supply systems directly. As Table 9.2 indicates, six of Canada's ten largest municipalities follow the "traditional" municipal public utility model (Calgary, Montreal, Ottawa, Toronto, Vancouver, Winnipeg). Two others (Halifax and Edmonton) have chosen to corporatize water supply services. Hamilton has chosen to continue with municipal management after having had a ten-year P3 contract for water supply, and Halifax has opted for a P3 for certain elements of its wastewater treatment system. Toronto and Vancouver recently explored restructuring options (with Toronto debating the creation of a stand-alone corporation and Vancouver debating a P3 contract for water treatment), which resulted in heated public debate, after which both municipalities decided to continue with direct municipal management.

## Increasing Involvement of the Private Sector

For much of the twentieth century, water supply was an "invisible" resource; little attention was paid to the systems that keep water running through our taps. By the 1990s, things began to change. Infrastructure systems – many of which had been built at the turn of the century – were aging, and many had

Table 9.1

**Business models for water supply in Canada**

| Business model | Who owns infrastructure? | Who operates infrastructure? | Legal status of operator | Legal framework | Who owns the shares? | Example |
|---|---|---|---|---|---|---|
| Government utility: direct management | Municipal or regional government | Municipal or regional administration | Government department | Public | n/a | Vancouver |
| Municipal board or commission | Municipal government | Commission or board | Public agency | Public | n/a | Peterborough |
| Cooperative | Users/ Cooperative society | Users or delegated authority | Cooperative society or corporation | Varies | n/a (or users) | Rural Alberta, Quebec, and Manitoba |
| Crown corporation | Government or utility | Utility | Usually defined by special law | Public or corporate | Government | Saskatchewan (SaskWater) |
| Corporatized utility | Government or private company | PLC as permanent concessionaire | Corporation | Corporate | Local/prov. government | Edmonton |
| Government utility: delegated management | Government or private company | Government and/or temporary private concessionaires | Corporation | Corporate | Private shareholders | Hamilton |
| Direct private utility | Private company | Private company | Corporation | Corporate | Shareholders or investor-owned | White Rock (BC) |

TABLE 9.2

**Water supply business models in large Canadian municipalities**

| Municipality | Agency | Business model |
|---|---|---|
| Calgary | Calgary Waterworks | Municipal utility |
| Edmonton | EPCOR Water Services | Public corporation |
| Halifax (water) | Halifax Regional Water Commission | Public corporation |
| Halifax (wastewater) | Halifax Regional Environmental Partnership | P3 |
| Hamilton | City of Hamilton | Municipal utility (formerly a 10 year P3) |
| Montreal | Public Works, City of Montreal | Municipal utility |
| Ottawa | Drinking Water Services Division, City of Ottawa | Municipal utility |
| Toronto | Toronto Works and Emergency Services | Municipal utility |
| Vancouver | Greater Vancouver Regional District | Municipal utility |
| Winnipeg | City of Winnipeg Water and Waste Department | Municipal utility |

not been properly maintained (CWWA 1997; Infrastructure Canada 2004; Sierra Legal Defence Fund 2004). Growing populations and increasing per capita demand put strain on existing systems. The availability of grants from higher levels of government that fund water supply infrastructure development has decreased in recent years. Many municipalities lack financing for necessary capital-intensive infrastructure replacements, and politicians lack the political will to increase water prices to economically sustainable levels.

At the same time, consumers' expectations are increasing. The importance of protecting drinking water quality was brought home to the Canadian public by water contamination incidents in Walkerton, Ontario, and North Battleford, Saskatchewan (O'Connor 2002a, 2002b; Prudham 2004; Woo and Vicente 2003). Many water supply systems required considerable investments in order to continue to meet the appropriate standards for drinking water. Environmental pressures are increasingly important, with increasing demands, at times backed up by new legislation, to reduce the environmental impacts of water use. In some regions, such as the Prairies, this is complicated by the increasing unpredictability of water resources due to severe droughts and unusual weather patterns – a possible signal of climate change impacts on water supply (Schindler 2001).

The political landscape is changing as well. In many countries, including Canada, governance frameworks have evolved significantly over the past two decades. Roles previously allocated to governments are now increasingly and controversially categorized as more generic social activities that can be carried out by other actors, such as private companies (Pierre 2000). Some observers characterize this trend as a unidirectional shift toward "distributed governance" (Rogers and Hall 2003), in which government responsibilities and functions are increasingly devolved to market (and, in some cases, community) actors (Pierre 2000; Rogers and Hall 2003). Mike Harris's administration in Ontario, for example, oversaw municipal amalgamation, deregulation, and the introduction of new legislation that enabled greater private sector involvement in the water supply sector (Bakker 2005; Bradford 2003; Prudham 2004).

## The P3 Model

Confronted with significant challenges in maintaining water supply systems, in a political climate conducive to involving the private sector in the provision of public services, some municipalities in Canada have entered into P3 contracts for water supply. To date, these contracts are concentrated in medium-sized communities and currently represent a small proportion of Canadian water consumers (Table 9.3). This pattern of distribution of P3s is due, in part, to the greater resources, expertise, and financing that larger municipalities may generate; in such cases, recourse to the private sector via P3s may not be considered necessary. Local political factors are also important. Over the past three years, the cities of Montreal, Toronto, and Vancouver have considered restructuring (a P3 in the case of Montreal and Vancouver, and corporatization in the case of Toronto). In all cases, restructuring proposals aroused significant public opposition – typically spearheaded by a coalition of environmental groups, organized labour, and other civil society groups. In all three cases, restructuring proposals were scaled back or shelved in response to public opposition.

## Potential Advantages and Disadvantages of P3s

There is limited information about P3 performance in Canada to date, as private companies usually make only limited performance information available,

Table 9.3

**Examples of public private partnership contracts in water supply in Canada**

| | Type of contract | Operator | Start Date | Duration (years) |
|---|---|---|---|---|
| Goderich (ON) | Management contract: water treatment | United States Filter Corporation | 2000 | 5 |
| Moncton (NB) | BOT: water treatment facility | United States Filter Corporation | 1998 | 20 |
| Canmore (AB) | Management contract: water treatment | EPCOR | 2000 | 10 |
| Hamilton (ON) | Management contract: wastewater treatment | Philips Environmental, then Azurix, then American Water Services | 1994 | 10 (now ended) |
| Halifax (NS) | BOT and management contract: wastewater treatment | Consortium (including United Water, Ondeo) | 2002 | 30 |
| London (ON) | O&M contract | Azurix, then American Water Services | 2001 | 10 |

and independent benchmarking protocols are not systematically applied to water utilities in Canada (unlike in other jurisdictions). It is thus difficult to assess the performance of private companies under different contracts. One exception is the case of Hamilton, Ontario, which has one of the earliest and largest P3s to date in Canada (Box 9.1).

As with any business model, P3s have advantages and disadvantages that vary significantly, depending upon the structuring of individual contracts. However, we can make some generalizations about the relative strengths and weaknesses of P3s. If properly structured, P3 contracts may offer several advantages. Necessary expertise, which may not be available in-house, may be obtained on an ongoing basis. Cost savings may be made through efficiency gains, although this depends upon the nature of incentives built into the contract. Increased flexibility in procurement, day-to-day management, and employment practices is another advantage often cited by proponents of P3s, although this varies, depending upon the legislative and regulatory frameworks in place in any given jurisdiction. Access to finance is another potential advantage: a variety of P3 contracts allow municipal governments to

BOX 9.1

## A public private partnership in Hamilton

In the early 1990s, Hamilton was facing a large accumulated deficit in infrastructure investment in water and wastewater treatment capacity; approximately 50 percent to 60 percent of its water and wastewater systems were fifty to one hundred years old, with significant implications for drinking and environmental water quality in the region, particularly in Hamilton Harbour (Hamilton 2001). In January 1995, the Regional Municipality of Hamilton-Wentworth signed a contract with a local company, Philip Services Corporation, delegating the management of the operation and maintenance of the city's water and wastewater treatment facilities, pumping stations, and reservoirs. The contract was not publicly tendered and Philip Services was the sole bidder. The contract value was assessed at approximately $187 million and, when signed, was one of the largest delegated management contracts for water services in North America.

Four different operators have managed the water supply system since the initial contract was signed in 1994. Philip Services operated under the terms of the contract for four years and, after the bankruptcy of its parent corporation, was sold in 1999 to Azurix Corporation, a newly created subsidiary of Enron. Subsequently, Enron sold Azurix to a US-based water services company, American Water. Shortly after completion of the sale, American Water announced that it would be taken over by a German multi-utility, RWE, one of the largest water services corporations in the world. The turnover in operators has been a source of debate in Hamilton. Questions were raised regarding the lack of competitive bidding for the original contract, whether the contract remained legally binding, and the possible financial implications for the city if it were to cancel the contract. Concerns have also been raised about water quality incidents, labour relations, and water and wastewater tariffs that had risen above the rate of inflation during the 1990s. In 2004, the city decided to take its water supply system back under public management.

The Hamilton case illustrates some of the ways in which P3s can lower accountability and transparency. No public participation was initiated by the city. The municipality chose not to publish its contract with the private operators (unlike other cities, such as Moncton and Goderich, which have made their contracts publicly available). Council meetings were closed-door sessions, in part, to protect "commercial confidentiality." Residents of Hamilton had limited means of finding out about the P3 process and had limited information with which to evaluate the performance of the private operators. Poor governance intensified an already heated political debate over the P3 contract.

---

SOURCE: Bakker and Cameron (2003, 2005).

delegate water supply to a private operator who provides project financing. From the perspective of governments, this strategy sometimes has the advantage of reducing apparent pressures on government budgets. The reduction in the government's borrowing requirement does not, however, necessarily imply lower bills for consumers, nor does it necessarily imply cheaper financing: the Walkerton report found that commercial finance was generally more expensive than public finance.[4] Gaining access to financing through the private sector can provide short-term relief to government budgets, but it may mean higher bills for consumers in the long run. Private finance is not a panacea for the financing shortfall in the water sector.

Moreover, as the municipality of Hamilton discovered, good governance is necessary to ensure that the potential advantages of P3s materialize. If governance is poor – for example, weak oversight mechanisms, poorly structured incentives, or unclear performance targets – performance is likely to be poor, no matter what business model is chosen. In many countries, independent regulatory bodies at the municipal, provincial, or even national level have been created, based on the belief that, without robust regulation, neither public nor private water suppliers are likely to perform well; often, these regulators perform "benchmark" comparisons of utilities in order to assess whether prices are reasonable and whether performance is efficient and effective. This is particularly useful in the case of P3 contracts, which typically require skilful contract administration in order to ensure that contractual obligations are being fulfilled – for example, that quality targets are being met.

In Canada, there is no such regulatory framework: P3s are currently regulated "by contract," in which municipalities are responsible for monitoring and regulation. Municipalities often lack the expertise necessary to oversee complex contracts, particularly those granting a large degree of autonomy to the contractor, such as concession contracts. Moreover, contractors typically have a greater degree of experience in P3 contracts than do municipalities. This is particularly the case in the water sector, which is very concentrated: globally, there are fewer than a dozen private companies capable of handling large-scale municipal water supply contracts (Finger and Allouche 2002). These companies typically supply millions of customers and run contracts simultaneously in many countries. Given this imbalance in information and expertise, supervision of the contract and any renegotiation may occur on terms favourable to the contractor but less favourable to the municipality, although this is not so likely to occur when municipalities carefully compare options (including an improved status quo) before choosing to enter into a P3.[5]

Moreover, the municipality may not be able to adequately supervise the condition of infrastructure being managed by the private company. If contracts are not carefully structured, and if monitoring is ineffective, the private sector may not maintain the water supply infrastructure to agreed-upon standards. The alternative – detailed independent regulation – is potentially costly and may undermine some of the desired advantages of P3s, such as greater flexibility in management. Multilateral financial institutions such as the World Bank make significant funds available for the creation of regulators for P3s in developing countries, and other countries with more extensive experience with water privatization, such as England, have created economic regulators. No such system exists in Canada.

Another potential disadvantage is that P3 contracts may result in reduced transparency and accountability to consumers, particularly in the case of long-term contracts. For this reason, P3 contracts are often politically controversial. This may have unforeseen consequences. For example, consumers in England's Yorkshire region did not respond to calls to conserve water during the extreme drought of the mid-1990s because of their resentment of the for-profit model and their perception that conservation would boost company profits (Bakker 2000). Given the risks of reduced transparency and weakened accountability, the Walkerton Inquiry recommended that P3 contracts be made publicly available, and it suggested that companies report requirements to municipal councils at more regular intervals than they do under normal corporate practice.

## Debating Water Privatization: Is Water a Commons or a Commodity?

Resistance to the involvement of the private sector is often influenced by ethical concerns, which can be characterized as the view that water should be treated as a commons and a human right rather than as a commodity (Table 9.4). People who hold the view that water is a commons often assert that water is a resource essential for life – for both humans and the environment. From this perspective, collective management – whether by communities or the state – is not only preferable but also necessary: private ownership of the water supply will, it is argued, invariably conflict with the public interest. Those who advance the "commons" view assert that conservation is most effectively encouraged through an environmental, collectivist ethic of solidarity, which encourages users to refrain from wasteful behaviour. The real "water crisis" arises from socially produced scarcity, in which a short-term logic of economic growth twinned with the rise of corporate power (and, in particular, water multinationals) has "converted abundance into scarcity" (see, for example, Shiva 2002). Accordingly, private companies should be excluded from water management, which should be organized as a "water democracy" consisting of decentralized, community-based, democratic water management, under which water conservation should be politically, socio-economically, and culturally inspired rather than economically motivated.[6]

In contrast, the "commodity" view of water asserts that the private ownership and management of water supply systems (as distinct from water itself) is possible and indeed preferable. From this perspective, water is no different from other essential goods and utility services. Private companies, who will be responsive both to customers and to shareholders, can efficiently run and profitably manage water supply systems. Incentives for water conservation can be provided through pricing: users will cease wasteful behaviour as, with increasing scarcity, water prices rise. Proponents of the commodity view assert that water must be treated as an economic good, as specified in the Dublin Principles[7] and in the Hague Declaration.[8]

Proponents of privatization argue that, through efficiency gains and better management, private companies will be able to lower prices, improve performance, and increase cost recovery, thus enabling systems to be upgraded and expanded. This is seen as critical in a world in which one billion people lack access to safe, sufficient water supplies. Privatization (the transfer of ownership of water supply systems to private companies) and private sector "partnerships" (the construction, operation, and management of publicly owned water supply systems by private companies) have, it is argued, worked well in other utility sectors.

Opponents of privatization point to successful examples of public water systems as well as to research that indicates that private sector alternatives are

Table 9.4

**The commons versus commodity debate**

| | Commons | Commodity |
|---|---|---|
| Definition | Public good | Economic good |
| Pricing | Free or "lifeline" | Full-cost pricing |
| Regulation | Command-and-control | Market-based |
| Goals | Social equity and livelihoods | Efficiency and water security |
| Manager | Community | Market |
| Access | Human right | Human need |

not necessarily more efficient, and are often much more expensive for users, than are well-managed public sector systems (see, for example, Estache and Rossi 2002). They assert the effectiveness of democratic accountability to citizens as opposed to corporate accountability to shareholders – an argument less easy to refute following the collapse of Enron, which, by the late 1990s had, through its subsidiary Azurix, become one of the world's largest water multinationals. From this perspective, the involvement of private companies is incompatible with guaranteeing a citizen's basic right to water. This is because private companies – answerable to shareholders and with the overriding goal of profit – will manage water supply less sustainably than will their public sector counterparts.

## Beyond the Public/Private Divide?

The debate over private sector involvement in water supply in Canada is highly polarized. Political controversy inevitably surrounds the introduction of the private sector into water supply management. Debating this issue should not sidetrack Canadians from the goal of ensuring long-term sustainability of water supply. Other countries have adopted a pragmatic approach and selected the best mix of the various models – public, private, and cooperative – to suit their needs (Box 9.2).

Improving water utility governance in Canada requires creative thinking about the best mix of options for different communities, along with a reasoned process of professional analysis and well informed public debate to decide between those options. Of course, the question of whether to involve the private sector in water supply management should never be considered in isolation and should always involve fundamental questions of sustainability and good governance, such as public health, environmental protection, transparency, public participation, equity, efficiency, and effectiveness.

The question of whether or not to privatize is more than merely technical: it requires a political debate that focuses on our worldviews of water and of society. Other countries have engaged in such debate. In Uruguay, for example, a national referendum was held in 2004 on the questions of whether "water is a natural resource essential to life" and whether access to piped water and sanitation services are "fundamental human rights." Sixty percent of voters agreed with the statements, thus leading civil society groups to call for constitutional reform and the banning of private sector involvement in water supply.[9]

Canadians have not yet engaged in such a national debate, but many local debates are ongoing across the country. As participants are discovering, questions about the respective roles to be played by communities, states, and private corporations raise broader issues of environmental sustainability and deliberative democracy. In debating private sector participation in the water supply, we are also debating the relationship between markets, states, and the environment. If we are to move beyond what risks becoming a stale confrontation between market fundamentalists and ardent defenders of the state, then we must make space for this collective debate.

BOX 9.2

**Beyond the public/private divide: New models for water supply utility governance**

In Chile, private sector participation in water supply has been allowed, which has meant an increase in water bills; but the government has implemented rebates for water bills for poor families. The rebates are linked to family income and ensure that no family pays more than a certain percentage of income for water.

In the Netherlands, all municipal water supply utilities are corporatized: utilities operate as commercial enterprises, charge full market rates, recover all of their costs, and are highly efficient. While operating like private companies, each water utility is overseen by a board made up of members from different government agencies. To prevent privatization, the federal government passed legislation prohibiting the sale of water company shares on the stock market.

In Porto Alegre, Brazil (home of the World Social Forum meetings), the publicly owned water supply utility has managed to extend services to nearly all residents of the city, assisted by a multi-level tariff structure that links bills to income and subsidizes poorer customers. Despite the dominance of socialist politicians in the city government, the utility outsources 40 percent of its business to local, private entrepreneurs, thereby saving money for consumers while supporting the local economy.

The Welsh water supply utility (with over 3 million customers) was privatized through flotation on the London Stock Exchange in 1989. However, water bills had risen well above the rate of inflation since 1989 due to the perceived increased risks of investing in the then private company. In 2001, the company's managers decided to end the experiment with privatization, and the company was converted into a not-for-profit utility, owned by its members and limited by guarantee. The company's AAA bond issue was successful, and it was able to significantly reduce water bills.

## NOTES

1 A market failure occurs when a market fails to meet the assumptions of standard neoclassical economic models, implying that it will not efficiently allocate goods and services. For example, market failures occur when property rights are not clearly defined or are unenforceable, when prices do not incorporate full costs or benefits ("externalities"), when information is incomplete, and/or in a situation of monopoly.

2 A natural monopoly is in effect whenever supply by one firm entails lower costs than does supply by more than one firm. Railroads and utility networks such as gas and water are classic examples of natural monopolies.

3 A "public good" is the term applied to a good that is non-excludable and non-rivalrous. In non-technical terms, this means that it is impossible to exclude some people from enjoying the good (the classic example is national defence) and that one person's enjoyment of the good does not diminish any other person's enjoyment of that good (e.g., my viewing the sunset does not diminish your ability to view the sunset [non-rivalrous]). In the case of water, the public health benefits of clean water supply are public goods.

4 The review of this issue commissioned by the Walkerton Inquiry found that, "in general, the financial capability of a municipal government and its ability to incur debt at favourable rates means that the cost of capital often tips in favour of public-sector based financing for water and sewerage projects" (Joe et al. 2002). Canadian municipalities have fewer options for financing than do their American counterparts as US municipalities can use municipal bonds as a financing mechanism (often supported at the state level by mechanisms to insure bonds and/or provide matching public funds).

5 The Walkerton report, for example, lists three options that municipal governments should consider when reviewing their systems: a municipal department, a municipal agency similar to a public utility commission (or board), and a municipal corporation (O'Connor 2002a, 2002b).

6 One example of this approach is the P7 Declaration on Water of 1997. The P7 (now P8) annual conference was convened for the first time in June 1997 by the Green Group in the European Parliament as an alternative to the G7 (now G8) Summit. Representatives from the world's poorest countries attend the conferences, which focus on the structural causes of and solutions to poverty.

7 The 1992 International Conference on Water and the Environment set out what became known as the Dublin Principles: fresh water is a finite and vulnerable resource, essential to sustain life, development, and the environment; water development and management should be based on a participatory approach, involving users, planners, and policy makers at all levels; women play a central part in the provision, management, and safeguarding of water; water has an economic

value in all its competing uses and should be recognized as an economic good. The Dublin Principles have been adopted by numerous international, multilateral, and bilateral agencies, including the World Bank.

8 The Ministerial Declaration of The Hague on Water Security in the 21st Century was made following the interministerial meeting known as the Second World Water Forum in 2000. See http://www.worldwaterforum.net.

9 In 2005, the Council of Canadians launched a campaign for the inclusion of water as a human right in Canada's Charter of Rights and Freedoms.

## REFERENCES

Bakker, K. 2000. Privatizing Water, Producing Scarcity: The Yorkshire Drought of 1995. *Economic Geography* 76 (1): 4-27.

—. 2003. From Archipelago to Network: Urbanization and Water Privatization in the South. *Geographical Journal* 169 (4): 328-41.

—. 2004. *An Uncooperative Commodity: Privatizing Water in England and Wales.* Oxford: Oxford University Press.

Bakker, K., and D. Cameron. 2003. *Setting a Direction in Hamilton: Good Governance in Municipal Restructuring of Water and Wastewater Services in Canada.* Program on Water Issues, Munk Centre for International Studies, Working Paper No. 1. (November).

—. 2005. Changing Patterns of Water Governance: Liberalization and De-regulation in Ontario, Canada. *Water Policy* 7 (5): 485-508.

Barlow, M., and T. Clark. 2002. *Blue Gold: The Battle against Corporate Theft of the World's Water.* London: Earthscan.

Bauer, C. 1998. *Against the Current: Privatization, Water Markets, and the State in Chile.* Boston: Kluwer.

Benidickson, J. 2002. *Water Supply and Sewage Infrastructure in Ontario, 1880-1990s: Legal and Institutional Aspects of Public Health and Environmental History.* Walkerton Inquiry Commissioned Paper 1. Toronto: Ontario Ministry of the Attorney General.

Bradford, N. 2003. Public-Private Partnership? Shifting Paradigms of Economic Governance in Ontario. *Canadian Journal of Political Science* 36 (5): 1005-33.

Brubaker, E. 2002. *Liquid Assets: Privatizing and Regulating Canada's Water Utilities.* Toronto: University of Toronto Centre for Public Management.

Co-operatives Secretariat. 1998. *The Co-operative Alternative To Public Service Delivery.* Ottawa: Co-operatives Secretariat, Government of Canada.

—. 2001. *Co-operatives in Canada 1999.* Ottawa: Co-operatives Secretariat, Government of Canada.

CWWA [Canadian Water and Wastewater Association]. 1997. *Municipal Water and Wastewater Infrastructure: Estimated Investment Needs 1997-2012.* Ottawa: CWWA.

Estache, A., and C. Rossi. 2002. How Different Is the Efficiency of Public and Private Water Companies in Asia? *World Bank Economic Review* 16 (1): 139-48.

Finger, M., and J. Allouche. 2002 *Water Privatisation: Trans-National Corporations and the Re-Regulation of the Water Industry.* London: Spon Press.

Goubert, J.P. 1986. *The Conquest of Water.* London: Polity Press.

Hamlin, C. 1990. *A Science of Impurity: Water Analysis in Nineteenth Century Britain.* Berkeley: University of California Press.

Hamilton, City of. 2001. *100 Year Report: Infrastructure Asset Management Strategy.* Hamilton, ON.

Hassan, J. 1998. *A History of Water in Modern England and Wales.* Manchester: Manchester University Press.

Infrastructure Canada Research and Analysis. 2004. *Assessing Canada's Infrastructure Needs: A Review of Key Studies.* Ottawa: Infrastructure Canada.

Joe. J., J. O'Brien, E. McIntry, M. Fortin, and M. Loudon. 2002. *Governance and Methods of Service Delivery for Water and Sewage Systems.* Commissioned Paper 17, The Walkerton Inquiry. Toronto: Queen's Printer for Ontario.

Johnstone, N., and L. Wood, eds. 2001. *Private Firms and Public Water: Realising Social and Environmental Objectives in Developing Countries.* London: Edward Elgar.

Jones, E., and D. McCalla. 1979. Toronto Waterworks: 1840-1877 – Continuity and Change in Nineteenth-Century Politics. *Canadian Historical Review* 60: 300-23.

Luckin, B. 1986. *Pollution and Control: A Social History of the Thames in the Nineteenth Century.* Bristol: Adam Hilger.

Melosi, M. 2000. *The Sanitary City: Urban Infrastructure in America from Colonial Times to the Present.* Baltimore: Johns Hopkins University Press.

NRC [National Research Council]. 2002. *Privatization of Water Services in the United States: An Assessment of Issues and Experience.* Washington, DC: National Academies Press.

O'Connor, D. 2002a. *Report of the Walkerton Inquiry.* Part I: The Events of May 2000 and Related Issues. Toronto: Queen's Printer for Ontario.

—. 2002b. *Report of the Walkerton Inquiry.* Part 2: A Strategy for Safe Drinking Water. Toronto: Queen's Printer for Ontario.

Pierre, Jon, ed. 2000. *Debating Governance.* Oxford: Oxford University Press: 138-66.

Prudham, S. 2004. Poisoning the Well: Neoliberalism and the Contamination of Municipal Water in Walkerton, Ontario. *Geoforum* 35 (3): 343-59.

Rogers, P., and A. Hall. 2003. *Effective Water Governance.* Global Water Partnership Technical Committee, TEC Background Papers, No. 7. Sweden: GWP.

Schindler, D. 2001. The Cumulative Effects of Climate Warming and Other Human Stresses on Canadian Freshwaters in the New Millennium. *Canadian Journal of Fisheries and Aquatic Sciences* 58 (1): 18-29.

Shiva, V. 2002. *Water Wars: Privatization, Pollution and Profit.* Boston: South End Press.

Sierra Legal Defence Fund. 2004. *The National Sewage Report Card: Grading the Sewage Treatment of 22 Canadian Cities,* Report No. 3.

Silva, G., N. Tynan, and Y. Yilmaz. 1998. Private Participation in the Water and Sewerage Sector: Recent Trends. In *Public Policy for the Private Sector*. Washington, DC: World Bank Group, Note 147: 1-8.

Woo, D.M., and K.J. Vicente. 2003. Sociotechnical Systems, Risk Management, and Public Health: Comparing the North Battleford and Walkerton Outbreaks. *Reliability Engineering and System Safety* 80 (3): 253-69.

# 10
# Liquid Gold? Water Markets in Canada

*Theodore M. Horbulyk*

Water evokes many emotions. To varying degrees, water is mythical, symbolic, and politically polarizing. Canadians may see water as an inheritance, a birthright, an asset, a resource, or a commodity. However, an important characteristic of water resources is that they have not historically been "commodified" very effectively. In Canada, relatively few water allocation decisions are decided directly by markets and market forces, although that situation is beginning to change.

## Economic Pressures on Water Allocation

Water is highly influenced by powerful economic forces, even when there are no direct markets for water or for some of the services water resources provide. Although there is active political discussion about the appropriate limits on government versus market roles in shaping a market economy, the effects of market-based decision processes are widespread and almost inescapable. This has important implications for the allocation of water. Consider a scenario where, in the future, water – including any of its quantity, quality, reliability, or environmental services attributes – becomes relatively scarce, regionally or nationally, in either Canada or the United States. This scenario is not unlikely: in Canada there are no specific and effective processes for spontaneously balancing regional or continental water resources. Therefore, it is entirely likely that, in coming years, water will, from time to time and place to place, be seen as relatively scarce or relatively abundant. This situation may come to be viewed as either temporary or permanent and will be associated with the impression that water has become relatively "cheap" or "expensive." This may, for example, emerge as a "water quantity" phenomenon, but it may also relate to the quality or reliability of any aspect of each nation's water resources. Given the fact that water plays a key role as a productive resource, or input, in making other goods and services, we can expect economic forces to react to regionalized or nationwide shortages according to the type and degree of integration present elsewhere in the economy.

In increasingly integrated economies, firms or investors from either region or country can acquire water by investing in the assets to which it may be linked. For instance, outside investors may purchase irrigated land that has integrated water rights, or they may purchase ownership positions (directly or indirectly through anonymous stock market transactions) in firms with valuable water rights or water access. Investors may move their water-intensive production facilities away from water-scarce and toward water-abundant areas, such as in expanding or relocating irrigated agriculture, agrifood processing, pulp and paper manufacturing, hydro or thermal electricity generators, and so on. Thus, Canadians should expect to see relative shortages or scarcity of water in any region of North America result in changes to the patterns of ownership and location of water-using production facilities.

Next, consider how changes in the relative scarcity or abundance of consumer goods can be linked to issues of water resource use and management in integrated economies. New attention is being paid to the concept of "virtual," or "embedded," water: the water that is necessary in order to produce commodities. For example, seventy times more water is used in the production of the world's food supply than is directly consumed "as water" by the world's householders. Future shortages in global food markets may start to put upward pressure on the prices of water-intensive commodities such as food grains, and this will translate into increased competition for water resources in food-producing countries such as Canada. In a world that will be adjusting to global climate change or to imbalances in population growth and economic development, increasingly open and active commodity trading systems and trade agreements could increase the rate at which these commodity shortages are transmitted to competing users of domestic water resources. These linkages work in both directions. Changes in world commodity trade can increase competition for domestic water resources, while, at the same time, domestic water management can influence Canada's trade volumes and international competitiveness.

As an example of external economic pressure on water resources, consider the water storage for hydroelectric generation in southern Alberta. Historically, water was impounded in the summer months and released to generate electricity for local markets in the winter months, when seasonal energy demand was greatest. However, recent summer seasonal energy demand has surpassed that in winter, even in local electricity markets. Furthermore, the expansion of the continental distribution grid has raised the relative influence – via market forces – of, for example, California's energy demands and US energy regulators in deciding how Alberta's dams will be operated. Thus, if the operators of hydroelectric storage facilities see high spot market prices for electricity in the summer months, and if there is sufficient electricity transmission capacity available to reach those markets, then these operators may spill more water from storage reservoirs through their generators. In this example, electricity market forces elsewhere on the continent have the effect of raising Alberta's river flows and of making more surface water available downstream in the summer months, when that water can provide much higher value to irrigators and environmentalists alike. More to the point, historically, there have been no water markets or similar enabling institutions to which either the irrigators or the environmentalists could turn to achieve equivalent transactions directly within their own province.

Trade-related competition for water resources will not only be related to irrigation and manufacturing; water-based tourism, fishing, recreation, and travel are water uses that are also affected by so-called "trade in services." Thus, it is likely that some of the increased competition for water resources might be experienced as increased demand for ecological and ecosystem uses of water. If international trade agreements increasingly include environmental safeguards that restrict water uses in other countries, the result may be added pressure on Canada's water.

## The Creation of Water Markets

One means of responding to these competing demands for water is to create water markets. Various US and Canadian jurisdictions have been developing markets for water or for various types of water rights, or they have been introducing other forms of pricing and market-based instruments. Especially where transactions and administrative costs can be kept low, and where market information can be conveyed easily (such as via the Internet), there is considerable potential for markets to anticipate temporary or permanent imbalances in water supply or demand.

How do water markets actually work? Various types have existed throughout history in such countries as Australia, Chile, Spain, and the United States (see Haddad 2000; Young and McColl 2005), and innovative approaches continue to be proposed and evaluated. In the simplest markets, resource managers make an annual prediction of water availability and then hold a water auction; in reality, however, this rarely occurs. Surprisingly, water markets do not always exchange volumes of water at all but, rather, the right or entitlement to use specific volumes of water under certain conditions.

In some cases, the "creation of a water market" might consist largely, and quite simply, of passing legislation describing the conditions under which governments will authorize the transfer of existing individual water rights. In these simple water markets, private buyers and sellers seek each other out, independently or with the assistance of intermediaries, and negotiate individual water sales. Historically, there may not have been any provision for these rights to move among water users, especially in exchange for monetary compensation. Indeed, even in jurisdictions where water markets already exist, many governments continue to issue rights or licences for other forms of natural resource use (such as timber harvesting, cattle grazing on public lands, and mineral extraction) that are not freely transferable among users in markets.

More sophisticated water markets might be promoted by the creation of specific market institutions, agencies, and structures that are to be operated by the government or by the private sector (usually working under government sanction). Clifford, Landry, and Larsen-Hayden (2004) use the term "water bank" to describe any of a diverse set of institutionalized processes used to transfer water entitlements among users via market processes. These authors survey water banks in twelve western US states and catalogue the banks' varied and evolving roles as, among others, clearing houses, electronic exchanges, brokers, market makers, and price setters.

In creating markets for water, there is usually considerable latitude for governments to prescribe the structure of the market and the nature of the "goods" to be traded. Indeed, a potentially important dimension of new water markets involves the specification of which parties the government will allow to participate. In practice, various jurisdictions have approached this issue differently. One might expect considerably different market behaviour and outcomes in, for example, a market that allows trades only among irrigators within one irrigation district than in a market that opens trading to any prospective purchaser. Some prospective purchasers might be potential users (e.g., those wanting to start a water-using business); other prospective purchasers might not be users in any consumptive sense (e.g., those who wish to preserve water in situ for environmental uses).

In some cases, the water rights offered for sale in a market transaction might correspond to a point and time of water use considerably different from that intended by the purchaser. Prevailing water laws or policies might only allow one to exercise the newly acquired right at the alternate location under specific new conditions, or it could require further payments for storage and distribution. Where there is competition for storage and distribution facilities, it may be preferable to have separate markets for water storage, for users' access to delivery capacity in a particular (natural or constructed) watercourse, and for use of the water itself. Under such a system of water markets (recently contemplated for adoption in Australia) water users would have to participate in a complete series of water market transactions before any water purchased could be used.

What impacts do markets have? Economists have historically examined and explained the role that markets and market-based processes can play in allocating all types of goods and services and in distributing the gains and losses that are associated with each transaction. Markets often display such advantages as bringing together willing buyers and sellers in a manner that can be flexible and highly responsive to changing conditions. Markets provide a basis for societies to organize their production, consumption, and trading activities, all at a far lower cost than might be achievable through an alternative system based on centralized planning or on individualized rations, quotas, or targets. Some markets play an important role in providing incentives to firms and to individuals to develop new sources of supply and to reduce usage or waste of resources. Markets can provide incentives to take risks, such as those involved in developing new technologies. Observation of market transactions can generate valuable information to consumers and to policy makers alike regarding the relative scarcity of particular goods and services, and it can provide valuable forward-looking signals about the expectations and beliefs of large numbers of market participants.

Recently, a number of economists described the specific application of markets to environmental goods and services, including markets for airborne emissions and for water (AIA 2005; Horbulyk 2005; Portney 2003; Woodward 2005). Almost always, these discussions are quick to describe the important limitations of markets, such as those that occur when the goods and services in question affect specific third parties (those not directly involved in the market transaction). This is the case for water resources, given that water is a multiple-use resource that is integral to human and environmental health. Another limitation arises when the services are diffused widely over groups of people who cannot effectively act with a single voice so as to generate market bids for goods and services that they value highly. In the case of water-related transactions, it could be costly, difficult, or impossible to observe accurately, to monitor, and to enforce all of the many transactions that could emanate from some new types of markets – such as markets for "pollution credits" defining allowable emissions from nonpoint sources (King 2005).[1]

Thus, the ability to generate improved outcomes through the use of markets is likely to rely heavily upon new and effective monitoring and enforcement practices on the part of government. Experiences with water markets in Australia and various US states support this idea (Young and McColl 2005). If potential gains are to be realized, then considerable effort on the part of governments is likely to be required in order to define appropriate water market rights, processes, and regulatory oversight. Dellapenna (2005) argues that the required degree of public intervention might be so large that the resulting arrangements hardly qualify as a market at all and, in fact, could better be viewed as an enhanced form of public management.

Another concern that can arise with water markets is that they may have the effect of capitalizing, or monetizing, historical or newly created entitlements to use "public" resources. Some people will see it as unfair that others gain direct financial advantage. In other words, if water markets create "liquid gold," then, regrettably, it may often be someone else's. A counter-argument is that it is exactly this opportunity for achieving individual gain that provides direct incentive to reduce wastage and to move water to higher valued uses. In some cases, putting a highly visible price tag on achieving such a change of practices only serves to highlight how poorly the resources were managed in the first place.

To the extent that people view water resources as a public resource and common inheritance, there will be philosophical opposition to creating market-based schemes that assign private ownership to water use "property rights." Either in perception or in reality, water resources have historically been part of some public "commons," and the sale of any entitlements may be seen as eroding this public asset. However, this type of public opposition may also be present in situations where water has not effectively been part of a public commons for quite some time. Here, the impetus for market-based approaches may be that patterns of private and public water use and priorities are well established – even if not formalized as water rights or appropriative allocations. Water may currently be allocated in ways that rigidly follow historical patterns but that, in a changing economy, are no longer seen as desirable. In these cases, support for the creation of water markets might be based on a desire to open up or to free up, once again for the public good, the process by which access to water resources is decided.

## Why Have Markets for Water Resources Become More Popular?

Why have markets for water resources, in particular, gained popularity in recent years? A number of authors have provided insight into this question. Zilberman (2005) argues that the pattern of water policy reforms in the United States and elsewhere is driven by the mix of specific pressures being exerted on politicians, resource managers, and users. In his view, real or anticipated scarcity is what is responsible for the establishment of systems of water trading, while, at the same time, the increasing financial cost of water system expansion has led to various forms of privatization or public-private partnerships. Increased concern about the environment has led to new emphasis on surface and groundwater quality, along with improved support for payments for watershed services. Concerns about fairness and social justice have led to pressure to regulate the prices at which water is sold or to provide subsidies, especially in the context of developing countries.

Clearly, there are connections among these developments. The adoption of water trading mechanisms generates price signals and market revenues that facilitate the role of private developers or operators. Zilberman (2005) emphasizes the combined roles of accumulated institutional knowledge and emerging crises as determinants of when and how rapidly transitions to market allocation might occur, citing the effect of the 1990s droughts in California on the rapid promotion and acceptance of water markets there.

Howitt and Hansen (2005) identify a number of precursors to the introduction of water markets (such as the existence of infrastructure to move and to store water) and to well-defined property rights to water that are both enforceable and transferable. These authors argue that trading is more likely to proceed through markets where it can occur with few adverse physical or financial effects on third parties, either because such effects are not present or because they are well regulated. In a data-intensive study of fourteen western US states between 1999 and 2002, Howitt and Hansen also identify the relative importance of short-term "leases" of water versus permanent water "trades." Each type of transaction may allocate future risks differently across buyers and sellers, and it may involve considerably different regulatory processes and burdens. The data show that, for twelve of the fourteen states, leases (including those used to enable various forms of water "options") are the principal form of water market transaction.

## Alberta's New Experience with Water Markets

To what extent are these precursors and determinants of US water market development relevant to the Alberta experience? Historically, Alberta, like the other provinces, had no experience with water markets. Residents make use of the numerous rivers that transect the province from west to east and that largely determined patterns of settlement and urbanization. Because of the gradient on the downslopes of the Rockies, many of these rivers and tributaries have previously been dammed for the purposes of hydroelectric generation and flood control, providing a valuable source of interseasonal surface water storage.

To make use of these resources, Alberta, like many western US states, has employed a system of appropriative water rights based on seniority. In Alberta, ownership of groundwater and surface water rests with the government, but rights to abstract and use that water are granted under such legislation as the Water Act (Alberta) and the Irrigation Districts Act (Alberta). Historically, water users could apply for water licences to use surface water or groundwater, and there were other provisions by which riparian landholders and other rural householders could use limited amounts of water without such a licence. Any licence issued was linked to the specific parcel of private land where the water could be used, and users had seniority based upon the time when they applied for the licence (this is known as the "first in time, first in right," or "prior appropriation," approach).

This system has come under increasing strain as water use has continued to grow, exacerbated by recent droughts. The Alberta Institute of Agrologists (AIA) (2005, 5) describes how growing water scarcity in southern Alberta caused the government to impose a moratorium on issuing any new licences in specific water basins. Once a landowner's new demands for water could not be met through an application process, it was clear that something had to be done. Without some reform of water policy, patterns of water use and of broader economic development could have remained locked into the historical water use pattern, without the flexibility to address changing needs.

It is significant, in Alberta's case, that the policy reforms that were implemented chose to respect and perpetuate the seniority-based system of appropriative water rights. Another jurisdiction might have chosen to address the underlying issues of water scarcity by introducing a system of proportional sharing in times of shortage or by introducing a scheme based on the relative priority of end uses of water. This would ensure, for example, that residential needs would be met before industrial ones (or vice versa). Alberta policy reforms chose not to rely directly upon water-pricing strategies to constrain or to redirect water use. Even with the existing water rights system, a new pay-per-use scheme would almost certainly have curtailed water use by an amount sufficient to meet current and future needs. However, with no extra water to allocate, with the endorsement of the established water rights system, and given the reluctance to rely upon restrictive pricing schemes, Alberta took the short and obvious step of adopting water markets as a way of avoiding being locked into historical water allocations.

Since 1999, Alberta has allowed various types of market transactions to reallocate diverse rights to water use. For the majority of commentators, the most significant development in Alberta's water use has been the recent introduction of permanent water transfers from one place and type of use to another (even though, by early 2006, less than ten permanent water transfers had been authorized). Space limitations do not allow a full description of the systems of water rights and trading currently used in Alberta, but key points are covered in AIA (2005) and Nicol (2005). Only since 2000 has the Water Act defined a process for permanent transfers of water licences, such as those that are held by diverse agents like irrigators, irrigation districts, cities, towns, and others. The buying of water licences is only open to those who already hold a licence, although one can become eligible by purchasing land that has a water licence attached. So far, this requirement appears to have kept many of those who would buy water rights to promote non-consumptive uses (such as purchases by a citizen's trust to enhance instream flows) out of the market. AIA (2005, 15) reports that Ducks Unlimited owns or controls more than 300 individual water licences, although this group actually takes delivery of licensed water to enhance waterfowl habitat and wetlands.

Nicol (2005) uses a series of case studies to describe six of the first transactions involving the permanent transfer of water licences away from the property to which each was tied historically. Through surveys and interviews, she explores the motivations, processes, and outcomes associated with each permanent transfer. It is clear that, as is often the case with a new regulatory procedure, these participants incurred considerable expense to set new precedents and protocols for those who might follow them.

Since 1999, the Irrigation Districts Act (Alberta) has authorized the temporary transfer (or "lease") of water rights held under an irrigation district's licence and exercised by individual irrigators within each district. Nicol (2005) studies some 222 temporary transactions that occurred in 2001 within the St. Mary River Irrigation District in southern Alberta, carefully examining the characteristics of buyers and sellers.

Outside of irrigation districts, the Water Act (Alberta) contains a provision to reassign water temporarily between one licensed user and another. These temporary transfers may have considerably lower regulatory costs than do permanent transfers. However, these temporary assignments are restricted to pushing the buyer's level of water use back up to the pre-existing licensed levels. These assigned water amounts cannot be so large as to augment the levels of use for existing or new users. In this specific detail, the assignment provisions currently available in Alberta are considerably more limiting than are the lease provisions studied by Howitt and Hansen (2005) in fourteen western US states.

How will Alberta's water markets develop in the future? Alberta's formative water markets do not yet feature prominent "water banks," "trading exchanges," or other marketplace infrastructure. Anecdotal evidence describes a number of individuals within the irrigation industry who are willing to act as market intermediaries, matching willing buyers and sellers and helping to propose the terms of a deal that might satisfy both. At least one private sector electronic trading exchange stands ready to commence water trading operations as soon as market activity can support it.

Computational modelling tools have been designed and calibrated for the southern Alberta watersheds in order to simulate the kinds of trading that might occur in times of drought. An example is a situation in which there are relatively flexible systems of water leases in the region (Horbulyk and Lo 1998; Mahan, Horbulyk, and Rowse 2002). However, there is a large gulf between the magnitude of water trading that has been seen in Alberta to date and what such models predict could be beneficial even within a single irrigation season. Of course, a common feature of many new markets for water, as with other environmental goods and services, is the relatively low volume of initial trades (except during times of crisis).

Recognizing this, Alberta has committed to considering an even broader set of policy reforms in order to build on these new markets, hoping to eventually make use of other forms of economic instruments to target the increased efficiency and productivity of water use (AIA 2005). Urquhart (2005), for instance, argues forcefully for choosing increasingly integrated approaches to resource management and implementing it wherever possible at the watershed or basin level.

Although it is too early to assess the ultimate success of Alberta's water markets, it seems that the ability to undertake various types of market-based transactions has provided a "pressure-release valve" for a system that had become gridlocked. Early evidence suggests that permanent water trades have been used rarely, that other forms of short-term trading are far more popular. In the future, one might expect to see greater use of options contracts, such as those based on contingent entitlements. These can be a cost-effective way for large consumers to adapt to future supply variability. Yet use of such markets may introduce policy debates about foreign ownership of water rights or about the desirability of having conservationists acquire and reallocate water supplies.

## Questions for Debate

The pressure that market forces increasingly place on water resources will impose choices – sometimes very controversial choices – on Canadians. As discussed at the beginning of this chapter, even where local decisions were made to limit direct markets for water resources, it seems unlikely that individual jurisdictions could effectively isolate them from the many market pressures that work indirectly through the production and trade of goods and services. Accordingly, a number of questions present themselves in this debate: Are Canadians satisfied with the methods by which rights to diverse water resources are defined and allocated? If historical, current, or future rights could be reassigned – with or without full compensation, temporarily or permanently, whether by market mechanisms or otherwise – what restrictions or controls might Canadians wish to see imposed on the possible outcomes? How can the governance of water resources be improved at all levels? There have, for example, been historical concerns about accountability and funding for water supply, treatment, and sanitation infrastructure. More recently, public-private partnerships (so-called "P3 initiatives") have been implemented in some jurisdictions, apparently even before broader terms of governance and accountability have been well established or understood. Similarly, the advent of market-based instruments may call for the development of specialized water courts and/or for new forms of water market regulation. Does the apparent opportunity to harness the power of market forces for water allocation create pressure to improve water governance? Or, conversely, is improvement in water governance capacity now facilitating the use of water markets?

To what extent, and through which processes, are Canadians prepared to anticipate and to resolve domestic and bilateral issues of interjurisdictional cooperation over water resources? Domestically, potential disputes are not only limited to those between specific provinces and the federal government but may also involve multiple provinces, First Nations, and cities or regions. Bilaterally, for example, there may be a need to revise the issues and processes covered by the International Joint Commission and the Great Lakes Charter Implementing Agreements.

What do Canadians view as the appropriate role of civil society and community groups in the governance of water resources? Who, for example, should be allowed to participate in water markets? A relatively recent trend in the United States involves the growth of small-scale, community-based, local watershed protection groups, along with much larger basin-wide water organizations. How can the potential value of such groups be harnessed? And how can their information needs be met? For example, in the United States, widely collected local stream flow and hydrology data are publicly and continuously available to all via the Internet. In Canada, data are less widely collected and less readily available.

## Conclusion

Water not only flows over and under provincial and Canada-US borders but is also embodied in, and influenced by, the growing trade in goods and services. If water use decisions were ever isolated from the dictates of market forces, this is no longer the case. Greater integration within and across Canadian and American markets is applying additional market pressure, both directly and indirectly, on the allocation and management of water resources. This greater integration may also provide an opportunity for increased cooperation in the management of transboundary water resources as well as an opportunity to share experiences in such areas as improving governance and community participation. Early evidence suggests that diverse forms of water markets and the implementation of market processes will be an ever-present part of this policy debate. As Zilberman (2005) notes, historically, short-term crises have served as powerful and effective catalysts, leading to rapid and profound changes in water policy and practice. To the extent possible, citizens across the continent will want to prepare for these crises by having their facts at hand and their homework completed, especially when faced with water policy choices that are highly controversial and not easily reversible.

## ACKNOWLEDGMENTS

The author acknowledges research funding support from the Alberta Ingenuity Centre for Water Research (http://www.aicwr.ca) and from the NCE Canadian Water Network (http://www.cwn-rce.ca). Portions of this chapter first appeared as Horbulyk (2004).

## NOTE

1 Nonpoint sources refer to small, diverse, and diffuse contributors to contaminant loading in a watershed, such as hundreds of individual farm fields and pastures that might release phosphorus into waterways. A sewage outfall or a large factory's discharge pipe, by contrast, would be considered a point source of phosphorus or other contaminants.

## REFERENCES

Alberta Institute of Agrologists (AIA). 2005. *Environment for Growth: People to Water, Water to People*. Edmonton: AIA. http://www.aia.ab.ca/policy.

Clifford, P., C. Landry, and A. Larsen-Hayden. 2004. *Analysis of Water Banks in the Western States*. Publication 04-11-011. Olympia: Washington State Department of Ecology. http://www.ecy.wa.gov/biblio/0411011.html.

Dellapenna, J.W. 2005. Markets for Water: Time to Put the Myth to Rest? *Journal of Contemporary Water Research and Education* 131: 33-41. http://www.ucowr.siu.edu/updates/131/08_dellapenna.pdf.

Haddad, B.M. 2000. *Rivers of Gold: Designing Markets to Allocate Water in California*. Washington, DC: Island Press.

Horbulyk, T.M. 2004. *Canada-U.S. Water Issues*. Briefing note for the Centre for Global Studies conference, entitled "Canada and the New American Empire." Victoria: University of Victoria.

—. 2005. Markets, Policy and the Allocation of Water Resources among Sectors: Constraints and Opportunities. *Canadian Water Resources Journal* 30 (1): 55-64.

Horbulyk, T.M., and L.J. Lo. 1998. Welfare Gains from Potential Water Markets in Alberta, Canada. In *Markets for Water: Potential and Performance,* ed. W. Easter, M.W. Rosegrant, and A. Dinar, 241-57. Boston: Kluwer Academic.

Howitt, R., and K. Hansen. 2005. The Evolving Western Water Markets. *Choices: The Magazine of Food, Farm, and Resource Issues* 20 (1): 59-63. http://www.choicesmagazine.org/2005-1/environment/2005-1-12.htm.

King, D.M. 2005. Crunch Time for Water Quality Trading. *Choices: The Magazine of Food, Farm, and Resource Issues* 20 (1): 71-75. http://www.choicesmagazine.org/2005-1/environment/2005-1-14.htm.

Mahan, R.C., T.M. Horbulyk, and J.G. Rowse. 2002. Market Mechanisms and the Efficient Allocation of Water Resources in Southern Alberta. *Socio-Economic Planning Sciences* 36: 25-49.

Nicol, L. 2005. Irrigation Water Markets in Southern Alberta. MA Thesis, University of Lethbridge.

Portney, P.R. 2003. Market Based Approaches to Environmental Policy: A "Refresher" Course. *Resources* 151: 15-18. http://www.rff.org/rff/Documents/RFF-Resources-151-Marketapproaches.pdf.

Urquhart, I. 2005. Alberta's Land, Water and Air: Any Reasons Not to Despair? In *The Return of the Trojan Horse: Alberta and the New World (Dis)Order*, ed. T.W. Harrison, 136-55. Montreal: Black Rose Books.

Woodward, R.T. 2005. Markets for the Environment. *Choices: The Magazine of Food, Farm, and Resource Issues* 20 (1): 49-51. http://www.choicesmagazine.org/2005-1/environment/2005-1-10.htm.

Young, M., and J. McColl. 2005. Defining Tradable Water Entitlements and Allocations: A Robust System. *Canadian Water Resources Journal* 30 (1): 65-72.

Zilberman, D. 2005. Emerging Water Policy Trends: An International Perspective. Paper presented at the 80th Annual Conference of the Western Economic Association International, 6 July, San Francisco.

# 11

# Trading Our Common Heritage? The Debate over Water Rights Transfers in Canada

*Randy Christensen and Anastasia M. Lintner*

In recent years, the Sunshine-Coast Regional District, which is located about eighty kilometres and a ferry ride away from Vancouver, has needed additional water. Like many municipal districts, it was eager to increase its tax base, so when approached by developers eager to expand on the booming housing market, the district immediately said yes – and then scrambled to find water.

The district turned its gaze to nearby Hotel Lake, a scenic but tiny lake described by one resident as constituting not much more than a "puddle on a rock." Local residents had long been concerned about the existing overuse of the lake, which is just over a kilometre at its longest, less than half a kilometre at its widest. As with many water sources in Canada, government officials had continued to grant water rights to Hotel Lake until conflict between competing uses was inevitable. Local residents were eager to protect the drinking water source, and conservationists and the federal fisheries agency worried about an endangered population of sockeye salmon in a stream that depended upon outflow and seepage from the higher elevation lake. All were concerned about the impacts of potential water diversions proposed by the district.

In 2003, the district had applied for a new licence to abstract water from Hotel Lake. Due to local concerns, the provincial government put the district's application for a new licence on hold and ordered a study of potential impacts. Instead of paying for a study that might not support its application, the district paid $85,000 to an existing owner of water rights to Hotel Lake, thereby obtaining a source of new water while avoiding an increase in the quantity of legally allocated water from the lake. In 2004, the provincial government approved this "water rights transfer,"[1] despite the fact that the water rights had not been used for thirty years and that the maximum volume allowed by the water right could result in a 70 percent increase in actual water use in the lake.

Fearing lowered water levels, decreased water quality, and the possibility that unregistered domestic users would lose access altogether, local residents and conservationists filed a legal challenge. The residents' appeal was successful in getting the transfer put on hold, pending further studies.[2] The resort to litigation will surprise few who are familiar with water use management. Water rights transfers are one of the most controversial water governance issues, and they are a microcosm of one of the most controversial debates of all: the extent to which water use should be managed as a commodity and water rights treated as private property. Where water rights transfers are allowed, litigation soon follows.

The Hotel Lake fight is one of the first of its kind in Canada, but it is safe to say it won't be the last. The growing demand for water – both inside and outside Canada – will cause increased conflict between existing users as well as a growing recognition of environmental needs. And those who seek water in the future may discover it's all spoken for. As discussed below, water rights transfers will either play a key role in addressing this inevitable conflict or they will exacerbate it.

## What the Heck Is a "Water Rights Transfer"?

At the outset, it may help to clarify some similar-sounding terms. A "water taking" refers to the actual physical capture and use of water. Water takings are often categorized as being a withdrawal or a diversion. A withdrawal generally refers to a water taking that is returned to or kept within the same watershed. On the other hand, a diversion generally refers to a water taking that is removed from a watershed. Throughout this chapter, the term "water use" denotes any water "taking." Any volume of water may be involved: from small quantities of water on land directly adjacent to a river or lake to large-scale removal of water long distances from the water source, including highly controversial "out-of-basin" diversions and water "exports."

To undertake any of these activities, it is generally necessary to have legal permission to withdraw or divert the water, which is referred to as a "water right." Water rights are usually attached ("appurtenant") to land and are passed along with title to the land. "Water rights transfer" generally refers to the conveyance of water rights from one party to another for use in a different place and often for a different purpose. Alberta has the most mature water rights transfer market, followed by British Columbia. A "water market" (discussed in Chapter 10, this volume) may refer to something as simple as the ability of private parties to arrange rights transfers between themselves (subject to government approval) on terms they negotiate, which may include payment to the holder. More commonly, the term evokes situations in which government policies, institutions, and brokers encourage and facilitate transfers.

### Transferability and Ownership: The Interrelation

The question, "who owns the water?" is controversial and generates highly charged public debate. Canadian law generally recognizes water as a public resource but allows the establishment of private property rights to the use of water. For both pragmatic reasons and reasons of public policy, full "ownership" of water in lakes, streams, or in the ground is not allowed. On the pragmatic side, water moves through a continuous cycle – through rivers, streams, and the ground – in ways that have no regard for the boundaries of private property or national borders, so granting legal title to a litre of water in a river is simply impractical. In terms of public policy, water serves myriad human and non-human needs, including environmental needs, and this makes full private ownership inappropriate (because of the need to balance between competing uses).

Regardless of policy and political objectives that may be served or compromised by water rights transfers (discussed below), there is no avoiding the fact that allowing water rights transfers brings us a step closer to private ownership of water. *Black's Law Dictionary* defines "ownership" as "a collection of rights to use or enjoy property, including the right to transmit it to others." Accordingly, before launching full tilt into a discussion of water rights transfers, it is worth taking a brief look at historical attempts to balance water's public and private interests as well as at the broad outlines of water management in Canada.

## The Historical Context

Aboriginal custom did not create private rights to water (Aboriginal water rights are the subject of Chapter 15, this volume); rather, it sought to ensure that all forms of water usage recognized and respected one's spiritual connection with water. In the Anishinabe tradition, for example, women are the caretakers of the water and any contemplated "use" is subject to the responsibility to "ensure the survival of the seventh generation."

To assert that European settlement brought a new approach to water management is an understatement. The idea of applying property rights to water use arose out of attempts to protect water's public uses while allowing private parties to use it for their own purposes. Historically, public uses predominated in Canadian common law, whose origins are in British and Roman law. The Roman approach to water held that the primary values of rivers and seas are preserved when they are held in common, with protection of public values being predominant; however, it admitted that marginal improvement to overall welfare might occur when some limited private access was allowed (Epstein 1994).

Based on Roman law concepts, European and Middle Eastern legal systems have long accepted rivers as *public property* (Teclaff 1972). Among others, contemporary Spanish and French laws expressly acknowledge that water in rivers is public property (Nanda 1977). In the United States, many state constitutions declare water resources to be public property (Blumm 1989). On the other hand, Islamic law views water and "great rivers" as *common property*, with private rights being confined to small volumes of water within well-defined boundaries (Nanda 1977). Even in Britain, which has a system of private "riparian rights" linked to land ownership, the common good was prioritized by allowing landowners adjacent to lakes and streams the right to use water so long as overall quantity and quality – and the ability of other right-holders to use water – were not unreasonably impaired.

Currently, it would be difficult to argue that Canada manages water for the primary purpose of protecting common values. While all Canadian jurisdictions, except Ontario and Prince Edward Island, explicitly vest the ownership of water in the Crown (Christensen 2005),[3] most provincial governments manage water in order to maximize private commercial and/or industrial activity, which "requires" granting private parties "secure" access rights.

## Current Water Management in Canada

The primary responsibility for managing water in Canada rests at the provincial level, and the disparate approaches adopted have resulted in an incomplete, inconsistent, and often ineffectual patchwork of laws and regulations controlling water use (La Forest 1973; Percy 1988). There are five major approaches to water rights in Canada: prior allocation, public authority, riparian rights, civil code, and Aboriginal water rights. As prior allocation and public authority systems are the only ones that allow water rights transfers, they are the ones given most attention here.

### Prior Allocation

Water use legislation in western provinces (British Columbia, Alberta, Saskatchewan, and Manitoba) and, to some extent, Nova Scotia is based on the "prior allocation doctrine."[4] A licensee acquires the exclusive right to use water from the date of the licence application ("first-in-time, first-in-right"). The first-in-time, first-in-right approach to water use sets up a seniority system whereby, in times of shortage, the most senior licence holder gets his or her full allocation before any junior licence holder. Generally, the rights are appurtenant (legally attached) to a particular parcel of land. These systems may allow limited use of water without a licence for specified purposes (e.g., domestic). In many cases, although increasingly less frequently, water licences were issued "in perpetuity."

These systems generally employ the underpinnings of a concept known as "beneficial use," where "beneficial" requires water to be used for purposes determined to be consistent with societal objectives and "use" requires actual use – commonly called "use-it-or-lose-it." In these systems, the purposes for which licences may be granted are defined, and the licences may be lost for non-use. Prior allocation systems have been harshly criticized. Water managers often grant rights until water supplies are over-appropriated. There is little oversight of licences through approaches such as metering or reporting of usage. Use of the water is often free or at very low cost, and there is generally no ability to "hold back" amounts for environmental needs.

Water rights transfers are often recommended to remedy the particular deficiencies of prior allocation systems. Alberta, for example, substantially revised its water use legislation in the late 1990s to accommodate water rights transfers. Although it has not amended its legislation in this regard, British Columbia is now granting water rights transfers applications, relying on an expanded interpretation of a "transfer of appurtenance provision," which previously allowed a single owner to change the place of use of a water right. Manitoba and Saskatchewan prohibit water rights transfers but are discussing possible adoption.

### Public Authority Management

In the Yukon, the Northwest Territories, and Nunavut the "public authority management" governs decisions about water use, which is generally implemented through local water boards.[5] All uses (except domestic and emergency uses) require a permit. This regime also has a "use-it-or-lose-it" cancellation component as well as provisions for cancellation or amendment for public interest considerations (including low water). Transferability of licences is permitted in the northern jurisdictions. Transfer of licences in the northern territories is rarely, if ever, motivated by unavailability of water.[6] This chapter is concerned with water rights transfers as a tool for addressing situations in which all available water in a local area has been allocated. As this occurs in Alberta and British Columbia, water rights transfers in the northern territories will not be further discussed.

## Riparian Rights

Water use legislation in Ontario and the Maritimes is based, in part, on the "riparian rights doctrine." This is inherited from England, where, historically, an owner of land that borders on a water source (e.g., land at the edge of a lake or land over which a stream flows) enjoys certain water rights called "riparian rights." Primarily, the landowner is entitled to have access to water flow in its natural quantity and quality (La Forest 1973). The landowner is also entitled to limited rights of use. Use must be for ordinary or domestic purposes, which (provided that the use occurs on the land itself) are not limited in amount. Extraordinary purposes (including irrigation and manufacture) are permitted if the use is reasonable and the water is returned to the water source substantially unaltered in quantity and quality. Thus, riparian rights do not permit use on non-riparian lands. A landowner, for example, could not agree to divert water from his or her property to provide drinking water for a neighbour. As such, riparian rights of use are not "transferable." Also, as groundwater does not flow in open channels, riparian doctrine does not apply to it. Ontario has placed a statutory limit on riparian rights by requiring a permit for use when water is withdrawn from surface water or groundwater sources in amounts above 50,000 litres per day for industrial uses and water bottling. Transferability of these permits is not currently allowed, but it is foreseeable that, when water sources become fully allocated, demands to transfer permits may arise.

## Civil Code Management in Quebec

Quebec's legal system is solely based on a statutory "civil code." The Civil Code of Quebec recognizes surface water and groundwater as a resource whose use is "common to all." A recent water policy further elaborated that the right to have access to and to use water should be applied "in a manner consistent with its nature" and, further, that the "government has a responsibility to regulate water use, establish priority uses and preserve its quality and quantity, while taking the public interest into account" (Ministère du Développement durable, de l'Environnement et des Parcs 2002). Permits for water use are not granted by any single agency. Permits granted are in relation to the type of use and are governed by different ministries.[7] Water use permits are not transferable.

### Aboriginal Water Rights

All of the four legal approaches to water use in Canada are subject to claims of Aboriginal rights and treaty rights (discussed in detail in Chapter 15). Aboriginal customs (or customary law) governed the use of water prior to European settlement. Even after the assertion of Canadian sovereignty over Aboriginal peoples (first claimed by Britain in 1763), Aboriginal customary law continues to exist in tandem with Canadian law. To resolve the existence of both Canadian and customary law, Aboriginal rights have been interpreted within the Canadian legal system as including certain Aboriginal customs and practices.[8] In 1982, Aboriginal rights and treaty rights became constitutionally protected. Any Aboriginal rights and/or treaty rights (that were not extinguished prior to 1982) can no longer be interfered with (i.e., "infringed") by government.[9]

## Are Transfers Good or Bad?

Water rights transfers, depending on your view, will either resolve excess water demands and conflict between users or they will exacerbate them. As in many places around the world, in Canada, the current system of water rights allocation and management has been strongly criticized in recent years. This is in part due to increasing stresses on water resources (Chapters 1 and 2, this volume), new social priorities (such as environmental protection), and changing political commitments (such as the federal government's reduced funding for freshwater monitoring).

Some observers argue that current water rights allocation systems are fundamentally flawed, particularly because flowing water is treated as having no monetary value: rights are often obtained for little or no money and can't be sold. Furthermore, applicable fees, if any, are sometimes the same, regardless of the amount actually used.[10] One commonly proposed solution to these issues is to provide greater legal recognition of water rights and to allow rights holders more freedom to transfer rights between themselves. Usually, such transfers involve monetary payment and, in effect, constitute a market. Such market-based schemes for natural resources management are controversial and, when proposed in relation to water, are often absolutely polarizing.[11]

According to supporters, the ability to transfer water rights gives holders an incentive to use water efficiently because they are permitted to sell or lease any water they do not use. Transfers also make possible the reallocation of water rights from economically low value uses to higher value uses. Theoretically, water rights transfers may help avoid conflict as existing rights holders are not forced to give up or share water rights and, in fact, will only do so when motivated by a sufficiently attractive financial offer. Water rights transfers may also increase the water supply available to new users as water purchasers would have an increased pool of potential sellers. New users could gain access to water without incurring the expense (and environmental impacts) of a new water supply.

Water rights boosters can even make credible appeals to environmental protection: where streams are over-allocated, rights transfers provide the most politically palatable manner of reclaiming flows for environmental purposes.[12] While buying water rights in order to meet environmental needs might seem distasteful, supporters argue that this approach creates a realistic possibility of meeting environmental needs in water-stressed areas, whereas a plan for forced take-backs of water rights will be implemented around the same time as the devil takes up ice-skating. The Nature Conservancy (an international non-profit organization dedicated to preserving biodiversity) has purchased water rights in order to provide for environmental protection. In Nevada, for example, the Nature Conservancy (in conjunction with the State of Nevada, the US Fish and Wildlife Service, and the Nevada Waterfowl Association) has purchased almost 30,000 acre-feet of water rights since 1989 in order to restore internationally significant wetlands (Nature Conservancy 2006).

Further, when transfer approval is sought, water systems may be designed so as to allow for the imposition of environmental requirements. For example, as a condition of transfer approval, a regime may allow regulators discretion to require that a certain percentage of the water allocation be dedicated to environmental objectives. While such requirements may generate environmental improvements, they do serve as a disincentive to undertaking transfers.

According to critics, water serves social and environmental needs that are too important to be left to the whims of market forces. Most would agree that the preservation of a commercially valuable fish population, or an endangered one, is more socially useful than is the profligate watering of an exclusive golf course; however, under a market approach, the golf course is more likely to get the water. In other words, environmental interests are not able to compete in the market as it is currently structured. The benefits of environmental integrity accrue to society (and non-human interests) collectively, and an expectation that self-interested individuals will voluntarily pay for public benefits at the needed scale is absurd.

Even if environmentally minded individuals were able to mount a serious attempt to reclaim substantial water rights, this approach to water use would still be unsatisfactory. The determination of where water rights should be repurchased should be made from an ecosystem perspective and should not be subject to individual whims with regard to protecting a specific water source. Moreover, these environmental "buy-backs" would be complicated by the fact that it is currently impossible for markets to appropriately attribute monetary value to all of water's uses; indeed, critics maintain that the basic concept of assigning water a monetary value in a traditional market is fundamentally flawed. Water's true economic value, if determined properly, would reflect its "ecosystem value" – all of the services that water provides to ecosystems (such as habitat for fish). In the absence of a purchaser for ecosystem services, complex economic estimation methods have to be used to calculate monetary value. Determining the monetary value of ecosystem services using these methods is very contentious because there are many aspects of water's role in the ecosystem that are extremely difficult to convert to a monetary equivalent (e.g., the aesthetic value of a wetland). And, even if the true economic value of water's ecosystem services were determined, the current market system would not capture it.[13] The result is that water's market price does not reflect its true economic value.

This difficulty of calculating the true economic value of water raises troubling questions about a policy that encourages the reallocation of water to "higher value uses" – uses that are based on market outcomes in which water's most basic services are monetarily ignored. Allocating water to the highest bidder also has the potential to exacerbate social inequity. As a common saying in the water-short US southwest has it: "Water flows uphill toward money." Globally, where market-based approaches to water allocation are adopted, subsistence farmers have been among the first to be priced out of the market (Bauer 1998).

Critics also question whether water rights transfers can live up to their promise to increase efficiency and reallocate supply. They suspect that, on the contrary, they will increase conflict by encouraging more intensive use of existing water rights, discouraging the natural attrition of water rights, and changing how water is used, thus making the rights to it more financially valuable (and, therefore, worth fighting over).

Water rights transfers also have the potential to harm other users. Even though the quantity of water that is allowed to be used does not change, rights transfers often change the amount of water actually used, the location of that use, the timing of that use (most significantly, the time of year), and "return flows" (water returning to the source after use, which is common for uses such as irrigation). Concern over these potential harms may stop many proposed water rights transfers from occurring, making transactions infrequent and rendering the purported benefits of water rights transfers more theoretical than real. In jurisdictions where water rights transfers are closely reviewed in order to prevent harm to other users and the environment, their number is correspondingly limited (Sax et al. 1991). Similarly, British attempts to allow water rights transfers were abandoned when it was determined that European Union environmental standards could not be met (Bakker 2004).

Loosening oversight to facilitate water rights transfers carries a very real risk of undermining the certainty of others' water rights. Changing the purpose, place of use, or timing of water use can affect the ability of other licence holders' rights (e.g., making it less likely that a full licence allocation will be met where licencees depend upon an agricultural user's return flows). Releasing transferors from the obligation to prepare extensive, time-consuming, and costly flows studies (a potential barrier to transfers) increases the likelihood of harm. Thus, attempts to facilitate water rights transfers and water markets could result in a somewhat ironic situation in which the market would result in the devaluing of water rights.

"Commodifying" water also carries another risk: the encouragement of speculation. The nature of water – including the fact that the geographical distribution does not necessarily correspond to population density and demand, as well as the high transaction costs associated with identifying a potential buyer – makes the use of markets awkward. Water markets intended to encourage the more efficient allocation of water may, instead, encourage hoarding. Given the importance of water to the range of human and non-human needs described above and elsewhere in this volume, the tying up of water rights in anticipation of future profit may further discourage reallocation and undercut the competitive position of a region or province. Speculation may also create negative environmental consequences by encouraging the wasteful use of water simply in order to maintain a right in good standing for a future sale.[14] Most obviously, speculation may raise moral concerns, particularly for those who feel that water is a human right rather than a commodity.

## Do Markets and Water Mix Like Oil and Water?

The debate over creating markets for water mirrors debates that have occurred regarding the market's ability to manage natural resources and to achieve public objectives. Herman Daly, one of the most well-known environmental economists in the United States, says that every economy faces three challenges – allocation, distribution, and scale. And he further argues that the market's efficacy with regard to each issue differs and, when it comes to natural resources, is often low.

"Allocation" refers to the apportioning of resources among different demands. Resources in this sense can be physical (machinery or commodities), human (workers, their time and skills), financial, or environmental ("natural resources"). Because resources are limited, we must allocate them in such a way as to provide the goods and services that people want and can afford. With regard to water, this would involve decisions such as how much of an extracted water supply should be put to municipal use as opposed to industrial, agricultural, or other uses. Markets can provide important signals regarding the optimal use of any type of resources for which there are competing demands.

"Distribution" – apportioning resources and the goods and services produced among different people – is more problematic than allocation. The market will provide a distribution of goods; however, because market transactions are based on willingness and ability to pay – the highest bidders will obtain the goods – there is going to be inequality. This may be of less concern with regard to goods such as fine wine or jewellery, but it is deeply disconcerting when the issue is water needed for drinking, sanitation, or subsistence food production. Globally, the World Health Organization has repeatedly warned that over one billion people do not have access to sufficient amounts of water for basic health and hygiene purposes (WHO 2000). To allow market forces to determine the distribution of water is, to put it charitably, callous. And many critics would argue that it is simply unethical.

The third economic challenge is that of "scale": how large can an economy become before it begins to harm the ecosystem that sustains it? With respect to water extraction, this is most easily understood in terms of how much water may be taken from a source before ecological functioning is unduly compromised. The market offers no mechanism for deciding what a desirable scale might be or for achieving it. And, unless specifically tailored to do so, a market will do nothing to correct a problem of scale that has become too large (e.g., an over-allocated water source).[15] Most environmental problems related to water use are "scale" problems that could be dealt with through a requirement to leave sufficient water for environmental purposes (along with related requirements to keep pollution and other disruptions of the water source to an appropriate "scale").

It is for these reasons that the proposed solution to water allocation disputes should be considered carefully and cautiously. Proposals that rely on strengthening private rights to water by adding tradability (a key component of private ownership) must be viewed in light of past experience. Specifically, water rights transfers are proposed as a solution to factors influencing inefficient water use – including individuals acting as if water has little or no monetary value and the problem of overallocation – which are endemic to the prior allocation system.

Are private water rights part of the solution or, as critics argue, part of the problem? Prior allocation has its genesis in assertions from businesses and investors that strong, secure water rights were needed if they were expected to take risks in establishing new enterprises (generally mining and agriculture, at least in the early days of colonial settlement). In response, governments granted water rights that were immune to regulation regarding water use efficiency or reallocation in light of changing societal or environmental needs (particularly as these rights were granted in perpetuity or with strong entitlements to renewals). Had the rights granted to private individuals been less strong (or "secure," to use the lingo of property rights advocates), the problems we face now might not have arisen at all. Canadian governments could have granted water rights subject to forfeiture for inefficient use or failure to adopt water-saving technology. In such a system, there would simply be no need to provide monetary incentives to water users to increase efficiency.

"Weaker" property rights, entailing greater constraints on users, would have avoided the problem that "strong" property rights have created.

If "strong" property rights are at the root of the water problems Canadians are now facing, what could strengthening property rights in water – by adding tradability – bring us in the decades to come? Canada lacks sufficient domestic experience to offer insights into how to gain the benefits and avoid the harm associated with water rights transfers. But other jurisdictions, such as California and Chile, do offer such insights.

## WATER TRANSFERS: THE EXPERIENCES OF CALIFORNIA AND CHILE

### California

California likely has longer and more extensive experience with transferring water rights than does any other jurisdiction in the world. As early as 1859, the California Supreme Court determined that water rights were "substantial and valuable property" that could be sold or "transferred like other property." Only one year later, the same court issued another decision that imposed serious restrictions on water rights transfers as the potential harm to other parties became apparent (*McDonald v. Bear River* and *Kidd v. Laird*).[16] This "no injury rule" has been interpreted to mean that one may only transfer a water right if it does not injure any legal user of water. Injury includes any change in water use that would harm those who have become reliant on unused water and return flows downstream. Water rights transfers are limited to the amount actually used, meaning that the person doing the transferring must actually give something up rather than simply capitalize on an unused paper right.

Under these historical rules, harm was avoided, but a limited number of water rights transfers were taking place each year (Sax et al. 1991). Sensing the potential to address some of the state's increasingly contentious water management issues, California took steps in the 1980s and 1990s to encourage more water rights transfers. These changes allowed short-term water rights transfers, water banking, and water rights transfers outside original areas of use, but it retained the prohibition against water rights transfers resulting in harm. During the first years after the changes, the state experienced significant conflicts, but these have now been reduced in number and water rights transfers are increasing, with hundreds of thousands taking place each year (Johns 2003). Quantities are reaching 1.2 million acre feet per year – almost a tenfold increase since the mid-1980s (Hanek 2003).

Environmentally, water rights transfers are looking good in California. In recent years, over one-third of water rights transfers in that state have occurred in an attempt to meet environmental needs (Howitt and Hansen 2005).[17] Leading environmental groups generally support the idea of water rights transfers, while occasionally opposing individual proposals. The State of California is even moving into being a market participant as well as a regulator, acquiring water rights for environmental purposes and for drought planning.

In short, water rights transfers in California have developed into a relatively non-contentious, frequently used tool that addresses problems of excess water demands and that also occasionally generates environmental benefits.[18] There is broad support for water rights transfers not only from those who embrace strong market initiatives (which is not surprising) but also from environmental groups (which is surprising).

Although there seems to be a growing consensus about the efficacy of water rights transfers within California's existing system, some have questioned the equity of the arrangements. Most of the water rights traded in California are for water provided through a heavily subsidized system of aqueducts that costs taxpayers over $400 million per year (Environmental Working Group 2005). Taxpayers subsidize the delivery of water to farmers and businesses who are able to reap the windfall of selling it at "full market value" – without any repayment of the subsides – often so that water can be put back or left in the stream for environmental purposes. Nonetheless, despite the rapid rise of the number of water rights transfers in California, some still argue for liberalizing water rights transfers even further and increasing the role of market forces (Mentor 2001). Those advocates might be envisioning a system more closely resembling that of Chile.

**Chile**

In 1981, the Chilean government enacted an extremely laissez-faire water law that privatized water rights, promoted free market forces and incentives in water use, and sharply reduced governmental regulatory powers in water management. Since then, the Chilean Water Code has been the world's leading example of a free market approach to water law and policy – a unique experiment in treating water rights not merely as private property but also as a fully marketable commodity (Bauer 1998). The 1981 Water Code is still in force today, protected by Chile's 1980 Constitution.

The adoption of free market reforms in 1981 has an interesting genesis. The Republic of Chile had adopted a civil legal system. The country's civil code governed water rights until 1951, when a separate water code was enacted. The water allocation system under the 1951 Water Code resembled that of the western United States, but water rights transfers were only allowed if the purpose of the water use remained the same as it had been before the transfer (Mentor 2001).

In 1973, the political situation in Chile changed dramatically. Under the control of General Augusto Pinochet, the Chilean armed services overthrew the socialist government of Salvador Allende. The military coup was a reaction to Allende's land reform efforts and to deteriorating economic conditions (Carrasco 1995). The military government adopted radical free market economic policies and curtailed the government's planning, regulatory, and proprietary roles over private industry and natural resources development. A group of US-trained, free market economists known as the "Chicago Boys" gained unprecedented influence over efforts to rewrite Chilean laws to further the government's economic policy (Carrasco 1995). The government's economic development model was export-oriented, and Chile's economy was opened to the world economy (Bauer 1998).

In 1981, the Chilean military government adopted a new Water Code that reflected its overall economic and political objectives (Bauer 1998). The Water Code increased the legal security of private water rights, thereby putting an end to questions about water rights ownership left by the agrarian reform. Government economists argued that the real boost in efficiency would come from price incentives and private trading. According to the Chicago Boys, market mechanisms would motivate users to save water in order to sell the surplus and to transfer water rights to higher-valued uses within agriculture (or other sectors of the economy).

Because the Chilean Water Code is such a paradigm for free market reforms, it has often been mentioned in international debates about water policy. The predominant view outside of the country is that the Chilean model of water management has been a success. The strongest proponents of this view have been economists at the World Bank, the Inter-American Development Bank, and related institutions, which have encouraged other countries to follow Chile's lead (Bauer 1998). These proponents – never ones to let facts (or lack thereof) get in the way of ideology – did little, if anything, to verify their conclusions.

Carl Bauer undertook the first in-depth analysis of the Chilean experience. He concluded that, in most parts of Chile, water markets were inactive and had a limited impact on the efficiency of water use and the reallocation of resources. While the promises of water markets were scattered and limited, the concerns about water markets were acutely felt. On issues such as social equity and coordinating multiple water uses (managing river basins, resolving water conflicts, and protecting river ecosystems and instream flows) the Chilean model demonstrated "serious weaknesses," and there are indications that speculation and hoarding is occurring (Bauer 1998). Bauer's examination leads us to conclude that the purported benefits of water markets in Chile were overpromised and have since been underdelivered. On the other hand, the risks of water markets have been fully realized.

## California, Chile, and Canada?

California's experience with water rights transfers has significant positive aspects, whereas Chile's gives rise to serious concerns. Will Canada come closer to Chile's experience or to California's? It is worth noting the similarities and differences among the three. In many ways, California's and Chile's water laws are similar to those in western Canada (the prior allocation system). All three have established licence or permit systems as the primary means for obtaining a water right.[19] All three systems respect private rights to use water while retaining state ownership of it (i.e., they distinguish between ownership of water rights and ownership of the water itself). In adopting these characteristics, all three have rejected riparianism as the legal framework for water rights ownership.

There are differences among the systems. California and western Canada allocate water according to a time-based priority system in which the claim of those with prior allocations of water is superior to that of junior water users. In Chile, water rights are divided into two classes, and the government equitably apportions water rights within each class (Ríos and Quiroz 1995).

There are other differences. California is diligent in ensuring that water users show that they are using water for beneficial purposes, whereas Chile does not do this. In fact, in Chile, once a water right is perfected, the owner may change the purpose of use without governmental approval. In western Canada, water must be used for defined purposes; in Alberta and British Columbia, water rights transfers require government approval. Furthermore, water rights in California may be lost if there is a prolonged period of non-use.

By contrast, Chilean water rights are not subject to forfeiture. In Canada, rights may be lost for non-use, although, in practice, this rarely happens.

Finally, California has adopted strong protections within its water rights transfer regime, but it also has a number of other forces that curb water management excesses. California guarantees robust citizen participation in water management decisions, something that is precluded in a number of Canadian jurisdictions. Unlike Canada, California has strong endangered species legislation (both at the state and federal levels) that is diligently enforced and that functions as the primary impetus for water rights transfers to environmental needs. California has also embraced other broad protections of the public interest, such as the "public trust doctrine."[20] The lesson for Canadians is that water rights transfers will only work well if they occur within a strong legislative framework – one that prioritizes the public interest and environmental protection, and that requires robust citizen participation in water management decisions. None of these conditions is currently being met in Canada.

Water rights transfers in California follow the same general approach as do those in other western US states. This cannot be said for British Columbia and Alberta. Alberta has adopted an explicit scheme of considering proposed water rights transfers that ensures the deliberation of relevant factors, such as harm to other users and the environment (see Alberta's Water Act). British Columbia's "transfers of appurtenance" are a different story. When applications are made, not only does the province fail to provide notice to the public generally, but it also fails to guarantee to provide notice to potentially affected licencees. In the *Hotel Lake* case referred to at the beginning of the chapter, the BC government took the position that harm from water rights transfers is impossible if the transfer does not increase the legal quantity of water authorized. On that basis, it refused to even consider effects on other users and the environment. The ministerial representative testifying in the Hotel Lake hearing went so far as to say that if an application for a water rights transfer was filled out correctly and the underlying rights had not been cancelled (regardless of whether they had been used), the application must be granted. Unlike in Alberta, in British Columbia, questions of environmental impacts, overexploitation, and overappropriation (due to the granting of too many rights for the actual quantity of water) are not currently being addressed in water rights transfer decisions.

In Canada, Alberta has some, but not all, of the protections that California has. There is at least a glimmer of hope that Alberta's experience may resemble California's. British Columbia, given current trends, will likely replicate the Chilean experience.

## Conclusion

From our perspective, the need to sustainably manage uses of fresh water is simply too important to preclude the use of any potential tool on ideological grounds. Canadian jurisdictions should remain open to the possibility of employing economic instruments for water use management, including water rights transfers.[21] However, it is important to take stock of the risks involved. Looking at the examples of California and Chile, it may be seen that there is a risk that other aspects of water governance may be so underdeveloped in Canada that increased use of economic instruments may impede rather than enhance sustainable water use management. To put it starkly, there is a danger that Canada's experience may replicate that of Chile rather than that of California.

The debate about water rights transfers also occurs within the context of a great unknown – the extent to which fundamental reform of water management systems will be undertaken. If the current systems remain in place, then it makes sense to move toward water rights transfers (in an appropriate manner, of course). California has managed to address a few of the most pressing water use management problems through water rights transfers. However, even if water rights transfers have improved water use management in California, the overall system remains highly unsatisfactory. It is possible that the small improvements brought about by water rights transfers have undermined the momentum to move toward real reform.

On the Canadian side of the border, there's no question of the need for real reform of water use. It is crucial to give priority to ecological function when determining water use management. Access to water through permitting should only occur after ecological integrity is ensured, after adaptive management has been implemented (so as to be able to continuously modify water uses in response to ecological, economic, and social "feedback"), and after the implementation of appropriate governance mechanisms at the appropriate geographic (watershed) and political levels. And there must at all times be clear accountability at higher levels of government (Brandes et al. 2005).

To sum up, water rights transfers should not just be viewed as some obscure debate among policy wonks or as a tool through which someone can obtain water. The decisions that Canada makes about whether and when to allow water rights transfers are central to the larger questions of how we manage water use, how we balance public and private interests, and how we achieve our collective social objectives (such as good water governance and environmental protection). Rather than proposing water rights transfers as the solution to our water allocation problems, we should focus on improving water and environmental governance before eventually considering water rights transfers as one potential – but limited – tool for water use management.

## NOTES

1 British Columbia's Water Act refers to this type of water rights transfer as a transfer of appurtenancy. Water rights transfers and appurtenancy are discussed below and in Chapter 10, this volume.

2 *"Hotel Lake," McClusky, et al. v. Assistant Regional Water Manager*, BC Environmental Appeal Bd. Decision 2004-WAT-0003(b) and 0004(b), issued 9 August 2005.

3 Ontario and Prince Edward Island have no positive statutory provision asserting the public ownership or vesting of water. However, their management approaches, as discussed below, are similar to those in several other Canadian jurisdictions.

4 This is a modification of a doctrine used in the western US states called prior appropriation. A similar approach was adopted in Australia as well.

5 In Nunavut, for example, the public authority resides in the Nunavut Water Board. As agreed to in the negotiated land claim (1993), the Nunavut Water Board oversees the use and pollution of all water sources. In the Northwest Territories, there are several boards that oversee the licensing of water use (Sahtu Land and Water Board, Gwich'in Land and Water Board, and Mackenzie Valley Land and Water Board). In addition to allocations, the Yukon Water Board also oversees the compensation process by which senior licences and exempted domestic uses are compensated for adverse effects anticipated from a new licence. There is no such process of compensation in other northern jurisdictions.

6 Personal communication, John Donihee, legal counsel for the Mackenzie Valley Land and Water Board, 16 August 2005.

7 For example, the Ministry of Natural Resources (and Hydro-Quebec) issue permits for power-related water use, the Ministry of Agriculture and Fisheries for agriculture-related water use, and the Ministry of Municipal Affairs for drinking water supply and infrastructure.

8 The interpretation of Aboriginal rights within the context of the Canadian legal system relies on the assertion of Canadian sovereignty, which did not have a basis in law at the time and thus may be invalid. A proper understanding of Aboriginal

water rights is well beyond the scope of this chapter. For a detailed discussion, see Kempton (2005).

9 To the extent that an Aboriginal and/or treaty right to water exists, Aboriginal customary law (such as the Anishinabe practice mentioned earlier) could govern water uses and take priority over all other uses (after ecological needs are met). In Yukon, First Nation riparian rights are associated with any Settlement Lands. If the Yukon Water Board authorizes a water licence that interferes with First Nations riparian rights, compensation must be paid to the First Nations in question. If First Nations riparian rights are substantially altered by an approved water use, the First Nations in question can apply to the Yukon Water Board for compensation.

10 See, for example, British Columbia, where licence fees are paid on the full licence allocation, regardless of use (Water Use Regulation, s. 7(10)).

11 Market-based schemes for the management of resources are increasingly in vogue. See, for example, Grafton (1996); Parliamentary Standing Senate Committee on Fisheries and Oceans (1998); and Branch, Rutherford, and Hilborn (2006) regarding "individual transferable quotas" for managing fish harvests.

12 Where transfers are allowed, non-profit groups such as the Oregon Water Trust obtain water rights to meet environmental objectives. A typical transaction might include supplying a rancher in an arid area with hay in return for her/him forgoing the irrigation of her/his own crop and leaving water instream. See http://www.owt.org. In the United States, government agencies often purchase water rights in order to meet environmental objectives.

13 When the true economic value of a good or service is not reflected in the market, the market fails to achieve the optimal outcome (e.g., maximize social well-being). This market failure is referred to as a negative externality. The real cost to society of "using up" ecosystem services is not paid by the private use of the water. This use of ecosystem services might be the result of water withdrawal or diversion (as discussed in this chapter) or pollution. When the social cost is greater than the private cost, the market solution requires either complete property rights or government intervention (such as a tax or charge, or quota, for water use). Water rights transfers, as they exist in Canada, are not sufficient to achieve the optimal outcome.

14 It has been reported that in the arid lands around Tucson, Arizona, some farmers have flooded fields to the point of non-productivity simply to maintain the water right for future sale to the growing urban area.

15 Markets may be useful in determining how reductions in scale should be achieved. For example, if it were determined that water usage needed to be reduced by 30 percent, markets might help to determine which rights should be forfeited (e.g., those willing to sell back the rights at the lowest price).

16 *Kidd v. Laird,* 15 Cal. 161 (Cal. Sup. Ct., 1860); *McDonald v. Bear River Co.,* 13 Cal. 220 (Cal. Sup. Ct., 1859).

17 Direct purchases, such as those made by state and federal entities in order to comply with federal environmental regulations (primarily augmenting stream flow to enhance fish runs), accounted for one-third of traded volume in 2001.

18 This is an area in which disputes related to California's failure to consider the impact of water rights transfers on groundwater use still occur. Another concern is the fact that some local rural economies have suffered as farmland is rendered idle when the water is sold (Hanak 2003).

19 The prior appropriation system in California originally recognized the establishment of water rights through diversion and use, without the need to apply to the government. Historical rights are now recognized within the administrative system, and new water uses require an application.

20 Briefly, the US doctrine states that, if the state holds legal title to resources, then it acts as a trustee for the benefit of the people of the state. As trustee, the state and its agencies are answerable to the courts in the exercise of their duty. Where private interests intersect with public claims, the former should give way to the latter (Sax 1970). A continuing duty is imposed on the state to supervise the exercise of water rights and to reconsider those rights when public trust values are endangered. This resulted in the divesting, without compensation, of water rights held by the City of Los Angeles (*Mono Lake* case).

21 Some other economic instruments that should be considered are revised water pricing and financial incentives (such as tax breaks for water efficiency investments).

## REFERENCES

Bakker, K. 2004. *An Uncooperative Commodity: Privatizing Water in England and Wales.* Oxford: Oxford University Press.

Bauer, C.J. 1998. *Against the Current: Privatization, Water Markets, and the State in Chile.* Norwell, MA: Kluwer Academic Publishers, 5, 21, 37-38, 52-56.

Blumm, M. 1989. Public Property and the Democratization of Western Water Law: A Modern View of the Public Trust Doctrine. *Environmental Law* 19 (12): 583, 576.

Branch, Trevor A., Kate Rutherford, and Ray Hilborn. 2006. Replacing Trip Limits with Individual Transferable Quotas: Implications for Discarding. *Marine Policy* 30 (3): 281-92.

Brandes, Oliver M., Keith Ferguson, Michael M'Gonigle, and Calvin Sandborn. 2005. *At a Watershed: Ecological Governance and Sustainable Water Management in Canada.* Victoria, BC: POLIS Project on Ecological Governance.

Carrasco, E.R. 1995. Autocratic Transitions to Liberalism: A Comparison of Chilean and Russian Structural Adjustment. *Transnational Law and Contemporary Problems* 5: 99, 104-6.

Christensen, Randy. 2005. *Groundwater Pricing Policies in Canada.* Toronto: Walter and Duncan Gordon Foundation.

Environmental Working Group. 2005. *California Water Subsidies*. http://www.ewg.org/reports/watersubsidies/execsumm.php.

Epstein, R.A. 1994. On the Optimal Mix of Private and Common Property. In *Property Rights*, ed. E.F. Paul, F.E. Miller, Jr., and J. Paul. Cambridge: Cambridge University Press.

Grafton, Quentin. 1996. Individual Transferable Quotas and Canada's Atlantic Fisheries. In *Fisheries and Uncertainty: A Precautionary Approach to Resource Management*, ed. D.V. Gordon and G.R. Munro, 129-54. Calgary: University of Calgary Press.

Hanak, Ellen. 2003. *Who Should Be Allowed to Sell Water in California? Third-Party Issues and the Water Market.* Public Policy Institute of California. http://www.ppic.org/content/pubs/R_703EHR.pdf.

Howitt, Richard, and Kristina Hansen. 2005. The Evolving Western Water Markets. *Choices* (American Agricultural Economics Association) 20 (4): 59.

Johns, Gerald. 2003. Where Is California Taking Water Transfers? American Society of Civil Engineers. *Journal of Water Resources Planning and Management* 129 (1): 1-3.

Kempton, Kate. 2005. *Bridge over Troubled Waters: Canadian Law on Aboriginal and Treaty "Water" Rights, and the Great Lakes Annex*. Toronto. http://www.anishinabek.ca/uoi/pdf/UINGL/kempton_greatlakes.pdf.

La Forest, Gerard V. 1973. *Water Law in Canada: The Atlantic Provinces.* Ottawa: Information Canada.

Mentor, Joe, 2001. Trading Water and Trading Places: Water Marketing in Chile and the Western United States. Conference paper: Globalization and Water Resources Management: The Changing Value of Water Conference, American Water Resources Association, Dundee, Scotland. http://www.awra.org/proceedings/dundee01/Documents/Mentor.pdf.

Ministère du Développement durable, de l'Environnement et des Parcs. *Québec Water Policy: Water – Our Life, Our Future*. http://www.mddep.gouv.qc.ca/eau/politique/index-en.htm.

Nanda, V.P., ed. 1977. *Water Needs for the Future*. Boulder: Westview Press: 46-54.

The Nature Conservancy. 2006. *Stillwater and Carson Lake Wetlands*. http://www.nature.org/wherewework/northamerica/states/nevada/preserves/art11311.html.

Nunavut. *Nunavut Waters and Nunavut Surface Rights Tribunal Act*, S.C. 2002, c. 10.

Ontario. *Ontario Water Resources Act*, R.S.O. 1990, c. O.40, as amended.

Parliamentary Standing Senate Committee on Fisheries and Oceans. 1998. *Privatization and Quota Licensing in Canada's Fisheries*.

Percy, David. 1988. *The Framework of Water Rights Legislation in Canada*. Calgary: CIRL.

Ríos Brehm, M., and J. Quiroz. 1995. *The Market for Water Rights in Chile*. World Bank Technical Paper No. 285: 1-2, 9.

Sax, Joseph. 1970. The Public Trust Doctrine in Natural Resource Law: Effective Judicial Intervention. *Michigan Law Review* 68: 471.

Sax, Joseph L., Barton H. Thompson, Jr., John D Leshy, and Robert H. Abrams. 1991. *Legal Control of Water Resources: Cases and Materials.* St. Paul: West Publishing.

Teclaff, L.A. 1972. *Abstraction and Use of Water: A Comparison of Legal Regimes*. New York: United Nations Department of Economic and Social Affairs.

World Health Organization. 2000. *Global Water Supply and Sanitation Assessment.* Geneva: World Health Organization and UNICEF.

PART 4

# Waterwise: Pathways to Better Water Management

GRAEME MACKAY/HAMILTON SPECTATOR

Despite increased attention to public health and water safety following the water contamination incident in Walkerton, Canadians remain concerned about water quality.

# 12
# A Tangled Web: Reworking Canada's Water Laws

*Paul Muldoon and Theresa McClenaghan*

Canada is facing a water crisis. This crisis is not simply a result of pollution, overuse, and depletion of water resources, but also of bad management, mainly due to the absence of a clear governance framework to oversee the protection, conservation, and good management of Canada's water resources. There are many reasons for this. Federal and, to some extent, provincial water governance capability in Canada is subject to political and leadership paralysis, in part because of the diverse views of the stakeholders within the policy debate. One view, often expressed by industry, is that water, like any other commercial asset, is a resource to be exploited, processed, and traded. Another view, often expressed by public interest groups, is that water is an inherently shared "social asset" vital to ecological and human health (Bakker 2003). The management "crisis" stems from the inability of governments to negotiate these views and to establish a robust governance framework for water (Boyd 2003).[1]

Making this task even more challenging is the fact that such a governance framework must integrate a broad array of water issues, ranging from drinking water protection and human health, to fisheries management and other economic interests based on water systems, to ecosystem sustenance and protection. And it must do so with regard to both water quality and water quantity. It is essential to allow for the diversity of situations, scales, and choices that will influence water governance decisions. The path forward requires us to develop a new water governance framework that embraces and builds upon Canada's complex jurisdictional situation – one that takes strength from diverse points of view rather than being paralyzed by them.

One of the obvious challenges, as explored by Owen Saunders and Michael Wenig in Chapter 6, is the fragmentation of jurisdiction over water resources. The Canadian Constitution divides the power to legislate over water resources among federal, provincial, territorial, and First Nations governments.

No one level of government can provide a complete framework for protecting water resources. Not unlike other areas of environmental law and policy, this fragmentation of jurisdiction has led to regulatory gaps, policy vacuums, and management by crisis. The stresses over water resources, along with the fundamental importance of water to all aspects of human and ecological systems, demand a new approach to water governance. The combined jurisdictions of the federal, provincial, territorial, and First Nations governments, along with the engagement of municipal government and an informed and diverse citizenry, present real opportunities to carve out a dynamic and resilient approach.

This chapter asks more questions than it answers. Our purpose is to launch a discussion that focuses on what a new governance framework for water in Canada might look like, recognizing that it would have to be developed with extensive input from the public, First Nations, all levels of government, and other stakeholders.

## Evolution of Canadian Water Policy

There is a rich and complicated history behind the formation of both federal and provincial water policy in Canada. We present only a brief outline of this history.

In 1909, the Crown, on behalf of Canada and the United States, entered into the Boundary Waters Treaty (BWT), which covered the waters from the "main shore to the main shore" of the lakes and rivers and connecting waterways (or portions thereof) that defined the international boundary between the United States and Canada (Bloomfield and Fitzgerald 1958; Cohen 1975). The treaty is important not only because of its content but also because it represented the distillation of over a century of debate and compromise on bilateral water policy issues. And, in some respects, it forged new ground.

The BWT protects the continued free and open navigation of all navigable boundary waters, extending to Lake Michigan. Each of the countries retains jurisdiction over use and diversion of waters on its own side of the international line, preserving the right to make claims in case of injury and to object to a use or diversion that interferes with navigation. Any new uses, diversions, or obstructions that interfere with natural levels or flows require the approval of the International Joint Commission (IJC), which was established by the BWT. The BWT also provides that the boundary waters and waters flowing across the boundary will not be polluted to the detriment of the natural waters on the other side of the boundary. It provides for an order of precedence for uses, with domestic and sanitary uses being first, navigation second, and power and irrigation third. Provision is made for referring to the IJC concerning matters of difference arising between the two countries.

Throughout the twentieth century, the IJC's approvals and considerations reveal the history of the development of Canadian water resources (Willoughby 1979; Spencer, Kirton, and Nossal 1981). The Columbia River, James Bay, and St. Lawrence River projects are a small sample of major hydroelectric and seaway developments. The IJC's references and studies document the long history of stresses on, and threats to, North American and, particularly, Canadian water resources.

Since the Second World War, each province has evolved its own legislative approach to water, and this has resulted in a large number of water regimes in Canada – regimes that are often quite different from one another. The last comprehensive federal review of Canadian water policy occurred in the early 1980s. At that time, the Progressive Conservative government formed a blue ribbon panel (Inquiry on Federal Water Policy) that travelled across Canada holding public hearings on the state and needs of Canadian waters. The report from that inquiry was widely accepted and remains largely relevant (Environment Canada 1985). Based on the findings of that report, the federal government developed its 1987 Federal Water Policy (Environment Canada 1987). This policy still provides the anchor for current approaches to water policy. It states that the "overall objective of the federal water policy is to encourage the use of freshwater in an efficient and equitable manner consistent with the social, economic and environmental needs of present and future generations" (Environment Canada 1987, 5). It then outlines five strategies and some twenty-five specific policy statements on a wide range of water-related issues.

In many respects, however, the 1987 Federal Water Policy does not address many of the challenges now facing Canadian water policy. Furthermore, most aspects of the 1987 policy have yet to be fully implemented. Of course, some of the strategic directions are quite general, and it is difficult to assess the progress made in their implementation. However, many of the directions were specific. Under "science leadership," the federal government committed to "develop and maintain, with the provinces and territories, water data and information systems directed to improving the knowledge available for managing Canada's water resources" (Environment Canada 1987, 9). Under "water pricing," the government committed to developing new water-efficient technologies and industrial processes (Environment Canada 1987, 10). Commitments were also made to implement integrated planning, tougher enforcement, and groundwater protection strategies as well as to engage in a "consideration" of federal safe drinking water legislation (Environment Canada 1987, 10-11, 19, 22). None of these initiatives has been fully realized, although each remains as relevant today as it was when the government committed to it in 1987.

Even where the Federal Water Policy did articulate a strategy and certain elements of it were implemented, gaps remain – and, with them, the need to update the policy to respond to current challenges. For example, the policy only superficially addresses issues pertaining to trade and water resources, private ownership of water, and export of water.

A good example of a gap in the Federal Water Policy occurred in 1995 when a small company named Nova Corp in Sault Ste. Marie applied to the Ontario Ministry of the Environment for a permit to take water by bulk carrier and export it to Asia. The permit was granted by the province but subsequently rescinded after a public and international outcry. Nova appealed the rescinding of the permit before an administrative tribunal; however, the matter was settled without a hearing, and the permit remained rescinded. The Nova case not only identified a gap in policy but also created a provincial, national, and international debate concerning the vulnerability of Canadian water for export. It was clear from this export application that there was little understanding of the ecological implications of exporting water by ships, pipelines, or other means. Moreover, there was little understanding of the trade implications that could result should the permit be granted. The fear was that, once a permit was granted to one company, similar permits would have to be granted to other companies, regardless of their domicile. Fear of losing sovereignty over water resources and the ability to protect ecological sustainability arose. As a result of the Nova case, the federal government discussed the need to renew and update its 1987 water policy. That exercise is still ongoing. In 2000, the government amended the International Boundary Waters Treaty Act with the intention of prohibiting out-of-basin transfers of boundary waters. It further committed to getting all provinces and territories to endorse an accord that would prohibit such transfers (International Boundary Waters Treaty Act). That accord has yet to materialize, although most provinces have now acted on their own to limit or prohibit exports.

Another example of a water policy gap is the Devils Lake diversion, which moves contaminated water from North Dakota via pipeline to the Sheyenne

River, which flows into Manitoba's Red River system. For the past several years, the Province of Manitoba has been fighting a multi-pronged battle to protect its Red River Valley water resources from the Devils Lake diversion. However, although the Canadian federal government supported Manitoba in its position, as did many US states and Canadian provinces, in the face of opposing US political interests, it was unable to get a reference to the IJC under the BWT. In August 2005, disappointingly, the US and Canadian governments issued a joint statement to the effect that the project would proceed, with monitoring and mitigation efforts. There appears to be no further attempt by either federal government to secure a reference to the IJC (DFAIT 2005).

A third example of a policy gap pertains to water quality, an issue of major concern in Canada (Pollution Watch 2004). The BWT prohibits pollution that will injure the waters – at least the boundary waters, of the other party. However, in practice, most point source water pollution occurs pursuant to provincially and territorially issued permits for industrial activity or municipal wastewater, and, in some circumstances and sectors, it is essentially unregulated by any level of government (and if it is regulated it has not been reviewed for a long time). The federal government does not have a comprehensive strategy to reduce water pollution; rather, it has set specific limits for a small number of sectors. In recent times, the federal government has paid little attention to programs that focus on, for example, Great Lakes protection. Despite increasing stresses on the ecosystem, Great Lakes programs are dormant, suffering from static or diminishing funding and the government's failure to develop new, innovative programs (along with its hesitancy to set new targets, timelines, and reduction goals) (Pollution Watch 2006; Botts and Muldoon 2005). Commencing in early 2006, the governments of Canada and the United States will begin their review of the Great Lakes Water Quality Agreement. The United States has been developing a regional strategy for Great Lakes restoration, including the introduction of two bills in Congress. The degree of Canadian activity concerning the Great Lakes is meagre by comparison, putting Canada in an awkward situation heading into the review (Botts and Muldoon 2005).

Upon reviewing 1987's Federal Water Policy, it is fair to say that the document is full of good intentions, and the necessity to either address current problems or overcome historical challenges is indicated. However, at this time, Canada lacks a federal water policy, never mind an overall policy that would integrate and coordinate federal, provincial, territorial, and First

Nations policy dimensions or provide effective leadership with regard to international issues such as the Devils Lake diversion.

In practice, the absence of a federal water policy means more than that there is no concise statement regarding federal government positions on any given water issue. What it means is that Canadian policy is continually governed in a reactive, crisis management mode. It also means that the Canadian public does not have the opportunity to debate and discuss policy options and opportunities. Furthermore, Canada cannot base trade positions on an existing policy framework. For example, arguing to limit water exports based on the conservation provisions of trade agreements requires that there be conservation policies underlining those arguments.[2] When it comes to fresh water, Canadians still believe in the "myth of abundance" (see Chapter 1; Sprague 2003). Despite the drinking water crises in Walkerton (O'Conner 2002a, 2002b) and North Battleford, the threats posed by climate change, deteriorating water quality in many areas across Canada, intensification of water use, challenges brought on by invasive species, stresses on groundwater use, existing and emerging border disputes, and so on (Pollution Probe 2006) – all of which demonstrate the need for robust water leadership – when it comes to defining a role for itself in Canadian water governance, the federal government has been left floundering.

## The Tangled Web of Jurisdiction

The lack of a current federal water policy is probably due to the fact that the issues are complex, have a direct impact on many economic and industrial interests, and may be unpopular to some stakeholders. It is also, no doubt, due to the fact that any federal policy must be correlated to and preferably integrated with provincial, territorial, and First Nations interests – a difficult task.

Both the federal and provincial governments have significant and shared constitutional powers over water, but it is the provinces that have the bulk of legislative power over property and land adjacent to and under water resources. First Nations rights, including rights of governance and control, are also significant and are still being interpreted and defined in negotiations and by the courts (see Chapter 15).

Among the roles that the federal government accepts in water protection and governance are responsibilities for navigation, oceans, fisheries, federal lands and waters (such as national parks and military bases), and First Nations lands and waters. Other roles include regulating toxic substances, promoting pollution prevention, and carrying out research. Provinces and territories have primary responsibility for regulating and protecting water quality, regulating drinking water systems, and making resource use and allocation decisions (Hughes, Lucas, and Tilleman 2003).

First Nations rights include access to and use of water resources (or the wealth and sustenance provided by those resources) and thus entail ensuring healthy water ecosystems that support healthy fisheries. Their rights may also include decision making with regard to the allocation of various aspects of water resources. The review of the extent of these rights is beyond the scope of this chapter (see McClenaghan n.d.; Royster n.d.).

Attempts to redefine federal-provincial and First Nations constitutional power through the Meech Lake Accord and the Charlottetown Accord did not expressly address environmental issues (and, where the latter did so, there is some debate as to whether it, in fact, moved the issue forward). The failure of these proposals is often cited as the rationale for concluding the January 1998 Environmental Harmonization Accord between the federal government and the provinces (except Quebec). This agreement attempts to define which jurisdiction should deal with what issue and how the federal and provincial jurisdictions should work together (in particular, on environmental standards, environmental assessment, and enforcement).[3] While the agreement is still in force, the commitment to it seems to have faded.[4]

Another important interjurisdictional governance mechanism that already exists is the Canadian Council of Ministers of the Environment (CCME) – a forum that brings together federal, provincial, and territorial governments.[5] The CCME sets priorities and then engages in developing approaches that may be adopted in various individual jurisdictions across the country. Aboriginal governments are not involved in the CCME. At the moment, freshwater issues such as water demand, water use management, water quality, and aquatic ecosystems are among the issues in which the CCME is engaged. Particular initiatives under way in Canada in 2005 involved municipal wastewater effluent, source-to-tapwater protection, water conservation and economics, and water quality guidelines, to name a few. As water and environment are not single "heads of power" (i.e., they don't fall under only one level of government),[6] and implementation of any approaches developed by the CCME will be up to the respective jurisdictions, the CCME approach to water governance is generally consensus-based. Although there are advantages to this approach, one of the main disadvantages is that most

CCME-developed approaches are often no more than expressions of the status quo. Although the CCME does recognize that environmental issues, including water protection and governance, cross political boundaries and that a cooperative approach is needed, it cannot provide the robust, innovative, and effective approach to water governance issues that Canada needs. Furthermore, its structure (which excludes First Nations governments), its approach (which sees public consultation as peripheral), and the mediocre results of many of its efforts combine to demonstrate that the CCME is not the answer to Canada's water governance crisis. First Nations communities are still largely without adequate drinking water. Provincial policies are still proceeding with, at best, intermittent and inconsistent federal participation. Federal participation is marooned just at the time when various water crises are intensifying, and water management, protection, and policy are in need of substantial revision.

One area that does work fairly well in terms of federal-provincial cooperation through the CCME is that of setting drinking water quality guidelines. Many provinces have adopted the Canadian Drinking Water Guidelines as their standards, although there are variations regarding both the adoption of additional parameters and the regulatory approach each of the provinces takes with respect to its drinking water standards. However, with regard to water exports, the approaches of the federal and provincial/territorial governments are not at all well coordinated. There are separate regimes, both federally and provincially. For example, the federal Boundary Waters Treaty Act prohibits certain bulk diversions or removals. At the provincial level, for example, Ontario's water taking regulation prohibits water transfers out of any of the three water basins defined in the province's regulations.[7] Another example involves the provinces of Ontario and Quebec, which are signatories to the recently concluded Great Lakes-St. Lawrence River Basin Sustainable Water Resources Agreement. Annex 2001 to the Great Lakes Charter constitutes an agreement among the Great Lakes provinces and states (one among the US states and one between the states and provinces) that proposes to provide decision-making standards regarding applications for certain diversions in the Great Lakes-St. Lawrence Basin.[8]

Another area that is working fairly well – albeit with major procedural delays – is the area of federal-provincial cooperation regarding infrastructure financing for municipal drinking water and wastewater systems. The federal government has made it a priority to contribute funds to infrastructure projects (through programs such as the Canada Strategic Infrastructure Fund), and, although the needs are far greater than can be provided for within the current scope of the funding programs, these initiatives are reaching many urban and small communities across the country. For example, many communities that have been operating only primary sewage treatment facilities have received federal and provincial dollars to build more modern sewage treatment facilities. On the other hand, many reports have demonstrated that the major issue with infrastructure is that of developing a viable, multigenerational plan for revenue generation and infrastructure replacement. Ontario, for example, is in the midst of exploring potential approaches to water governance with its Sustainable Water and Sewer Systems legislation (still not proclaimed as of February 2006, pending development of the relevant regulations). The release in July 2005 of the Water Strategy Expert Panel's report confirmed that, between 2005 and 2020, $30 billion worth of investment in water infrastructure is needed in Ontario alone. Therefore, there is a role for the federal government not only in targeting dollars to assist with some of the most egregious infrastructure needs but also in developing models of sustainable funding for such infrastructure.

It may well be that Canadian federalism is not structured in such a way as to allow for the formation of one unified water policy. What may be possible, however, is the development of a federal water strategy that is integrated and coordinated with provincial, territorial, and First Nations interests and that allows for the participation of engaged citizens. For example, the people of Nunavut are exploring the vision expressed in the Bathurst Mandate, which recognizes the equal importance and responsibility of individuals, of community collectives, and of governments in providing for strong, healthy communities (Nunavut 2005).

A Canadian water governance strategy may be able to reflect both the commonality and the diversity of Canada itself. Two of the benefits of working toward such a strategy include public debate and the opportunity for policy development.

## Principles of a Canadian Water Governance Strategy

The first step in developing a Canadian water governance strategy involves identifying principles with which to ground it. The most important policy objective underlying the strategy is the achievement of sustainability. Sustainability, in this context, refers to the maintenance, preservation, and protection of healthy ecosystems for present and future generations. Accordingly, even though the governance strategy should be directed toward water sustainability, it has an obvious and direct relationship with other policies, including biodiversity, land use planning, and climate change, among many others. The goal of sustainability would have to be discussed and further articulated. For example, the concept of intergenerational equity must be a component of the goal of sustainability.

There are many principles that support the goal of water sustainability. Without being exhaustive, the key principles to be developed in a new Canadian water governance strategy are likely to include:

- *Investment and financing:* infrastructure renewal and polluter pay principle.
- *Science and information:* full data and monitoring and enhancing scientific capacity.
- *Approaches to management:* precautionary principle, watershed planning, pollution prevention, and conservation.
- *Role of civil society:* right to know, public participation in decision making.

These principles must be manifest at both the provincial and federal levels. Although a number of other principles should be included, for present purposes, these provide a good example of the areas in which a Canadian water governance strategy is needed.

## The Federal Role

Even with Canada's fragmented jurisdiction over water, the federal government plays a very important role in water governance. Some areas of federal involvement include science, infrastructure renewal, conservation, a federal safe drinking water act, and Aboriginal water rights.

### Science

The federal government should provide the scientific foundation for decision making. Basic decisions dealing with emissions and releases of water, water flows, and water quantities are currently being made without recourse to adequate information. Improving the scientific basis for making such decisions should be one of the cornerstones of federal involvement. The government can either undertake this work itself and/or it can support provinces and territories as well as third parties (such as universities) in doing so. At the same time, it is critical to define the social and policy framework within which this science is situated. For example, is science being asked to contribute to precautionary decision making with respect to water resources? The criteria for decision making are as important as is the information base upon which decisions are made. Transparent policy debate should dictate where science is most likely to make a contribution to sustainable decision making.

### Infrastructure Renewal

While one of the governing principles of water governance should be sustainable pricing, generally at full cost, it must be recognized that, in many areas of Canada, there is an enormous deficit in terms of infrastructure needs, and this is often exacerbated by issues of distance, population, and scale. The federal government should develop funding mechanisms to ensure that resources are available to overcome these challenges. For instance, where feasible, federal grants may include requirements for the long-term financial health of a water system. Federal grants must also provide long-term assurance to First Nations and remote communities that they will be able to operate safe drinking water systems.

### Conservation

It is crucial that the federal government coordinate with the provinces, territories, and First Nations on a conservation strategy for Canadian water. Conservation goals should be unified, integrated, and non-contradictory, and they should be based on sound science. While many conservation measures may have to be implemented at the provincial level, there are many opportunities for federal involvement, including setting standards for water appliances, providing support for agricultural conservation, and assisting with the development of industrial best practices across the country.

### Pollution Prevention

The federal government has declared that pollution prevention is a federal concern, yet it has done relatively little to operationalize it.[9] The thrust of this principle is to examine why society generates pollutants in the first place (rather than trying to control them once they have been generated). Once this has been determined, the focus shifts to ways to redesign processes and products to avoid pollution. In effect, pollution prevention is an industrial strategy that promotes innovation, efficiency, conservation, and clean technologies. Federal leadership in this area is crucial for all industrial and commercial sectors.

### Federal Safe Drinking Water Act

Although various political parties have proposed federal safe drinking water legislation, none now exists. Accordingly, there is no coordinated framework within which to discuss drinking water policy across Canada. Equally important, there is no coordinated water policy pertaining to matters solely within federal jurisdiction.

## Aboriginal Role

Aboriginal governments may want to consider asserting a water governance model in accordance with their Aboriginal and treaty rights (and/or in accordance with Indian Act provisions, if applicable) in order to govern and protect their water resources. First Nations water governance takes many forms within Canada, and there are long-standing models elsewhere on the continent that may be examined. In New Mexico, for example, tribal governments have decision-making authority even over non-tribal activities if those activities affect the tribe's water resources.

Two current water policy initiatives demonstrate that Canada is far from achieving a harmonious mechanism with which to deal with First Nations water governance. In one example, that of the Annex 2001 negotiations referred to above, the Great Lakes provinces and states did not include First Nations governments in the negotiations. For example, it was only very late in the process that Ontario began to inform some of its First Nations communities and umbrella associations about the Annex 2001 negotiations. Several Ontario First Nations see enormous implications in the terms of Annex 2001 for their unceded Aboriginal rights and properties, including lake beds and the ability to govern and control the water resources upon which their communities depend, both economically and culturally.

In the other example, the watershed-based source water legislation introduced by Ontario in December 2005 (the Clean Water Act),[10] huge questions remain regarding how First Nations communities will be involved in developing watershed-based source protection plans and how their objections or wishes will be respected within the proposed committee structures. The question of integral First Nations involvement and governance ability is even more acute in the northern part of Ontario (i.e., most of Ontario), where First Nations communities are often the only or primary communities in the midst of vast watersheds.

## Provincial and Territorial Roles

In addition to federal and First Nations efforts to develop a Canadian water governance strategy, the provinces and territories need to integrate their laws, regulations, and programs to create an effective strategy. The goal is to ensure that there is a Canada-wide set of principles for the sustainable use of water. Although provinces may vary in their approaches to furthering such principles, the idea is to ensure the coordination of efforts and sharing of ideas (as well as data, science, and other such information). In the following section, we look at steps that might be taken to develop a truly Canadian water policy – that is, one that includes provincial, territorial, and First Nations as well as federal concerns.

## Toward a Canadian Water Governance Strategy

This chapter argues that there is a pressing need for a Canada-wide water governance strategy. While the CCME is supposed to provide leadership and to guide environmental policy, what is required, along with inclusion of the public, is a new mechanism for federal, provincial, territorial, and First Nations relations. The establishment of a mechanism to coordinate and implement Canada-wide water policies is obviously no simple task. However, the failure to do so is not an option, as today's crises could result in long-term, large-scale disruptions of the ecosystem. There are several urgent reasons for implementing a new water regime in Canada. First, increasing use of waters within Canada demands a coherent policy. Canadians are still world leaders in terms of quantities of water used (and abused). Second, a number of reports and studies verify that the world is becoming "water poor," suggesting that Canadian water will become a source of global envy. Canada must demonstrate leadership with regard to the appropriate management and conservation of this water not only to protect its own resources but also in order to export its knowledge and expertise (as opposed to the water itself). Third, Canadian water resources are under growing stress from pollution, invasive species, and intense use due to such activities as aquaculture, urban development, and climate change, to name but a few. At this point, due to the dismantling of many of the research and monitoring programs throughout the 1990s, it may be difficult to assess the state of Canadian waters. Fourth, the drinking water systems of many First Nations are in dire condition and need sustainable, long-term care in order to meet First World drinking water standards.

In developing a Canadian water governance strategy, it is essential to recognize that there are huge differences in water issues across the country. The issues of greatest concern in the Great Lakes-St. Lawrence region differ from those in, for example, Saskatchewan. The solutions that would work in large urban centres are not the same as those that would work in rural areas. Nevertheless, there would be important gains for decision making and ecological sustainability if a Canadian water governance strategy could be developed, outlining which proposed solutions are most applicable. A goal of strategy development should be to arrive at a framework that is both equitable and pragmatic across the country.

For example, the concept of watershed-based water resource stewardship, involving multiple levels of government, First Nations, and citizen engagement both for ecosystem and for drinking water protection, is gaining ground across the country. At minimum, this should be considered a best practice for all governments when it comes to designing water legislation, policy, and rules. However, the details of implementation may differ across the country, depending on the institutions already in place and the history and scale of the watersheds. What is needed is not identical solutions but, rather, equitable and useful mechanisms, resources, and institutional support for water governance decisions that will meet the needs of all Canadians. Achieving greater integration, taking advantage of the best innovations, sharing the best practices, avoiding reinventing wheels, providing leading edge research, and having access to federal resources are all essential to a good Canadian water governance strategy.

It is also critical that a new Canadian water governance strategy provide citizens with decision-making rights and that it not assume that water governance is the purview of "governments" alone. Citizens, NGOs, and other stakeholders can offer important insights into water governance and protection. However, those insights are often ignored, lost, or overlooked by our current inflexible governance and decision-making institutions.

What is needed is a strategy that not only allows for but also expects meaningful citizen and stakeholder participation. For example, Ontario's proposed watershed-based source protection planning legislation envisages broad citizen participation as well as extensive participation by landowners, stewardship organizations, business, and others. In British Columbia, the Fraser

Basin Council (2005) has been encouraging and noting the importance of Aboriginal-non-Aboriginal relationships in water and other resources through both formal and informal arrangements that demonstrate the importance of working together. In May 2005, the UN's Sixth Global Forum on Reinventing Government adopted the Seoul Declaration, which calls upon governments across the globe to increase transparency and participatory decision making in order to achieve sustainable development.

How, then, do we develop this Canadian water governance strategy? We need a process that is transparent, that provides for extensive public engagement, and that includes a negotiation process that includes First Nations as well as provincial and federal governments. Citizen engagement in the process of water governance, as well as the highly important roles of municipalities and watershed bodies, must also be embraced. A suggested first step would be the establishment of a public inquiry on water, which would be a joint federal-provincial-First Nations initiative. The inquiry would conduct research, hold public hearings, and then report on what would be the best institutional framework for such a strategy.

Among other things, a water governance inquiry should ask Canadians:

- How do you want water governance decisions to be made?
- Who should participate in water governance decisions and in protecting Canada's water resources?
- What values should guide these decisions?
- What institutions should provide oversight over water governance decisions?

Clearly, the task facing this inquiry would be substantial. Fashioning an effective water strategy is a challenging endeavour, but it is essential if we are to rework governance mechanisms to ensure that Canadian federalism works for, rather than against, better water management.

## NOTES

1 David Boyd (2003, 15) notes that, "unfortunately, the Canadian legal system's approach to water issues is heavily influenced by the enduring myth of endless water ... As a result, our laws and policies generally fail to provide the level of water security sought by Canadians."

2 For a review of some trade implications and water, see Great Lakes United and the Canadian Environmental Law Association (1993).

3 The 1998 Environmental Harmonization Accord has always been the subject of debate. For example, see House of Commons Standing Committee on Environment and Sustainable Development (1997). It has also been the subject of litigation: see *Canadian Environmental Law Association v. Canada* (Minister of the Environment) (1999).

4 For further background and context on federal-provincial environmental relations, see Harrison (2002).

5 The mandate and activities of the CCME are described at http:// www.ccme.ca.

6 "Head of power" is a legal term that denotes the administrative responsibility allocated to specific orders of government and/or government institutions.

7 The three water basins defined by Ontario Reg. 387/04 are the Great Lakes-St. Lawrence Basin, the Nelson Basin, and the Hudson Bay Basin.

8 The formal names for these agreements are the Great Lakes-St. Lawrence River Basin Sustainable Water Resources Agreement and the Great Lakes-St. Lawrence River Basin Water Resources Compact. They can be read at http://www.mnr.gov.on/mnr/water/greatlakes/index.html.

9 See Canadian Environmental Protection Act, Statutes of Canada c. 33, "Declaration" (Statutes of Canada 1999).

10 The proposed law was introduced on 5 December 2005 and was called the Clean Water Act, 2005 (Bill 43) (Legislative Assembly of Ontario 2005).

## REFERENCES


Bakker, Karen. 2003. Liquid Assets: How We Provide Water Depends on Whether We View Water as a Commodity or as a Public Good. *Alternatives Journal* 29 (2): 17-21.

Bloomfield, M., and Gerald F. Fitzgerald. 1958. *Boundary Waters Problems of Canada and the United States*. Toronto: Carswell.

Botts, Lee, and Paul Muldoon. 2005. *The Evolution of the Great Lakes Water Quality Agreement.* East Lansing: Michigan State University Press.

Boyd, David. 2003. *Unnatural Law: Rethinking Canadian Environmental Law and Policy.* Vancouver: UBC Press.

Cohen, Maxwell. 1975. The Regime of Boundary Waters: The Canadian-U.S. Experience. *Recueil des Cours* 146: 220-39.

DFAIT. 2005. *Joint Canada-U.S. Declaration on the Devils Lake Diversion Project.* Ottawa: Department of Foreign Affairs and International Trade. http://w01.international.gc.ca/minpub/Publication.asp?publication_id=382873&language=E.

Environment Canada. 1985. *Currents of Change: Final Report.* Inquiry on Federal Water Policy. Ottawa: Environment Canada.

—. 1987. *Federal Water Policy*. Ottawa: Environment Canada.

Fraser Basin Council. 2005. Website Survey of Aboriginal–Non-Aboriginal Relationships. http://www.fraserbasin.bc.ca/programs/documents/SurveyReport_FN.pdf.

Great Lakes United and the Canadian Environmental Law Association. 1993. *NAFTA and the Great Lakes: A Preliminary Survey of Environmental Implications.* November.

Harrison, Kathryn. 2002. Federal-Provincial Relations and the Environment: Unilateralism, Collaboration, and Rationalization. In *Canadian Environmental Policy: Context and Cases*, ed., D.L. VanNijnatten and R. Boardman, 123-44. Don Mills, ON: Oxford University Press.

House of Commons Standing Committee on Environment and Sustainable Development. 1997 (December). *Report on the Harmonization Initiative of the Canadian Council of Ministers of the Environment.* Ottawa.

Hughes, Elain, Alastair Lucas, and William Tilleman. 2003. *Environmental Law and Policy*, 3rd ed. (Chapter 2). Toronto: Emond Montgomery.

McClenaghan, Theresa. N.d. Why Should Aboriginal Peoples Exercise Governance over Environmental Issues? *UNB Law Journal* 51: 211-30.

Nunavut. 2005. *The Bathurst Mandate Pinasuaqtavut: That Which We've Set Out to Do.* Nunavut Government website: http://www.gov.nu.ca/bathurst.htm.

O'Connor, Hon. Dennis R. 2002a. *Report of the Walkerton Inquiry.* Part 1. The Events of May 2000 and Related Issues. Toronto: Queen's Printer.

—. 2002b. *Report of the Walkerton Inquiry.* Part 2: A Strategy for Safe Drinking Water. Toronto: Queen's Printer.

Pollution Probe. 2006. *Toward Sustainable Water Policy in Canada.* Preliminary Discussion Points for the Pollution Probe Water Policy in Canada National Workshop Series. Workshop No. 1, 6-7 February, Winnipeg, Manitoba.

PollutionWatch. 2004. *Shattering the Myth of Pollution Progress in Canada: A National Report.* Environmental Defence and the Canadian Environmental Law Association. December. http://www.pollutionwatch.org.

—. 2006. *Partners in Pollution: An Assessment of Continuing Canadian and United States Contributions to Great Lakes Pollution.* Toronto: Environmental Defence and the Canadian Environmental Law Association. http://www.pollutionwatch.org.

Royster, Judith V. n.d. *Environmental Federalism and the Third Sovereign: Limits on State Authority to Regulate Water Quality in Indian Country.* http://www.ucowr.siu.edu/updates/pdf/V105_A4.pdf.

Spencer, Robert, John Kirton, and Kim Richard Nossal, eds. 1981. *The International Joint Commission Seventy Years On.* Toronto: University of Toronto Press for the Centre for International Studies.

Sprague, John B. 2003. Myth of Abundance: The Common Claim That Canada Boasts the World's Largest Supply of Fresh Water Is False. *Alternatives Journal* 29 (3): 28.

Willoughby, William R. 1979. *The Joint Institutions of Canada and the United States.* Toronto: University of Toronto Press.

# 13
# Are the Prices Right? Balancing Efficiency, Equity, and Sustainability in Water Pricing

*Steven Renzetti*

At a recent workshop sponsored by the federal government's Policy Research Initiative, a number of speakers addressed the issues of municipal and provincial water pricing in Canada.[1] Several observations emerged from that workshop. The first is that there are a number of reasons why it is important to get water prices "right." These include encouraging an efficient allocation of water, improving water quality, providing water suppliers with adequate revenues, and encouraging innovation and conservation. The second is that Canadian municipal and provincial governments rarely get water prices right and that this creates a wide variety of problems, including over-consumption, water use conflicts, deteriorating infrastructure, declining water quality, and stifled innovation in water-conserving technologies. The third is that the current challenge is how to move beyond the already well-documented criticisms of Canadian water-pricing practices and to implement water-pricing reforms.

Before reforms can proceed, however, it is necessary to have an idea of what these "right" water prices should look like. A rather daunting feature of water-pricing reforms is that they need to address a number of possibly divergent policy goals. For too long, economists have advocated changes to public sector prices while giving short shrift to environmental and equity concerns as well as to the political nature of rate-setting. By the same token, environmental and social activists have proposed reforms that have not addressed the financial concerns of water utility managers.

If water prices are to be fair, promote efficient water use, and protect water quality – that is, if we are to get them "right" – then they should be:

- *Financially sound:* prices should be designed so that the agencies that supply potable water suppliers and treat sewage earn enough revenue to pay all their bills.

- *Efficient:* prices should inform water users of the full social costs of their water use so that users can make efficient water consumption decisions.
- *Environmentally sustainable:* prices should promote water conservation and, more generally, contribute to sustaining aquatic ecosystems.
- *Equitable:* there is a growing consensus that access to potable water is a basic human right and, thus, that the cost of providing clean water should not pose an unacceptable burden on low-income households.

It is important to note that it is unlikely that reforming water prices alone will ensure improvements in municipal water systems and improved provincial water allocations; rather, reforming water prices is one part of a broad effort to improve water management in Canada. The modest purpose of this chapter is to investigate how to reform water pricing at the municipal and provincial levels so as to satisfy the above-listed policy goals. Because Canada's Constitution grants provinces primary jurisdiction over natural resources, water prices differ not only at the municipal level but also at the provincial level, where some provinces charge for access to self-supplied water. I examine both municipal water supply pricing and provincial raw water pricing. In both cases, I consider the current state of pricing and ask what is needed to reform water prices in order to make them more financially sound, efficient, environmentally sustainable, and equitable.

## Municipal Water Supply Prices

### Current State of Municipal Water Prices

A recent report from Environment Canada (Burke, Leigh, and Sexton 2004) summarizes the state of municipal water pricing and highlights some of the shortcomings of the country's water prices. Across the country, there is a remarkable diversity of forms of water prices. In particular, the report indicates that 37 percent of households face a flat rate charging system (i.e., they pay a fixed monthly fee independent of water use), while 39 percent face water prices that are constant (i.e., the unit price does not vary with the amount consumed). A small proportion of Canadian households (13 percent) face water rates where the unit price decreases with the amount consumed, and an even smaller proportion (10 percent) face water rates where the unit price increases with the amount consumed. The most striking of these figures is the 37 percent of households whose payments for water

TABLE 13.1

**Average monthly water charges by rate type, 1999**

| Province/Territory | Percentage of population with metered service | Average monthly charge for flat rate ($/month)[1] | Average monthly charge for volumetric rate ($/month)[2] |
|---|---|---|---|
| Newfoundland | 0 | 18.52 | –[3] |
| PEI | 1.0 | 24.21 | 39.33 |
| Nova Scotia | 89.9 | 17.05 | 32.34 |
| New Brunswick | 46.4 | 29.60 | 36.96 |
| Quebec | 15.4 | 16.78 | 14.34 |
| Ontario | 82.2 | 27.73 | 30.00 |
| Manitoba | 96.6 | 30.73 | 48.42 |
| Saskatchewan | 98.7 | 32.84 | 37.49 |
| Alberta | 73.8 | 30.82 | 42.84 |
| British Columbia | 24.1 | 26.52 | 22.73 |
| Yukon Territory | 43.1 | 49.61 | 42.41 |
| NWT | 98.4 | 62.80 | 87.20 |

1 Flat rate means that the household pays a fixed monthly fee that does not vary with the amount of water consumed.

2 Volumetric rate means that the amount paid per month varies with the amount of water consumed.

3 There were not enough rate structures in the province to determine this value.

SOURCE: Burke, Leigh, and Sexton (2004), various tables.

service are not connected in any way to the amount that they consume. The impact of flat rate water charges is highlighted by the fact that water use was 70 percent higher when consumers faced these types of rates than it was when they faced volume-based rates. Table 13.1 compares the average monthly water charges faced by residential consumers by rate type (flat rate versus volume-based rates) in 1999 across the provinces. It is interesting to note that most provinces have higher monthly charges under volumetric rate structures and that the highest monthly charges are observed in the Prairie provinces and the north.

The Environment Canada report also allows us to look at how municipal water prices have been changing recently. The evidence is decidedly mixed. The good news is that real prices[2] have risen in all provinces between 1991 and 1999. The not-so-good news is that, over this period, the rate of increase in prices has barely exceeded the rate of inflation in New Brunswick, Quebec, Ontario, Saskatchewan, and Alberta. This suggests that most water agencies are not aggressively promoting water conservation or pursuing improved financial viability through higher water prices.

## Assessment of Municipal Water Supply Prices

### Financially Sound?

Any self-sustaining agency must receive sufficient revenues in order to be financially viable. These revenues may come from user fees, grants and subsidies, and other sources. A fundamental shortcoming of the current state of water pricing in Canada is that the revenues generated by water prices are insufficient to meet all of the operating and capital needs of water suppliers. In part, this has been because municipal water agencies have received subsidies from provincial and federal governments and transfers from municipal governments. In a widely cited report, the National Round Table on the Environment and Economy (1996) estimated that unmet water and wastewater infrastructure needs in Canada were between $38 billion and $49 billion in 1996, and capital costs for the following twenty years would be in the order of between $70 billion and $90 billion. At the same time, only 50 percent of the cost of maintaining and operating water infrastructure was actually being met through cost recovery from users of the systems.

### Efficient?

In order for water use to be economically efficient, consumers must be made fully aware of the costs of their water use decisions. Specifically, the price paid for water must equal its marginal cost (i.e., the incremental cost of the water and the construction and maintenance of the water collection and distribution network that supplies the water). Given the nature of the technology of water collection, treatment, and distribution, it has been demonstrated that this marginal cost of supply is likely to vary with distance, time of use, and type of use. For example, the farther a household is from the storage facilities, the more expensive it is to supply. In addition, supply spikes during peak periods (within the day peak use occurs in the early morning and early evening; within the year, peak use occurs during summer months). Finally, some user groups are more costly to serve because of the nature of their water demands.

This set of observations strongly suggests that efficient water prices should be differentiated by quantity, user group, distance, and time. Prices are commonly differentiated by user group, but rarely can these differences be tied to cost differences. Further, as is indicated above, a small proportion of water suppliers do have prices that vary according to the quantity of water used but, again, there is little evidence that these rates are tied to the marginal cost of supply. It is very rare, however, to find water prices that vary according to distance within an urban area or even according to the time (within the day or within the year) of consumption. Renzetti (1999) studies a sample of Ontario water suppliers and finds that prices charged to residential and commercial customers are found to be only one-third and one-sixth of the estimated marginal cost for water supply and sewage treatment, respectively. For example, the average price to residential customers is $0.32 per cubic metre, while the estimated marginal cost is $0.87 per cubic metre. The implications of this mispricing include inflated consumption levels, overextended distribution networks, and weakened incentives to develop water-conserving technologies.

**Environmentally Sustainable?**

There is an important and poorly understood connection between two seemingly unrelated facets of water supply: how agencies carry out their cost accounting and the impact that these agencies can have on aquatic ecosystems. The idea is that a necessary precondition for reforming water prices is to first ensure that the full costs associated with water use are enumerated. These costs need to include not only the opportunity cost of purchased inputs (labour, capital, materials, etc.) but also the value of raw water supplies and the costs of the environmental and human health damages arising from diminished water quality.

Renzetti and Kushner (2004) examine the implications of incomplete cost accounting by considering the operations of a representative agency responsible for water supply and sewage treatment. The water agency serving the Niagara region (population 410,000) in southern Ontario supplies over 80 million cubic metres of potable water and treats over 85 million cubic metres of sewage. In 1998, it recorded total operating costs of $22.5 million for water supply and $41 million for sewage treatment. These figures reflect the agency's accounting for the costs of labour, materials, energy, and capital repairs. Unfortunately, the region's water supply and sewage treatment operations impose other costs on society, which are not reflected in the agency's cost accounting. For the purposes of this chapter, the most important unaccounted cost stems from the water agencies' use of raw water and its impact on local water quality.[3] When these environmental and other unaccounted for costs are included in conventionally measured costs, it is estimated that the agencies' costs could rise by $10.4 million to $34.9 million, or 16 percent to 55 percent of recorded costs. It is important to point out that the regional water agency in Niagara is not unusual in this regard, as it uses cost accounting methods that are commonly employed by water and sewage treatment agencies across Canada.

This idea of promoting environmental protection by changing public agencies' accounting rules might sound far-fetched, but it is already being implemented in Europe. Under the European Union's Water Framework Directive, in the near future, all water supply and sewage treatment facilities must measure the economic value of their use of raw water supplies, along with any changes to aquatic ecosystems, and incorporate these estimated costs into water and sewage prices (Chave 2002). Further, it has been suggested in Ontario that the extra revenues earned by expanding the definition of water supply costs could be used to support water agencies' source water protection efforts (Technical Experts Committee 2004).

**Fair?**

To many observers, the fairness of water policies and, in particular, water prices is as important as is the efficiency of those policies. Unfortunately, assessing the fairness of Canadian water prices is difficult for a number of reasons. First, there is relatively little information on the relationship between water-related expenditures and different household income levels. It is generally believed that water-related expenditures fall as a share of income as income rises (Wasny 1986). This is important because it would indicate that efforts to raise prices could have a bigger impact on low-income households than they do on average- or high-income households. Unfortunately, the fact that many low-income households live in apartments where individual water use is not metered and, thus, not billed complicates our understanding of this relationship. Second, there is no consensus among policy analysts or, more generally, society on the extent to which prices for public services such as water supply should differ from the cost of supply in order to address equity concerns (e.g., see the discussion in Bocking 2002). On the one hand, it can be argued that it is most appropriate to set prices at their respective marginal cost so as not to distort supply and demand decisions and to deal directly with the fundamental causes of poverty or an undesirable distribution of income by other policy means. On the other hand, it can also be argued that access to potable water is a human right and that employing efficient or full-cost pricing may impinge on that right by harming those households who are least able to afford water.

Despite these difficulties, there is reason to believe that current water pricing in Canada harms many low-income households. Non-volumetric charges and constant rate structures are the predominant forms of water prices in Canada and, unfortunately, may do the least good for low-income households. If a low-income household (such as a single senior citizen) uses a relatively small amount of water, then switching away from a flat rate to a volumetric rate will likely result in a smaller water bill for the household, since the former charge is typically based on the municipality's "average" level of consumption. A similar problem arises when households face a constant price for water. An important limitation of constant price structures is that they usually do not generate a revenue surplus that can be used to subsidize water rates at low levels of consumption. Finally, as noted above, the revenues generated by today's water prices are inadequate to support needed infrastructure repairs and improvements. The reduction in system reliability that results from this underinvestment is most likely to affect poor water users most heavily as they do not have the financial resources to protect themselves from system interruptions (e.g., by buying bottled water).

Even the presence of more complex price structures can work against low-income households. Burke, Leigh, and Sexton (2004) point out that over three million Canadian households live in towns and cities with volumetric water rate structures that include a minimum monthly charge. The problem is that this charge reflects "a volume of water that is greater than the normal range of residential water use" (4). These observations suggest that, in contrast to other services (such as electricity and natural gas supply), where there has been significant concern over the fairness of rate structures, the impacts of alternative rate structures on low-income households has not been thoroughly considered. Fortunately, much of what has been learned about the design, for example, of "lifeline" rate structures (i.e., subsidized rates for low levels of consumption) in telecommunications and electricity can be applied to water supply. For example, concerns over the impacts of rate increases in Washington, DC, led to a number of proposals for altering the way in which rates would be raised so as to minimize the impact on the city's poor households. These included lifeline rates, waiving the fixed monthly charge, and replacing quarterly billing with monthly billing. A report prepared by the US Environmental Protection Agency provides an excellent review of the relative benefits and administrative costs of these options (US Environmental Protection Agency 2002).

## Provincial Raw Water Prices

### CURRENT STATE OF PROVINCIAL RAW WATER PRICING

The second area of interest is pricing for access to raw water at the provincial level. While water pricing at this level has not received the same degree of attention as has municipal water pricing, it is still important because of the role of provincial governments in determining the allocation of very large volumes of water across sectors. Under the Canadian Constitution, provincial governments have almost full primary jurisdiction over the allocation of water (exceptions include transboundary waters and waters on First Nations lands). All provinces have enacted permit systems to regulate large-scale water withdrawals. Most firms, energy producers, mines, and water agencies that withdraw large volumes of water must first obtain a provincial permit to do so. For the most part, these permits are issued on a "first-come, first-served" basis after differing degrees of scrutiny of the impacts of potential withdrawal on pre-existing permitted withdrawals and local ecosystems. While the scrutiny of potential environmental impacts has increased recently, Nowlan (2005, 9) observes that "water laws in Canada historically have not placed a premium on conservation." As other chapters in this volume point out, Canada's water policies and laws have historically been directed at promoting water use (e.g., through the support of interbasin transfers) rather than curbing it. Permits are usually issued for a specific number of years and are not transferable. This last feature stands in contrast to the extensive reliance on water permit trading and leasing that occurs in water-short areas of the United States.

### ASSESSMENT OF PROVINCIAL RAW WATER PRICING

**Financially Sound?**

The levels and objectives of the prices charged for permits to withdraw water are perhaps the most unsatisfactory features of provincial efforts to regulate water use. Seven of thirteen provinces and territories levy some type of one-time or recurring fee for these permits, while the remaining provinces (including Quebec and Alberta) do not levy any fee at all. Of those jurisdictions that do charge, fees are very low: ranging from $0.01 to $143 per 1,000 cubic metres of permitted annual water withdrawals (Nowlan 2005). This wide variation in fees implies that there is also a wide variation in the contribution those fees make to supporting government programs. The current system of permit issuance obviously generates no revenue in those provinces without fees (such as Quebec). Ontario has very recently introduced limited administrative fees that are only meant to recoup some of the costs associated with assessing applications for water permits (Ontario Ministry of the Environment 2004). By way of contrast, the fees from water use permits earned the BC government nearly $400 million in 2004 (British Columbia Ministry of Finance, 2005).

**Efficient?**

A second problem regarding pricing for access to raw water at the provincial level is that few provinces have clearly stated what these fees are supposed to accomplish (Dupont and Renzetti 1999). In particular, there is no reason to believe they contribute to the efficient water abstraction decisions made by self-supplied water users. This is because the fees are not designed to reflect the opportunity cost of withdrawn water (i.e., the value of the permitted water in its alternative uses). Furthermore, the rights to withdraw water are not allocated according to the relative net benefits of proposed water use nor are they allocated with a clear understanding of the opportunity cost of the proposed water use. While the criteria to assess an application for a permit to withdraw water vary across the provinces, none of the provinces assesses the value (or tax revenue or jobs or any other indicator of economic significance of the proposed water use) that water use might generate for it. Moreover, if there are changes in the relative values of alternative allocations of water, then permits cannot be traded after being issued. In effect, permitted water withdrawals are frozen in place during the life of the permit, despite any changes to local economic or environmental conditions. This inability to trade water use rights in the face of changing demands for water has created serious imbalances in Canada and even more so in Europe and parts of the United States. In situations like these, holders of historical water rights (typically farmers) are unable to sell or lease their rights to water-short cities and businesses, despite the latter's willingness to compensate rights holders for their water.

**Environmentally Sustainable?**

There are a number of features of the process used to assess applications for direct water abstractions that are meant to protect the surrounding aquatic environment. For example, Nowlan (2005) points out that provinces are increasingly expecting applicants for water use permits to demonstrate what impact, if any, their proposed withdrawals will have on neighboring water bodies. The interest in this chapter, however, is a narrower one: do the fees charged for these permits also support this effort to protect the environment?

In order to answer this question, let's consider what provincial water prices would have to look like if they were to promote environmental protection. Most important, the prices would have to inform water users of the estimated costs of any impacts that their water use has on the environment or other users. Thus, if a proposed water withdrawal lowers groundwater levels or increases costs and risks to downstream users (due to reduced quality of return flows or due to greater likelihood of inadequate stream flows), then prices should reflect these impacts. Similarly, if a proposed water withdrawal is related to a use of water characterized by high consumption levels (such as water bottling), then this should be reflected in higher withdrawal fees. Regrettably, provincial water prices exhibit few of these features. For example, the costs arising from external environmental impacts associated with proposed uses have little or no bearing on fees (when fees are charged).

**Fair?**

Dupont and Renzetti addressed the issue of fairness of the pricing of access to water resources at the provincial level. At the time, we argued that,

> by allowing industry, public utilities and farming operations free access to water resources, a provincial government is undertaking an implicit and poorly understood redistribution of wealth. This is because, by failing to capture some of the economic value created by the application of water in production processes, a government de facto allows that value to be directed toward those individuals and groups which have gained access to the use of fresh water resources. This is reinforced by the fact that some water users return water to the basin in a degraded state, thereby imposing costs upon other users. (Dupont and Renzetti 1999, 365)

It seems reasonable to argue that, as the owner of a scarce and productive natural resource, the Crown is entitled to share in the economic value created by the application of water in industrial processes. This has certainly been the position taken by provincial governments with respect to the use of other renewable natural resources (e.g., forests). As indicated above, a number of other provincial governments already levy some type of charge for direct water withdrawals. Thus, it would seem fair to expect users to pay a reasonable fee in order to secure the use of a natural resource that contributes to their profitability. This point is made stronger when you note that the provincial government provides infrastructure (such as dams and flood protection) and services (such as protecting water quality and hydrologic information) that increase the value of water to users.

## An Agenda for Reform

This chapter began by asking whether it is possible to reform water prices in a way that contributes to four policy goals simultaneously. These goals are to make water prices financially sound, economically efficient, environmentally sustainable, and fair. Getting prices "right" in this way would be an important part of a larger strategy to reform water management in Canada. Other elements of this strategy (such as enhanced governance and transparency of decision making) are discussed elsewhere in this volume. Fortunately, there are some fairly clear directions for reforming water prices and, indeed, some of these reforms are being undertaken or investigated by provincial and municipal governments. The reforms being considered by Canadian governments, however, are too limited and most likely need to be redirected or extended in order to ensure that water prices promote efficiency, equity, and environmental sustainability.

### REFORM OF MUNICIPAL WATER SUPPLY PRICES

There are several things that can be done to improve municipal water prices. These include the following:

- *Enhanced cost accounting.* The government of Ontario recently passed legislation (Sustainable Water and Sewage Systems Act) that requires all water suppliers and sewage treatment facilities to revise their accounting procedures so that the full cost of their activities is recorded and reflected in prices. This is a valuable first step, and it is hoped that other provinces follow suit. Unfortunately, the definition of full cost that will appear in the regulations supporting the new law is unlikely to include external costs (such as the opportunity cost of raw water supplies or environmental or health damages arising from disinfection byproducts or sewage plant effluents). In contrast, the Water Framework Directive being enacted by the European Union does include these external costs in its definition of full costs.

- *Marginal cost pricing.* Assuming that a water supplier already meters water use, then an important step would be for it to adopt marginal cost pricing as a new basis for rate design. This may not be a simple undertaking. It will require changes in the way water agencies conduct their cost accounting and rate design efforts. Furthermore, it may lead to fluctuations in the agency's revenues. Nonetheless, past experience in water supply systems and other public utilities demonstrates the benefits of making this type of change. In the short run, adopting a summer surcharge is an attractive and feasible option that will inform consumers of the fact that water supply in dry summer months is significantly more expensive than it is during other time periods. It will also help to protect aquatic ecosystems when water supplies are most scarce. Another slightly more complex option would be to introduce zonal or distance-based water pricing.

An important question, however, concerns the impact of these types of reforms on low-income households. More generally, one might ask whether it is possible to reform municipal water prices as suggested above while not having a negative impact on low-income households. This is clearly a difficult task, but the experience of Los Angeles, in its efforts to reform water prices, shows that it is not insurmountable. Hall and Hanemann (1996) report that Los Angeles experienced recurring water shortages and faced the prospect of new sources of water that would be very expensive. This situation prompted the mayor to appoint a blue ribbon panel to investigate the reform of water prices as part of a broader strategy to address water supply problems. The panel was composed of community representatives, elected municipal officials, water utility managers, and academic researchers.

The panel studied the existing cost accounting and pricing rules and made several important recommendations. The first was to switch away from conventional public utility cost accounting and toward the economically more meaningful marginal cost accounting. The second was to adopt complex price structures that would better reflect the cost of supply to different parts of the city, different climate conditions, and even different lot sizes. At the same time, the panel remained cognizant of the potential impacts of rate reform on low-income households. Thus, it recommended that some of the excess revenue generated by the preferred increasing block rate structure be used to create a credit that would subsidize water consumption by low-income consumers.

Despite the complexity and controversial nature of the issues surrounding water-pricing reform, the panel's recommendations were largely adopted by the city council. It appears that there were a number of features that contributed to the success of the panel. These included strong political leadership, substantial public participation, the availability of technical expertise from water agency staff and academic researchers, a willingness to use information on user group characteristics to craft complex rate structures, and, finally, the panel's adoption of a broad perspective on the objectives of rate reforms (pricing reforms had to satisfy several objectives, including improved efficiency, equity, political feasibility, and revenue stability).

### Reform of Provincial Raw Water Prices

Once again, we want to consider whether there are possible directions for reforms that would make provincial raw water prices financially sound, economically efficient, environmentally sustainable, and fair. While the shortcomings of provincial water-pricing regulations are no less significant than are those of municipal water pricing, the appropriate directions of reform are somewhat less clear. The reasons for this are twofold. First, provincial governments must decide how they will allocate water in the future. Assuming that the goals I have set out are reasonable, then it is difficult to see how the status quo can be maintained. While the processes for assessing permit applications have been made more sophisticated in terms of their consideration of environmental impacts, they continue to do little to encourage technological innovation, promote water conservation, or contribute to government revenues. As a result, provincial governments should either move to reform their raw water prices and use these as the primary instrument for determining water allocation, or they should move to a more market-orientated system of using water use permits that are transferable and whose prices are determined through trading. Some provinces are moving in these directions. Alberta has tentatively moved towards introducing water markets, and preliminary research has shown this institutional structure to have significant potential to promote efficiency in water use while protecting instream water needs and third parties potentially affected by water trading (Horbulyk and Lo 1998; Nicol, Klein, and Bjornlund 2005). In contrast, both Ontario and Newfoundland have recently reformed their fees for water use permits.

Second, assuming that a province has decided to employ water permit fees as its principle policy instrument, it must decide upon the primary objective of those prices. If it is to generate revenue to support their water-related programs, then the task is to design prices that recoup some of the surplus generated through private sector water use without unduly distorting users' decision making. This could be done in a straightforward fashion with a lump sum application fee and a recurring annual licence fee. Unfortunately, this would do little to promote conservation, environmental sustainability, or improved fairness of water allocations. On the other hand, if the purpose is to promote efficient water use and enhanced protection of aquatic ecosystems, then provincial raw water prices must reflect the opportunity cost of water use (including the costs of environmental impacts). In this case, the intent would be to alter behaviour as confronting users with higher fees is meant to encourage conservation and innovation in water-saving technologies.

Dupont and Renzetti (1999) simulate the impacts of alternative forms of water permit fees in Ontario. For example, one simulation considers the impact of a "two-part" pricing approach that has an annual fee ($2,500 for commercial and industrial users and water utilities and $300 for farms) and a volumetric charge of $0.003 per cubic metre for most self-supplied water users. The estimated impacts of introducing such a pricing structure included an expected drop in aggregate water withdrawals of approximately 5 percent, an average increase in the production costs for permit holders of 0.10 percent, and an increase in annual government revenues of approximately $88 million. The average household would experience approximately a 0.2 percent increase in its cost of living as a result of increases in water and electricity prices. Assessing the impact of reforming provincial raw water prices on low-income households would require further research. Furthermore, these types of fees could be tailored to more closely reflect local water supply-demand balances or to protect vulnerable ecosystems by varying the volumetric charge regionally or temporally (e.g., the fee could rise in summer months to reflect the higher opportunity costs of withdrawals under diminished river flows and lake levels).

## Conclusion

There are many dimensions of Canadian water policy that are world class. Canada leads the world in policies related to flood damage reduction, integrated watershed management, and designing community-based efforts to improve water quality. Unfortunately, the same cannot be said for Canadian water prices. Simply put, water prices at the municipal and provincial levels are in bad shape, and they fail in each of the four areas set out at the beginning of this chapter. They do not generate the revenues needed to support water agencies; they do not inform consumers of the full costs of their water use decisions; they do not contribute to protecting environmental ecosystems; and they do not satisfy basic principles of fairness. These shortcomings have not arisen out of a lack of technology or know-how; rather, they reflect not only the water supply industry's historic disinterest in using water prices for any purpose other than raising revenue but also a general provincial and federal neglect of this important part of water management. Despite this, things are beginning to change. There is an increased awareness of the importance of water prices, and isolated first steps are being taken toward getting them right. This chapter shows that reforming water prices to satisfy financial, economic, and environmental policy goals – while avoiding adverse impacts on low-income households – is both desirable and feasible. Further, these reforms can be part of a larger effort to improve governance, promote efficient water allocation, and protect aquatic ecosystems.

### Notes

1 A synopsis of the workshop is available at http://policyresearch.gc.ca/page.asp?pagenm=freshwater_symp. A number of the papers from the workshop are published in *Canadian Water Resources Journal* 30, 1 (2005).

2 "Real prices rising" means that prices have gone up faster than has the rate of inflation.

3 The other main source of error in cost accounting concerns the way in which the agencies record their capital costs. See Renzetti and Kushner (2004) for details.

## REFERENCES

Bocking, R. 2002. Pricing as an Instrument for Water Conservation. Chapter 15.2 in proceedings of Water and the Future of Life on Earth Workshop, 22-23 May, Morris J. Wosk Centre for Dialogue, Simon Fraser University, Vancouver. http://www.sfu.ca/cstudies/science/water.htm.

British Columbia Ministry of Finance. 2005. *Budget 2004: Estimated Revenue by Source.* http://www.bcbudget.gov.bc.ca/2004/bfp/bgt2004_part2.htm.

Burke, D., L. Leigh, and V. Sexton. 2004. *Municipal Water Pricing 1991-1999.* Ottawa: Department of the Environment. http://www.ec.gc.ca/water/en/info/pubs/sss/e_price99.htm.

Chave, P. 2002. *The EU Water Framework Directive: An Introduction.* Colchester, UK: IWA Publishing.

Dupont, D., and S. Renzetti. 1999. An Assessment of the Impact of a Provincial Water Charge. *Canadian Public Policy* 25 (3): 361-78.

Hall, D.C., and W.M. Hanemann. 1996. Urban Water Rate Design Based on Marginal Cost. In *Marginal Cost Rate Design and Wholesale Water Markets: Advances in the Economics of the Environment,* ed. D.C. Hall, vol. 1, 99-122. London: JAI Press.

Horbulyk, T., and L. Lo. 1998. Welfare Gains from Potential Water Markets in Alberta, Canada. In *Markets for Water: Potential and Performance,* ed. K.W. Easter, M. Rosegrant, and A. Dinar, 241-57. Boston: Kluwer Academic Press.

National Round Table on the Environment and the Economy. 1996. *State of the Debate: Water and Wastewater Services in Canada.* Ottawa: Canadian Council Of Ministers of the Environment.

Nicol, L., K. Klein, and H. Bjornlund. 2005. Irrigation Water Markets: Evidence from the Murray-Darling Basin in Australia and the South Saskatchewan River Basin in Alberta. Paper presented at Reflections on Our Future annual conference of the Canadian Water Resources Association, 15-17 June, Banff, Alberta.

Nowlan, L. 2005. *Buried Treasure: Groundwater Permitting and Pricing in Canada.* Report prepared for the Walter and Duncan Gordon Foundation, Toronto. http://www.gordon.org.

Ontario Ministry of the Environment. 2004. Water Taking and Transfer Regulation (O. Reg. 387/04). http://www.e-laws.gov.on.ca/DBLaws/Source/Regs/English/2004/R04387_e.htm.

Renzetti, S. 1999. Municipal Water Supply and Sewage Treatment: Costs, Prices and Distortions. *Canadian Journal of Economics* 32 (2): 688-704.

Renzetti, S., and J. Kushner. 2004. Full Cost Accounting for Water Supply and Sewage Treatment: Concepts and Application. *Canadian Water Resources Journal* 29 (1): 13-23.

Technical Experts Committee. 2004. Watershed-Based Source Protection Planning. Report to the Ontario Minister of the Environment. http://www.ene.gov.on.ca/envision/water/spp.htm.

US Environmental Protection Agency. 2002. Rate Options to Address Affordability Concerns for Consideration by District of Columbia Water and Sewer Authority. Philadelphia. http://www.epa.gov/water/infrastructure/pricing/Affordability.htm.

Wasny, G. 1986. Household Water Expenditures in Major Metropolitan Centres in Canada. *Canadian Water Resources Journal* 11 (4): 33-45.

GRAEME MACKAY/HAMILTON SPECTATOR

Water conservation measures often rely on informal sanctions of community members and neighbours.

# 14 Moving Water Conservation to Centre Stage

*Oliver Brandes, David Brooks, and Michael M'Gonigle*

As with actors in a long-running play, Canadians speak from a well-worn script when it comes to discussing our uses of water. The script is performed automatically. When, as individuals, we want water, we simply say: "Turn on the tap." When, as a community, we want more houses or industry or agriculture, we simply tell our politicians: "Build more dams." Against a backdrop of thundering rivers, endless lakes, and towering glaciers, this national performance comes naturally. But the times, they are a-changin', and the old lines don't work as well as they used to.

A new script for water management is emerging. Instead of focusing on getting more and more, new technology and innovation are moving toward sustainable water management. Communities are demanding "no new dams" and are inspiring water managers to begin thinking creatively about how to deliver services, not just water. Society is beginning to recognize that the largest source of new water for Canadians won't be new at all but, rather, more efficient use of the water we already have. Recycling and conservation hold the potential to reduce use even further. Just as Amory Lovins (1985) taught us to think in terms of "negawatts" as we met new electicity demands by reducing electricity use, so we must learn to think in terms of "negalitres" as we meet new water demands by reducing water use.

But the success of this new script is by no means assured. The old approach, with its theme of "man over nature," is supported by a powerful myth concerning the abundance of Canada's fresh water. Society may be unwilling to move from a familiar past that promises abundance toward a new reality that demands its participation in the challenging dialogue required for a sustainable future.

Despite the myth of abundance and the power structure that supports it, we do not believe that Canada's old approach to fresh water will be viable for much longer. As emphasized in previous chapters, Canada may rank high in per capita availability of water, but most of that water is found where Canadians are not (e.g., flowing to the Arctic) or is unavailable because it exists only as a slowly renewable stock in aquifers or lakes. Beyond this geographic mismatch, a diverse set of management "threats" conspire to challenge our illusion of limitless water (Environment Canada 2004). These include:

- *Infrastructure limits:* freshwater supply, wastewater collection, and treatment costs are outpacing available community resources (with estimates of unmet water and wastewater infrastructure ranging from $23 billion to $49 billion (NRTEE 1996; Environment Canada 2003).
- *Safety concerns:* the provision of water treated to drinking water standards is taxing management capabilities (as demonstrated in Walkerton and North Battleford).
- *Wild cards:* climate change (with attendant depletion of glaciers and longer periods of drought), ecosystem deterioration (including the widespread loss of wetlands), and new forms of pollution (such as waterborne endocrine disrupters, among others).

To support the ecological services that underpin our quality of life, Canadians must limit their demands for water and start protecting existing water sources. Responding to these threats requires changing behaviours and attitudes. But with change comes opportunity. The new script will be about innovation and new relationships with our surroundings. Opportunities include using technologies that ensure that our toilets use half the water they now use (or even no water at all); recycling our "waste" water and turning it into a source for other uses; and, in the urban area, turning expansive water-guzzling lawns into oases of native drought-resistant plants. Ultimately, our communities must be designed *for* water conservation not just retrofitted when we approach water limits.

This transformation does not just happen; fortunately, there are transitional steps. First, we gain time by changing from supply management to demand management. Then, we begin to address deeper concerns about the fundamental nature of the demands themselves. This moves us to a new stage – the "soft path," where water is viewed primarily as a service rather than as a product. Water conservation then moves to centre stage, triggering a fundamentally larger societal transformation. This transformation is about a new way of dealing with water, nature, and *ourselves* – ultimately, it is a form of participatory management called "ecological governance." Ecological governance is the overarching framework for sustainability – in our metaphor, the theatre – where ecosystem health and processes are carefully considered at all levels of decision making, both upstream and downstream and throughout the watershed.

As shown in Table 14.1, the move from supply-side management to the soft path follows a spectrum of water management approaches that move us from compulsively *getting* more resources to thoughtfully *governing* resource uses. Such innovation is already under way in places as diverse as Australia, South Africa, Europe, Israel, Florida, and California. One need only think of the energy efficiency revolution of the 1970s to recognize a familiar pattern. Taken together, these sequential yet overlapping steps represent a strategy that ensures a paradigm shift from wasteful to sustainable consumption of fresh water in Canada.

## A New Script: Changing from Supply to Demand Management

Although attitudes and policies are slowly changing, as witnessed by various local efforts to be "water wise," the old paradigm of supply management still dominates. This approach is predicated on large, centralized infrastructure with physical components (dams, reservoirs, pumping stations, and networked pipes) and corresponding institutional components (large-scale, hierarchical government agencies, rigid legal systems, and engineering firms). Supply management flourished in a time when dams were viewed as symbols of modernization. The Hoover Dam still draws tourists to celebrate the first "megadam" – or what India's first president referred to as "temples of modernity."

The Hoover Dam was built at the beginning of an era of engineering megaprojects that were fuelled by the promise of better lives and eternal progress. The dominant belief was in "supply management": the idea that nature was to serve our needs and that we could ensure this not by adjusting to the rhythms and cycles of the seasons but, rather, by channelling nature so that we could have water wherever and whenever we wanted it.

Underpinning supply management is the philosophy that water is not a limiting factor to growth and that the only real constraints are technological ability and the financial cost of the infrastructure needed to harness, store, and deliver water. Future growth is modelled on extrapolations from past and current consumption patterns, which, in turn, set in motion engineering plans to increase capacity to meet anticipated future needs. This historic

TABLE 14.1

**A spectrum of water management approaches**

| Policy | Dominant discipline | Range of policy choice | Fundamental question | Planning process | Outcomes |
|---|---|---|---|---|---|
| Supply management | Engineering | Low: infrastructure based | How can we find new resources to meet future projected needs for water based on past trends in water use and population growth? | Planners model future growth, extrapolate from historic patterns of consumption, plan for an increase in capacity to meet anticipated future needs, then locate and develop a new source of supply to meet that need. | Construction: dams, pipelines, canals, wells, desalination |
| Demand management | Economics | Medium: short-term cost-benefit driven | How can we reduce demands for water to conserve the resource, save money, and reduce environmental impacts? | Planners model growth and account for a comprehensive efficiency and conservation program to maximize use of existing infrastructure. Increasing capacity would be a final option as part of a least-cost approach. | Water as an economic good; technical fixes; efficiency gains |
| Soft Path | Social sciences with physical limits | High: based on self-reflective political processes | How can we deliver the *services* currently provided by water in new ways that recognize the need for long-term "system" changes to achieve social sustainability? | Planners describe a desired sustainable future state (or scenario) based on economic and demographic variables and then "backcast" to devise a feasible and desirable path to that future with sustainability built into the economic, political, and sociocultural choices made along the way. | Ingenuity, new options, habits, and patterns of use |

SOURCE: Adapted from Brooks and Wolfe (2004) and Brandes et al. (2005).

pattern of ever-increasing capacity – typically underwritten by government subsidies that reinforce the belief that water is not a valuable resource – does little to provide incentives for efficiency, let alone to encourage conservation. Though supply management has given Canadians a wealth of water access, the financial costs of new water supply are almost doubling every decade (Serageldin 1995), and preliminary evidence is that the environmental costs are mounting even faster (Postel and Richter 2003). This cycle of use-build-use-build embeds past water habits (and waste) into future planning even as the reality of ecological limits becomes apparent.

Demand management offers an alternative approach, which seeks simultaneously to save money, conserve water resources, and reduce environmental impacts. Demand management has gained recognition in a number of resource fields, including energy and transportation. With new technologies and institutions, we can use less water but still achieve the same benefits from household, industrial, commercial, and agricultural water use. Low-flow fixtures, drip irrigation, reuse and recycling technologies, consumer education, and pricing and regulatory initiatives can all help to "manage demand."

Developing (and maintaining) new power plants, highways, and pipelines is costly. Avoiding new infrastructure expenditures and reducing consumer demand offers real potential. Most water policy experts believe that cost-effective savings of 20 percent to 50 percent of water use are readily achievable (Brandes and Ferguson 2003; Brooks and Peters 1988; Postel 1997; Tate 1990; Vickers 2001). Cost effectiveness is often masked because users get water at less (typically much less) than its full cost. As a result, their incentive to improve water use efficiency is limited. Even "full cost accounting" ignores the non-monetary environmental effects of water supply and disposal. If these costs were taken into account, the cost effectiveness of efficiency improvements would be even greater.

A recent state-wide study in California profiles a compelling example of the potential savings possible with demand management. The Pacific Institute (Gleick et al. 2003) reports that total commercial, industrial, residential, and institutional water use could be cost-effectively cut by at least 30 percent using "off-the-shelf" technologies. This improvement can be achieved more quickly and cleanly than can any new supply project now under consideration. They emphasize that "the potential for conservation and efficiency improvement in California is so large that even when the expected growth in the state's population and economy is taken into account, no new water-supply dams or reservoirs are needed in the coming decades" (Gleick et al. 2003, 1).

There are examples of community-wide demand management in Canada, and they are slowly becoming more common. For example, Cochrane, Alberta, reduced water consumption by 15 percent by giving away toilet dams, low-flow showerheads, and faucet aerators, with the result that the city was able to defer a multimillion-dollar pipeline. Port Elgin, Ontario, avoided a $5.5 million expansion of its water treatment plant by spending $550,000 to install 2,400 residential water meters and by establishing an intensive conservation program (Boyd 2003, 51; CWWA 2005). Many smaller-scale examples are also emerging. For instance, the C.K. Choi building at the University of British Columbia harnesses multiple demand management technologies that, together, significantly reduce water use.

A whole galaxy of water efficiency and conserving measures exist. Some demand management measures, including composting toilets, drip irrigation,

BOX 14.1

**The C.K. Choi Building, University of British Columbia**

Significant water savings are realized through a series of features. Composting toilets installed in this project do not require water for flushing (and therefore eliminate sanitary waste). City water is generally only required for the low-flow lavatory faucets (spring-loaded to further reduce waste) and kitchen sinks. Irrigation of site planting material is provided solely from collected rain water (stored in an 8,000-gallon subsurface cistern) and recycled grey water from the building. Projected water usage is approximately 300 gallons per day – a reduction of over 1,500 gallons of potable water per day.

and greywater reuse and recycling are recent technologies. Others, such as cisterns and rain barrels, are much older but can have a significant impact on water use. Demand management measures are not only about technologies but also about innovative practices, such as xeriscaping[1] and "water star" product labelling. "Carrots" and "sticks," such as conditional funding, rebates, tax benefits, mandatory plumbing requirements, and pricing schemes that penalize excessive water use, are also among the measures used to manage demand. Table 14.2 provides a list of some of the water demand management measures that are commonly used in North America.[2]

## A Fresh Approach to "New" Water

Making conservation the next best source of "new" water requires regulators and water resource managers to assess the risks associated with new technologies and to build demand management infrastructure into future systems. This entails long-term planning and a comprehensive approach, all of which require specific and extensive government leadership and action.

Fundamentally, demand management necessitates an integrated approach that considers surface water and groundwater, source and wastewater, and energy. The process must involve water users in planning the appropriate mixture and timing of measures to ensure that they suit local conditions. There is no silver bullet. A diversity of measures must be used strategically and comprehensively to reap the full benefits. For example, a rebate (incentive) for replacing water-inefficient appliances should be supported with a public education program to help instil a conservation ethic in water users. Such a program should, in turn, be supported by a new water rate (pricing) structure that increases water rates according to volume used or time of use (where "smart meters" are installed).

Demand management is a major step forward, but it is only the first step. Most importantly, it does not pay close attention to the differences between efficiency and conservation. In the simplest terms, efficiency is a means and conservation an end. In most cases, efficiency will allow for some conservation, but it may also "serve as permission to consume" (Moezzi 2002). Think of the simple task of watering lawns: efficiency dictates the use of low-flow sprinklers, but with ever more lawns to water, sprinklers become just a better way to keep doing something we should no longer be doing in the first place. In contrast, conservation suggests planting greenery that does not require watering at all.

Table 14.2

**Water demand management measures**

| General categories | Specific examples | |
|---|---|---|
| Sociopolitical strategies | Information and education | |
| | Water policy | |
| | Water use permits | |
| | Landscaping ordinances | |
| | Water restrictions | |
| | Plumbing codes for new structures | |
| | Appliance standards | |
| | Regulations and by-laws: | • Turf limitation bylaws |
| | | • Once-through cooling system bans |
| Economic strategies | Rebates for more efficient technologies (e.g., toilets, showers, faucets, appliances, drip irrigation) | |
| | Tax credits for reduced use | |
| | Full-cost recovery policies | |
| | High-consumption fines and penalties | |
| | Pricing structures: | • Seasonal rates |
| | | • Increasing block rates |
| | | • Marginal cost pricing |
| | | • Daily peak-hour rates |
| | | • Integrated sewer and waste water charges |
| Technological strategies | Metering | |
| | Landscape efficiency | |
| | Soil moisture sensors | |
| | Watering timers | |
| | Micro and drip irrigation | |
| | Cisterns | |
| | Rain sensors | |
| | Efficient irrigation systems | |
| | Soaker hoses | |
| | Leak detection and repair | |
| | Water audits | |
| | Pressure reduction | |
| | System rehabilitation | |
| | Efficient technology: | • Dual flush toilets |
| | | • Low-flow faucets |
| | | • Efficient appliances (dishwashers/washing machines) |
| | Recycling and reuse: ranging from cooling and process water to greywater for toilets or irrigation, to treating and reclaiming wastewater for reuse. | |

SOURCE: Adapted from Brandes and Ferguson (2003, 40).

To translate the new script from a selection of individual demand management efforts into a complete shift in social behaviour, it is necessary to implement a set of measures. Universal metering and conservation-based pricing are important cornerstones to any demand management program. As other authors in this volume have demonstrated, water exhibits negative price elasticity, which means that, as prices go up by a certain percentage, rate of use falls – but not necessarily by the same percentage.[3] As anyone who has eaten at an "all-you-can-eat" buffet can attest, a flat rate leads to overconsumption. A comparison of Canadian cities shows reductions in water use of up to 70 percent when water is priced by volume rather than by a flat rate (Brandes and Ferguson 2003; Campbell 2004; Cantin, Shrubsole, and Alt-Ouyahia 2005). Indeed, for maximum conservation benefits, prices should *increase* with the volume taken (increasing block rate) so long as this can be done fairly, without penalizing poorer families.

Social marketing is another component of a comprehensive approach to managing demand. Conventional education programs that focus merely on disseminating information fail to address the true barriers to behavioural change. Social marketing invests time and effort at the outset in order to address these barriers from program design to implementation (McKenzie-Mohr 2004). Ontario's Region of Durham, for example, took this approach in its outdoor water efficiency program. The program started in 1997 with the region employing summer students in a community-based social marketing program to work with homeowners to reduce residential lawn watering. The result was a 32 percent reduction in peak water demand over a three-year period (Maas, 2003, 16).

Water reuse and recycling technologies also present opportunities to reduce water use and to promote water sustainability. They are particularly useful in regions with physical supply limitations, such as the Okanagan Basin in British Columbia and the dry grasslands of southern Alberta. Typical applications include using treated municipal wastewater to irrigate non-food crops, urban parkland, landscaping, and golf courses. Though water recycling is not always the cheapest option when only direct accounting costs are included, lifecycle costing may show long-term economic benefits. About 80 percent of the cost of typical sewage systems goes to collection and only 20 percent to treatment. Over the longer term, we can save money by capturing and recycling stormwater and wastewater locally. In addition, by diverting treated wastewater to irrigation, would-be pollutants become valuable fertilizers, rivers and lakes are protected from contamination, crop production increases, and reclaimed water becomes a reliable, local supply. Unfortunately, conventionally trained sanitary engineers emphasize the linear approach to managing water and sewage – use, collect, treat, and then dispose of waste. And the benefits of closing the cycle – use, collect, treat partially, and then use again – go unrealized (Postel 1997, 128, 134).

At present, water reuse and recycling in Canada are practised on a relatively small scale, and the sophistication varies regionally, depending on the availability of water supplies and regulatory flexibility (Schaefer, Exall, and Marsalek 2004). Renzetti (2005) shows that the net rate of water reuse or recirculation in Canadian manufacturing has hardly changed between 1981 and 1996. Reuse and recycling can be a powerful component of demand management, especially in areas where water demands are rapidly increasing or where there are conflicts among users. Israel, for example, treats 70 percent of its urban wastewater, which is then used for agricultural irrigation (Gleick 1998).

## Overcoming the Implementation Gap

Options such as conservation-based pricing, social marketing, and reuse-recycling act synergistically. As prices reflect a "truer" cost of the resource and citizens understand the importance of using alternatives, recycling-reuse options become more economically and socially feasible, spurring innovation and technological advance. This relationship creates a cycle of opportunity.

Despite the success stories outlined above, demand management remains on the periphery of Canada's current water management strategies. Glen Pleasance of the Canadian Water and Wastewater Association's Water Efficiency

BOX 14.2

**Vernon, British Columbia: Closing the loop**

The City of Vernon in British Columbia has embraced reuse-recycling as both a water treatment method (through its reclamation process) and as an alternative source of supply. Annually, it uses one billion gallons of treated wastewater to irrigate 2,500 acres of agricultural, recreational, and forestry land. The result of this program is that, since 1977, there have been virtually no discharges of the city's municipal effluent into Okanagan Lake, and water demands that might have been withdrawn from local water sources have been reduced.

SOURCE: Cohen, Neilson, and Welbourn (2004); Waterbucket.ca (2005).

Committee notes that "fewer than 20 percent of Canadian municipalities have established demand-side management plans" (Cantin, Shrubsole, and Alt-Ouyahia 2005, 3). A study in Ontario notes that, although individual demand management measures are being implemented, supply management still predominates. For example, only one in five of the Ontario municipalities studied had a DSM plan or strategy in place (de Loë et al. 2001, 57, 66). This lack of planning reflects the persistence of the myth of water abundance and results in the entrenchment of a host of barriers, such as outdated plumbing codes, largely free use of groundwater, and artificially low water prices[4] that block more widespread implementation of water demand management (Brandes and Ferguson 2004, 44; Nowlan 2005).

Progress to limit our supply-side reliance has been slow. A more comprehensive approach to managing demand is only possible with all stakeholders – government, business, and civil society – working to make the necessary structural changes. As a first step, we must recognize that demand management is the key to unlocking society's potential to address long-term prosperity and water sustainability. As Sandra Postel (1997, 191) notes, demand management is a "last oasis ... large enough to get us through many of the shortages on the horizon, buying time to develop a new relationship with water systems and to bring consumption and population growth down to sustainable levels."

## Expanding the Script: Water Soft Paths

Demand management can bring about significant *incremental* improvements in water management. Ultimately, however, it will not lead to a sustainable future that ensures enough water for ecosystems and for human beings. As Sachs (1998) states, "An increase in resource efficiency alone leads to nothing unless it goes hand-in-hand with an intelligent restraint on growth ... efficiency without sufficiency is counter-productive; the latter must define the boundaries of the former." Water soft path approaches offer a more holistic and comprehensive approach to water management than what we have now. They begin from the premise that sustainability is essential, and they differ fundamentally from conventional (or hard path) water planning in several ways. The most important difference between the two approaches is the way water is perceived: the soft path approach perceives water not as an end product but, rather, as a service or a means of accomplishing certain tasks, such as cooking, sanitation, or food production (Brooks 2005).

### Asking Why Instead of How

Aside from ecological requirements for water, only a few uses – most obviously drinking – require water as the end product. Within a soft path approach the objective is to satisfy demands for services rather than to supply water per se. The demand management approach focuses on "how" – how to do the same with less water; whereas the soft path focuses on "why" – why use water to do this in the first place? The question is not *how* to improve the efficiency of crop production with new irrigation systems but, rather, *why* irrigate instead of using rainwater harvesting and other techniques that multiply the effectiveness of rain-fed agriculture? Or, even more fundamentally, *why* grow crops in areas where the ecosystems are not suited to them?

Other changes follow from this shift in orientation. Most important, while demand management approaches focus on making technological and some institutional changes in order to improve water use efficiency, soft path approaches focus on broader changes that conserve natural resources and alter how water efficiency calculations are made. The demand management approach can indicate how to produce a tonne of paper with less water, for example, but it cannot suggest how the pulp and paper industry could reduce water use by decreasing production. Demand management can indicate how to produce food with less water, but it cannot suggest a reduction in meat consumption that would save significantly more water.

Clearly, the soft path involves much more than choices about efficiency: it also involves social choices and cultural values (Wolfe and Brooks 2003). Soft paths provide the macro complement to the microanalysis provided by demand management. Demand management focuses on short-term operational choices for specific projects and systems; soft paths focus on long-term planning for entire societies. Demand management can indicate what will or will not "pay off" but cannot predict water use in a future "conserving" society. This is where soft path analysis and soft path strategies are most instructive.

### From Approach to Policy

Demand management and soft paths are complementary. Together they can serve both our economy and our environment. The methodology for the soft path analysis is relatively well developed, and it can be summarized in terms of four underlying principles – treating water as a service, ensuring ecological sustainability, conserving quality as well as quantity, and "backcasting" in order to develop policy.

Unlike the standard benefit-cost (or cost-effectiveness) analyses commonly used in demand management, the soft path approach to analysis does not claim to be "objective." Soft path analyses and strategies are explicitly normative, establishing the values of sustainability and equity as social goals and promoting the changes in technology, institutional structures, and personal consumption patterns needed to achieve these goals.

The soft path is a long-term approach to water management. It takes time for change to take root; it especially takes time to rebuild physical and institutional infrastructure to permit local reuse and recycling and to promote changes in demands for water-intensive services. And such change

BOX 14.3

**Soft path principles for water management**

*Treat water as a service rather than as an end in itself*

Except for a few relatively small human uses, such as drinking and washing (and water reserved to support ecosystems), a soft path approach does not view water as a final product; instead, it views water as a means of accomplishing specific tasks, such as sanitation, attractive yards, or agricultural production. When water is viewed as a service, managers don't focus exclusively on traditional technologies and infrastructure: they also promote education and social marketing, local reuse and recycling, urban redesign for conservation, and different modes of farm management. Changing practices and behaviour will offer a wide range of ways to reduce water use while maintaining desired services.

*Ensuring ecological sustainability*

Soft paths recognize ecosystems as legitimate users of fresh water; therefore, environmental constraints are built in from the start in order to limit the amount of water withdrawn from natural sources and to establish conditions on the quality of water returned to nature. Every soft path must be tested for its effects on the environment, and any option that puts sustainability at significant risk must be rejected.

*Conserving quality as well as quantity*

High-quality water is critical to human health, but water quality requirements vary with end-use. A contaminant that is toxic for one use may be benign or even beneficial for another. We don't want animal waste in our drinking water, for example, but we eagerly seek it for gardens and farms. Yet, in most of Canada, we irrigate with drinking water. For both economic and physical reasons, it is almost as important to conserve the quality of water as it is to conserve its quantity. We only need small quantities of high-quality water (mainly for household purposes and special industrial tasks), but we need huge quantities of low-quality water (mainly for irrigation on farms and cooling at generating stations and industrial plants). Soft path policies are designed from the start to match the quality of water supplied to the quality required by the end-use. The key is to cascade water systems, ensuring that wastewater from one use becomes input for another use – from a washing machine to a garden, or from an industrial cooling system to make-up water for boilers.

*Looking ahead by working backwards*

Soft paths require a set of policy changes and program plans that will, over time, move society toward water sustainability. How that route is built comprises another unique characteristic of the soft path. Traditional planning starts from the present and projects forward toward the future. Soft path planning does just the reverse. Through a technique called "backcasting" – in contrast to "forecasting" – a sustainable and attractive future state concerning water sources and uses is defined. We then work backward from our vision of this future state to identify policies and programs that will connect us to it.

SOURCE: Brandes and Brooks (2005).

requires community action and engagement, which can only begin with a vision of a water future that is both sustainable and equitable. The environmental and social costs – in habitat loss, decline in security of supply, watershed degradation, and infrastructure expense – of continuously increasing our demand for water are too great to ignore.

## A New Theatre: Toward Ecological Governance

Water policy cannot be neatly separated from the rest of our economy. The process of moving to efficiency and to conservation-based approaches to water management must include recognizing that this will discomfort entrenched interests and will displace workers in some industries. However, the longer we remain fixated on past practices and status quo options, the more intense this upheaval will be. Regardless of long-term economic, social, and environmental benefits, sustainable water management will only emerge within the context of larger social changes and fundamental political commitments. Beyond the decisions and rules made by government, other participants – in particular, business and civil society – will be critical if we are to move ecological principles from the periphery to the core of decision making and, ultimately, make water management truly sustainable.

Confronting inertia and the many institutional barriers that block widespread water conservation is part of what can be called "ecological governance" – a form of governance in which ecological sustainability is incorporated into the very fabric of government, industry, and civil society.

Ecological governance addresses the way individual and collective decisions are made within watersheds and, more broadly, within the ecosystems that support all human life. As demand management becomes more long term, integrated, and comprehensive, people will begin to interact differently with the landscape and so create a soft path that embodies a different relationship to water. In the process, we find ourselves shifting from "managing the natural environment" in response to unsustainable human demands to "reshaping human governance" to fit within ecological limits.

The soft path approach to water management is part of a larger debate about the future of government regulation, in which people are asking how policy and practice can be more responsive to the innovative potential now demanded by citizens and regulators. In response, a new approach, often referred to as "democratic experimentalism," is emerging: "innovative regulatory strategies that seek to combine local deliberative experimentation with central coordination and rolling improvements in performance standards" (Karkkainen 2002, 998). Ecologically based democratic experimentalism requires

BOX 14.4

**Principles of ecological governance**

Ecological governance includes:

- embedding ecosystem values and considerations within the decision-making processes (including economic calculations) of government, business, and the public
- taking uncertainty seriously so that future consequences are reflected in flexible processes that extend decision making beyond the electoral cycle, calculations beyond short-term returns on investment, and personal values beyond isolated consumer preferences
- experimenting with new approaches to legal regulation, resource management regimes, and jurisdictional designs that can inculcate ecological thinking into decision making
- enhancing the ability to make explicit sociopolitical choices at the macroeconomic level in order to "shift" the demand curves rather than just move along them.

SOURCE: Brandes et al. (2005).

governance systems to begin to acknowledge their role within the natural world and to develop new techniques to strengthen human-environment relationships. However, the provision of local responses that promote innovation and that allow for experimentation under provincial and/or national coordination allows for the possibility of making comprehensive social changes.

Central to this concept is the need for regional and national coordinating bodies to share their knowledge with those facing similar problems. This information pooling increases the efficiency of public administration and heightens its accountability by involving citizens in decisions that affect them. The experimentalist approach helps institutions to adapt and to grapple with persistent problems that limit sustainability (Dorf and Sabel 1998). It entrenches adaptive management – a process of learning-by-doing – and facilitates solutions that have been adapted to meet local needs. This allows for the flexibility needed in the face of uncertain markets and a changing climate (Moench, Caparian, and Dixi 1999).

The French water parliament system is a leading example of this type of "new" governance. In France, the government has shifted its focus from centralized control to facilitating local action. It has done this by devolving authority to a system of local management authorities who operate on a watershed level. These "water parliaments" take the watershed as the starting point for sustainable water management. Local authorities develop plans tailored to local conditions, while the government, having clear directives and objectives, guides and supports the process on a national level (da Motta et al. 2004, 37). In a country as large and diverse as Canada, this type of "nested" planning and coordination is a promising option for effectively governing our water resources (Brandes et al. 2005).

Commitments to such watershed approaches are emerging in Canada. Examples range from the conservation authorities in Ontario to more recent institutional arrangements, such as the Fraser Basin Council and various provincial water strategy commitments, including a clear "watershed approach" in Alberta and Quebec. These developments suggest the potential for the ongoing transformation of Canadian water governance. However, if this type of approach is to take root, then Canada needs to make a more serious commitment to it than it has done to date. Learning from the experiences of Australia, South Africa, and Europe could go a long way toward moving the Canadian effort forward.

## Conclusion

In the short term, a serious commitment to demand management will stave off the need for most new water supplies. That window must be used to initiate a conversation concerning our attitudes toward water and to begin adjusting our physical and institutional infrastructure toward soft path options that treat water as a service rather than as a product. Ultimately, we have to "shift our paradigm" – not a trivial task – from a focus on "getting" to one of "governing." And we must do this not just for water but for all natural resources. Complementary policies for water demand that management, water soft paths, and watershed-based institutions all fit comfortably within the framework of ecological governance. Ecological governance, thus, is the catalyst for innovation that would promote a system based on efficient, equitable, and sustainable alternatives to today's linear, extractive, and far from sustainable patterns of natural resource use.

Happily, and in contrast to much of the world, we in Canada have a remarkable natural endowment as well as the economic capital and the human capabilities to make changes to ecological governance in a gradual way. What remains to be seen is whether we have the sense and the political will to do so, for the time available is limited. To return one last time to our metaphor, the new theatre of ecological governance will have no audience: we will all be on stage.

### Notes

1. "From the Greek word *xeros*, meaning dry, Xeriscaping designs draw on a wide variety of attractive indigenous and drought-tolerant plants, shrubs, and ground cover to replace the thirsty green lawns found in most suburbs. A Xeriscape yard typically requires 30-80 percent less water than a conventional one, and can reduce fertilizer and herbicide use as well. One study in Novato, California, found that Xeriscaped landscaping cuts water use by 54 percent, fertilizer use by 61 percent, and herbicide use by 22 percent" (Postel 1997, 159).
2. Good resources for a detailed discussion of the many options available for reducing water use include Vickers (2001); Gleick et al. (2003); American Water Works Association, http://www.awwa.org; Water Wiser, National Water Efficiency Clearinghouse, http://www.waterwiser.org; Canadian Water and Wastewater Association, http://www.cwwa.ca; Canadian Council of the Ministers of the Environment (CCME), http://ccme.ca.

3 The elasticity of water to price is always negative, but the extent of the reaction varies with the use. Typically, for a home, the percentage decrease of water use for drinking, cooking, and cleaning will be less than the percentage increase in price, but the percentage decrease in use for irrigating lawns, cleaning sidewalks, and washing cars will be greater than the percentage increase in price.

4 The lack of appropriate withdrawal charges is one of the primary causes of artificially low retail prices for water. These low prices create a chain of overconsumption, starting with primary users such as industry, farmers, and municipalities. Low withdrawal prices eventually result in water utilities overcapitalizing. With the increased capacity of supply water, utilities then have an incentive to maintain low prices and high levels of use to ensure that capacity is maximized (Brandes and Ferguson 2004, 44).

## References


Boyd, D.R. 2003. *Unnatural Law: Rethinking Canadian Environmental Law and Policy*. Vancouver: UBC Press.

Brandes, O.M., and D.B. Brooks. 2005. *The Soft Path for Water in a Nutshell*. Victoria, BC: POLIS Project on Ecological Governance.

Brandes, O.M., and K. Ferguson. 2003. *Flushing the Future? Examining Urban Water Use in Canada*. Victoria, BC: POLIS Project on Ecological Governance.

—. 2004. *The Future in Every Drop: The Benefits, Barriers, and Practice of Urban Water Demand Management in Canada*. Victoria, BC: POLIS Project on Ecological Governance.

Brandes, O.M., K. Ferguson, M. M'Gonigle, and C. Sandborn. 2005. *At a Watershed: Ecological Governance and Sustainable Water Management in Canada*. Victoria, BC: Project on Ecological Governance.

Brooks, D.B. 2005. Beyond Greater Efficiency: The Concept of Water Soft Paths. *Canadian Water Resources Journal* 30 (1): 83-92.

Brooks, D.B., and R. Peters. 1988. Water: The Potential for Demand Management in Canada. Ottawa: Science Council of Canada.

Campbell, I. 2004. Toward Integrated Freshwater Policies for Canada's Future. *Horizons* 6 (4): 3-7.

Cantin, B., D. Shrubsole, and M. Ait-Ouyahia. 2005. "Using Economic Instruments for Water Demand Management: Introduction." *Canadian Water Resources Journal* 30 (1): 1-10.

Cohen, S., D. Neilsen, and R. Welbourn, eds. 2004. *Expanding the Dialogue on Climate Change and Water Management in the Okanagan Basin, British Columbia*. Final Report, 1 January 2002 to 30 June 2004. Ottawa: Climate Change Action Fund.

CWWA [Canadian Water and Wastewater Association]. 2005. Water Efficiency Experiences Database. http://www.cwwa.ca/WEED/Results_e.asp.

da Motta, R.S., A. Thomas, L.S. Hazin, J.G. Feres, C. Nauges, and A.S. Hazin. 2004. *Economic Instruments for Water Management: The Cases of France, Mexico and Brazil.* Northampton, MA: Edward Elgar.

de Loë, R.C., L. Moraru, R.D. Kreutzwiser, K. Schaefer, and B. Mills. 2001. Demand-Side Management of Water in Ontario Municipalities: Status, Progress, and Opportunities. *Journal of the American Water Resources Association* 37 (1): 57-72.

Dorf, M., and C.F. Sabel. 1998. A Constitution of Democratic Experimentalism. *Columbia Law Review* 98 (2): 267-473.

Environment Canada. 2003. *A Primer on Fresh Water: Questions and Answers.* 5th ed. Ottawa: Environment Canada.

—. 2004. *Threats to Water Availability in Canada.* Burlington, ON: National Water Research Institute, 128.

Gleick, P. 1998. Water in Crisis: Paths to Sustainable Water Use. *Ecological Applications* 8 (3): 571-79.

Gleick, P., D. Haasz, C. Henges-Jeck, V. Srinivassan, G. Wolff, K. Cushing, and A. Mann. 2003. Waste Not, Want Not: The Potential for Urban Water Conservation in California. Oakland, CA: Pacific Institute for Studies in Development, Environment and Security.

Karkkainen, B. 2002. Adaptive Ecosystem Management and Regulatory Penalty Defaults: Toward a Bounded Pragmatism. *Minnesota Law Review* 87: 943-98.

Lovins, A. 1985. Saving Gigabucks with Negawatts. *Public Utilities Fortnightly,* 19-26.

Maas, T. 2003. *What The Experts Think: Understanding Urban Water Demand Management in Canada.* Victoria, BC: POLIS Project on Ecological Governance.

McKenzie-Mohr, D. 2004. Community-Based Social Marketing. http://www.cbsm.com.

Moench, M., E. Casparian, and A. Dixi. 1999. *Rethinking the Mosaic: Investigations into Local Water Management.* Boulder, CO: Nepal Water Conservation Foundation and Institute for Social and Environmental Transition (ISET).

Moezzi, M. 2002. The Predicament of Efficiency. ACEEE (American Council for an Energy Efficient Economy). http://enduse.lbl.gov/Info/ACEEE-Pred.pdf.

Nowlan, L. 2005. *Buried Treasure: Groundwater Permitting and Pricing in Canada.* Toronto: Walter and Duncan Gordon Foundation.

NRTEE [National Round Table on the Environment and the Economy]. 1996. *State of the Debate on the Environment and the Economy: Water and Wastewater Services in Canada.* Ottawa: NRTEE.

Postel, S. 1997. *Last Oasis: Facing Water Scarcity.* New York: W.W. Norton and Company.

Postel, S., and B. Richter. 2003. *Rivers for Life: Managing Water for People and Nature.* Washington, DC: Island Press.

Renzetti, S. 2005. Economic Instruments and Canadian Water Use. *Canadian Water Resources Journal* 30 (1): 21-30.

Sachs, W. 1998. Bargaining for the Rest of Nature. *Aislin Magazine* 22.

Schaefer, K., K. Exall, and J. Marsalek. 2004. Water Reuse and Recycling in Canada: A Status and Needs Assessment. *Canadian Water Resources Journal* 29 (3): 195-208.

Serageldin, I. 1995. Toward Sustainable Development of Water Resources. Washington, DC: The World Bank.

Tate, D. 1990. *Water Demand Management in Canada: A State of the Art Review.* Social Science Series No. 23. Ottawa: Inland Waters Directorate Water Planning and Management Branch, Environment Canada.

—. 1997. Economic Aspects of Sustainable Water Management in Canada. In *Practising Sustainable Water Management: Canadian and International Experiences*, ed. D. Shrubsole and B. Mitchell. Cambridge, ON: Canadian Water Resources Association.

Vickers, A. 2001. *Handbook of Water Use and Conservation*. Amherst, MA: WaterPlow Press.

Waterbucket.ca. 2005. Conservation Success Stories. BCWWA (BC Water and Waste Association), Water Sustainability Committee. http://www.stewardshipcanada.ca/communities/waterbucket/mi_successstories/mi_successStories18.asp.

Wolfe, S., and D.B. Brooks. 2003. Water Scarcity: An Alternative View and Its Implications for Policy and Capacity Building. *Natural Resources Forum* 27: 99-107.

PART 5

# Water Worldviews: Politics, Culture, and Ethics

# 15
# The Land Is Dry: Indigenous Peoples, Water, and Environmental Justice

*Ardith Walkem*

My Uncle Bill would often talk with me about the beaver and Twaal Valley. Twaal Valley is the heartland of the Cook's Ferry Nlaka'pamux people, and the valley floor is marked by the intricate circles where we would build our *s?isktkn* (winter homes). Standing in the middle of the largest field is *Nq$^{w}$iycut(t)n*, with its ancient towering lodge pole pine that marks the area where Nlaka'-pamux people would gather each fall. We would trade, laugh, hold *xitl'ix* (a form of community court hearings), settle disputes, and make plans. *Nq$^{w}$iycut(t)n* literally translates as "place to dance"; it is the place that still holds the heart of the Cook's Ferry Nlaka'pamux. It is the place that our people were forced to leave during a cold winter when smallpox had killed nineteen of every twenty people while we were living communally in our winter homes (my grandfather remembered moving from there as a small boy). It is the place that we returned to rebuild our families. It is also the place where no Nlaka'pamux people live now.

The water source for Twaal sits in a very high mountaintop, with steep fall-offs in either direction. One direction is to the south, to Twaal. The water's southern flow was directed by a series of beaver dams built in a marshy knoll near the top of the mountain. The beaver dams gathered the water and eventually sent it downward to Twaal, sustaining the Nlaka'pamux. To the other side of the mountain is a valley settled by newcomers who established homesteads there before noticing that it was a dry valley. When newcomers thought to investigate possible water sources for the valley (which was periodically wet but not year-round), they eventually came to the mountaintop and its network of beaver dams. At first, they tried destroying the dams, but the beavers would quickly rebuild them and ensure the continual supply of water for Twaal. The newcomers then went in and systematically killed the entire beaver population on the mountaintop and rerouted the water to their dry valley. The volume of water flowing down to Twaal was no longer enough to sustain the Nlaka'pamux people, and, family by family, we were forced to move away. My Uncle Bill always said that we should find a way to bring the beaver back, to bring water to the land once again.

The experience of the Nlaka'pamux people of Twaal is far from unique. Many Aboriginal peoples have similar experiences of displacement from territories, conflicts over waters, and interference with lifeways and traditions as a result of newcomer activities that divert, pollute, or degrade waters. Indigenous peoples and territories have been subject to colonization as newcomers first came to extract wealth and resources and later stayed to establish settlements and impose foreign laws, governance, and values on indigenous territories (including waters) and peoples. Colonization has disrupted indigenous peoples' ability to sustain themselves on the land and diminished the ability of our territories and waters to sustain life. Indigenous cultures are closely tied to the lands and waters, and when waters are endangered, the very identity and survival of indigenous peoples is endangered.

## Legal Recognition of Indigenous Peoples' Rights to, and in, Water

Historically, Canada has simply denied that any indigenous territorial rights (including waters) exist. Land use and development has proceeded with little regard for the indigenous peoples who draw their lives from the lands and waters. Despite many years of colonization, the destruction of our territories, and the imposition of foreign laws and values, indigenous peoples have maintained awe and reverence for the life-giving force of water and, across generations, have continued to call for the return of indigenous laws and traditions so that we can protect our peoples, waters, and territories.

Indigenous peoples have increasingly turned, with mixed results, toward Canadian courts in an effort to find a legal means (within the Canadian colonial state) to protect water. There are several possible sources within Canadian law that deal with the recognition and protection of indigenous peoples' rights to, and in, water. These include reserve water rights and Aboriginal title, Aboriginal rights, and treaty rights.

### Reserve Lands and Water Rights

As settlement proceeded across the country, Canada set aside parcels of land ("reserves") for the use and benefit of indigenous peoples and required that they move onto these lands. Some reserves include an explicit allotment of water for domestic, agricultural, or other purposes (Bartlett 1986; Union of

BC Indian Chiefs 1991; Walkem 2003). Where a reserve allotment does not explicitly include water, it is arguable that the reserve should be understood to include a sufficient supply of water to allow the people to make full and beneficial use of the land, including water for domestic and economic purposes. In the United States, there is legal recognition, through the Winters Doctrine, of the fact that, when government reserved land for indigenous peoples, water supply sufficient to enable the people to sustain themselves was also protected (*Winters* and *Walker* cases[1]; Getches 1997).

Despite the fact that reserve lands are federal creations (under s. 91(24) of the Constitution Act of 1867), reserve water allocations fall under provincial or territorial water regimes. In some cases, provinces have either refused to honour reserve water allocations and have cancelled them outright or issued licences that reduce the water available to these lands (*Westbank* case[2]; Union of BC Indian Chiefs 1991). Provincial failure to honour water allocations included in reserve creation remains a contentious issue. In some Prairie provinces, water allocations were included as part of the reserves established under treaties, and these treaty promises have not been fully honoured (Bartlett 1986; Quinn 1991). The Peigan Nation of Alberta recently settled a lawsuit against Canada and Alberta, recognizing that the reserve established for the Peigan under treaty six also included a reservation of water (Glenn 1999).

In British Columbia, Canada used indigenous peoples' reliance on the inland and ocean fisheries to justify setting aside much smaller reserves than those set aside in the rest of the country (it was argued that a smaller land base was required because indigenous peoples could sustain themselves on the fishery) (Harris 2004; Smith 1985; Union of BC Indian Chiefs 1991; Ware 1983). Canada has subsequently reduced indigenous peoples' access to the fishery, denied that reserves created on ocean waters include the tidal foreshore and seabed where shellfish and other marine life are harvested, and consistently failed to protect the indigenous fishery from over-fishing or pollution (Harris 2005). Recognition of indigenous peoples' reserve water rights, including to fishing stations and the sea bed and foreshore, remains an outstanding issue (Kempton 2005; Starr 1985).

Lack of safe drinking water is an ongoing problem in many reserve communities (and non-reserve communities, such as Métis settlements). In approximately 20 percent of indigenous communities across Canada, the water supply is contaminated and poses significant health risks for indigenous peoples (Assembly of First Nations 2001). This contamination is caused by land uses that have proceeded without consideration for indigenous peoples.

These land uses include agriculture, mining, and logging of the watersheds from which indigenous communities draw their domestic water supply. The resultant contamination is exacerbated by a lack of adequate funding for proper water treatment facilities.

## Aboriginal Title, Aboriginal Rights, and Treaty Rights

Aboriginal title, Aboriginal rights, and treaty rights are protected under s. 35 of the Constitution Act of 1982, which reads: "The existing aboriginal and treaty rights of the aboriginal peoples of Canada are hereby recognized and affirmed." Constitutional protection limits government actions that infringe Aboriginal title, Aboriginal rights, or treaty rights to, or in, water. The constitutional protection discussed here only refers to those limited areas in which Canadian governments or courts are willing to recognize indigenous peoples' territorial interests. Indigenous peoples' understanding of our territorial rights and responsibilities (including our jurisdiction to act to protect waters) is that they are inherent, far broader than Canadian law currently allows, and do not require Canadian legal or political recognition in order to exist.

### Aboriginal Title

Canada's historic denial of the existence of Aboriginal title ended when, in the *Delgamuukw* case, the Supreme Court of Canada recognized that such title continued to exist and could not be extinguished by provinces.[3] Aboriginal title recognizes the relationship between indigenous peoples and their territorial homelands (a far broader area than that allotted to reserve lands). Aboriginal title is a communal interest, flowing from indigenous peoples' historic relationship with their territories (including waters) and reflects the fact that we have land tenure and resource management systems that have been in practice since time immemorial. A right to, and in, water itself is included as part of Aboriginal title. Oceans, lakes, rivers, streams, wetlands, ice, and permafrost are all included as part of Aboriginal title territories.

Aboriginal title encompasses the right to choose to what uses lands and waters can be put, and it includes an economic component that protects both traditional economies and evolving modern uses of resources (*Delgamuukw*). Most important, Aboriginal title recognizes indigenous peoples' right to be involved in all land and water use decisions that affect their territories (*Delgamuukw*). To prove Aboriginal title, indigenous peoples must show that they used and occupied an area, according to their own laws and traditions, at the time the Crown asserted sovereignty over that area (or, in the case of

Métis Aboriginal title, when the Crown asserted effective control) (*Delgamuukw* and *Powley* cases[4]).

Some indigenous peoples have initiated court cases to establish the existence of their Aboriginal title in an effort to protect their territories and have their own laws regarding land and water management respected as well as to ensure that their right to choose the uses to which their territories can be put is legally protected. The Haida Nation, for example, is currently involved in litigation claiming Haida title to all of Haida Gwaii (the Queen Charlotte Islands), including the deep sea and ocean bed surrounding the islands. The Haida hope that recognition of Haida title will result in the fuller recognition of Haida laws to protect Haida Gwaii by ensuring that all uses of lands, waters, and resources maintain the sustainability of Haida Gwaii into the future. Similar court cases have been initiated in other parts of the country (e.g., the Anishinabe are claiming title to areas of the waterbed of the Great Lakes).

In *R. v. Van der Peet*, the Supreme Court defined Aboriginal rights as "collective rights that contribute to the cultural and physical survival of Aboriginal peoples" and that include practices, customs, or traditions that are "integral to the distinctive culture" of any indigenous people.[5] Many activities protected as Aboriginal rights (including fishing, hunting, gathering, and spiritual practices) are closely tied to waters and rely upon a continuing supply of clean water.

Indigenous peoples may be able to challenge government actions that infringe upon Aboriginal rights that entitle them to a continued supply of pure water, uninterrupted in quality and flow. For example, logging a watershed may disrupt indigenous spiritual practices that centre around isolated bathing pools, while contaminating or diverting watercourses may interfere with Aboriginal rights to the fishery.

**Treaties**

Historic treaties are nation-to-nation agreements between indigenous nations and representatives of Canada. Treaty rights reflect the mutual intentions of the parties at the time that the treaty was entered into and are defined according to the texts of the treaties themselves, including any oral promises made between the parties. A treaty may protect hunting or fishing rights, and the water necessary to sustain those rights is also protected. For example, in *Claxton v. Saanichton Marina Ltd.*,[6] the Tsawout people were able to stop a marina development that would have disturbed the eel grass necessary to sustain their crab fishery, which is protected under the Douglas Treaty. Many treaties contain the provision that indigenous peoples will be able to continue to sustain themselves on the lands reserved to them, and this provision implies the full protection of the water necessary to fulfill the terms of the treaty (Walkem 2003).

Modern land claim agreements are modern treaties that usually involve the indigenous group, federal government, and a provincial or territorial government. Many land claim agreements deal with waters (e.g., the Nisga'a Final Agreement) and usually require that the indigenous group agree to recognize the Canadian governments' jurisdiction over water in exchange for a guaranteed water allotment and some form of co-management of watercourses.

Constitutional protection of Aboriginal title, Aboriginal rights, and treaty rights provides a vehicle through which indigenous peoples can challenge government actions, or authorization of third party activities, that pollute or degrade waters. As a result of legal decisions such as *Haida* and *Taku,*[7] governments are required to consult with indigenous peoples and to accommodate Aboriginal title, Aboriginal rights, and treaty rights. Consultation and accommodation (which must occur prior to a First Nation's going to court and proving the existence of these rights) require that the Canadian government seek the input of indigenous peoples and seriously consider their rights when it comes to decision making concerning government actions that affect them. For example, this could involve ensuring that plans are changed to allow for the protection of waters within Aboriginal title territories and/or that activities are cancelled if they pollute or destroy waters that support an indigenous fishery.

Despite the promise of constitutional protection of Aboriginal title, Aboriginal rights, and treaty rights, the ability of indigenous peoples to protect waters remains constrained. They are still required to go before Canadian courts to "prove" that their rights exist (a lengthy and financially draining process). Even if indigenous peoples are able to establish the existence of their rights, Canadian courts often engage in a balancing process (i.e., balancing indigenous peoples' constitutionally protected rights against the socioeconomic interests of newcomer society) that ultimately fails to adequately or fully protect these rights and limits the ability to protect water (*Van der Peet* case; *Haida)*. Many indigenous peoples have argued that Canadian courts are colonial institutions that represent Canadian society and interests. They argue that these institutions cannot be relied upon to protect indigenous peoples or the territories necessary to sustain indigenous cultures and traditions into the future (Alfred 1999; Barsh and Henderson 1997). Canadian court decisions largely protect Canadian government jurisdiction and law-making power, and, conversely, only protect indigenous practices or activities, but not indigenous jurisdiction or law-making power. Canadian jurisprudence and society are at a crossroads. Indigenous laws are recognized as forming part of the common law associated with Aboriginal title, Aboriginal rights, and treaty rights. In court cases such as *Van der Peet* and *Delgamuukw* the Supreme Court of Canada recognized that indigenous laws (held and carried forward through oral traditions) must be meaningfully incorporated into decisions made about indigenous peoples' territories. However, in practice, this recognition remains largely theoretical (existing only on the dry pages of legal judgments) and has yet to achieve its full or even partial potential. In Canada, despite the fact that indigenous peoples' laws and water rights have been recognized on paper (s. 35, Constitution Act of 1982 and in legal judgments), the legal landscape remains dry and arid, barren of any practical space for indigenous laws. Constitutional and legal recognition is hollow if it does not provide and protect jurisdictional space for indigenous laws on lands, waters, and territories that are constitutive of Aboriginal nationhood and existence.

## Indigenous Laws Matter

The absence of indigenous laws matters: for the waters, for indigenous peoples, for all peoples, for all life. The denial of indigenous laws has allowed activities that undermine the ability of waters and lands to sustain life into the future. Current Canadian water management policy is not working. As John Borrows (1997, 420) observes:

> There is a real need to reformulate how we plan and participate in the design and governance of human settlements. Increasing alienation from our natural and social environments has nearly overwhelmed our ability to effectively function in the places we choose to live. Cities and towns are being eviscerated through pollution, poverty, congestion, crime, and a loss of control over the political means to change this situation ... There are significant problems of desertification, deforestation, and soil erosion ... [W]e are witnessing fisheries' stock collapse, the draw-down and pollution of ground water, species extinction, the depletion of

stratospheric ozone, and the increase in atmospheric carbon dioxide.

Water teaches us that everything is cyclical. Living without regard for the cyclical nature of water has resulted in newcomer society acting as though there are no repercussions to logging watersheds, releasing pollutants and other contaminants into waters, or consuming water for industrial, agricultural, or domestic purposes beyond viable levels. Instead of curbing human activities to respect the natural course and flow of water, current water and land use decisions assume that human activities can be washed away or diluted. Missing from this equation is a fundamental understanding of the nature of water, of our human status as just one form of life within a world teeming with plants, animals, and fish – all of whom draw life from our shared waters.

Canadian resource management is largely human-centred and is based on the assumption that waters, lands, and resources exist to serve human needs and aspirations. Beliefs such as this reflect the misguided notion that the natural world can be brought under control (or brought within an acceptable degree of predictability) and that any damage caused by human activities can be contained and addressed through evolving technologies. Indigenous laws and traditions, in contrast, recognize the spiritual, physical, and emotional interconnectedness between people and other living things and appreciate the inherent value of water as a force in itself (Salmon 2000; Turner 2005).

The persistence of a decontextualized management system (which holds that the impacts of development can be assessed by reference to scientific knowledge that is removed from the land) has been a systemic barrier to the recognition of indigenous laws. Misconceptions (if not outright racism) prevent indigenous peoples' environmental knowledge (which includes laws for resource conservation and use) from being incorporated into existing management regimes. Notably, we often hear the myths that indigenous knowledge and resource laws are "anecdotal, nonquantitative, nonecological, narrowly pragmatic, irrational, [or] unsubstantiated" (Sherry and Meyers 2002, 346).

Indigenous traditions reflect a land ethic, or sense of place, that situates people within their territories and infuses indigenous laws with respect for our relationships with and mutual dependence upon the other life forms that share the ecosystem. A central feature of this land ethic involves the recognition that decisions cannot be made independent of context (based on scientific or economic assessments) but, rather, must be made on the land, with an eye to assessing how all life on that land will be affected by any decisions that might be made regarding its use. And one must recognize that it is human activities that must respond to the environment, not vice versa.

The impacts of current developments on water tend to be measured in terms that minimize the importance of indigenous peoples' relationship with, and reliance upon, water. According to Enrique Salmon (2000, 1, 332), "when ecologists, land managers, environmentalists, and conservationists speak and write about endangered species and their potential loss, they rarely mention the loss of human cultures that work to enhance their homelands." Measuring the impacts of developments only on certain segments of society, without considering the loss of indigenous peoples' lifeways, is a form of environmental racism.[8] Environmental racism has led to, and continues to perpetuate, land and water use decisions that interfere with indigenous peoples' abilities to sustain their existence into the future.

The lives of indigenous peoples are intricately tied to the land and to the waters. As those who live closest to the land and rely most heavily upon it, indigenous peoples strongly feel the effects of water depletion, pollution, or other changes. Water is the lifeblood of the land and of the indigenous peoples and cultures that rely upon it and its waters:

> The traditional economy of Indigenous Peoples is closely intertwined with water, and when water is degraded, polluted or unavailable, all aspects of our physical, social, cultural and economic well being are affected ... The traditional foods which sustain Indigenous Peoples require healthy and pure water, including salmon, trout, wild rice, moose, deer, geese, berries, roots, ducks, deer, clams, whales, caribou, corn, beans, squash, ptarmigan, lobster, herring, eel, seaweed, as well as other fish, plants and animals. When water becomes polluted, the traditional food sources that our peoples have relied upon for centuries become polluted, and the survival of our cultures and societies is endangered. (Walkem and Schabus 2004)

Unfortunately, instances in which indigenous peoples' cultures and livelihoods have been affected by indiscriminate water use and development are too numerous to list. When water sources are depleted or poisoned, fish or other aquatic life contaminated, or a watershed logged, indigenous peoples lose part of their cultures and traditions, which are tied to those waters.

In the James Bay area of Quebec, a large portion of land was flooded to generate hydroelectric energy, and the repercussions for the Cree and Inuit peoples have been immense. Entire watercourses developed toxic levels of mercury due to a chemical reaction between decaying organic matter and submerged rocks, causing fish and waters to become toxic (Notzke 1994; Moses 1999). Accidental flooding during the course of operations of the hydroelectric facility caused further damage by drowning a caribou herd of ten thousand. These impacts are not borne by the general Canadian or Quebec populations – who benefit from a steady electrical supply – but by the Cree and Inuit whose lives are linked to the waters, land, fish, and caribou.

The Inuit in the Arctic increasingly live with the effects of pollution released outside of their territories and cycled to the North through the natural hydrological cycle. The bioaccumulation of toxic contaminants such as DDTs, PCBs, and POPs in the waters and traditional Inuit foods (including polar bear, seal, fish, and caribou) is causing severe health concerns (Brown 2001). An additional threat is posed by global warming, which is melting the polar ice caps and permafrost and threatening traditional Inuit lifeways.

Indigenous communities in the Great Lakes Basin have witnessed the transformation of this freshwater body from a life-giving and life-maintaining force to a source of serious ecological threat. Increased incidence of cancers, birth defects, diabetes, and multiple chemical sensitivities among indigenous peoples in the Great Lakes area has been linked to industrial and domestic contaminants in the water basin. The bioaccumulation of toxins and pollutants entering the food chain has resulted in these toxins being present in the breast milk of indigenous mothers who consume traditional foods gathered directly from the water or from animals and plants that depend on the water. In response to the existing and transgenerational health threats these activities pose, Anishinabe women have organized "water walks" along the shores of the Great Lakes both to uphold their responsibility to give voice to the water and to raise awareness. Indigenous peoples have formed the organization known as EAGLE (Effects on Aboriginals in the Great Lakes Environment) in order to collect traditional ecological knowledge and to monitor the effects of Great Lakes pollution.

Open-netcage salmon farming, on both the Pacific and Atlantic coasts, highlights the failure of Canadian water management policy to account for the long-term health of the waters, fish, and aquatic life as well as the lives of the indigenous peoples who rely upon ocean resources. In British Columbia, there is mounting evidence of the damage that fish farms cause to human health as well as to other marine life through the release of pesticides, fungicides, fecal matter, PCBs, and other contaminants that directly flow into the ocean waters – not to mention the large-scale (hundreds of thousands) escapement of farmed salmon (Bruce and Walkem 2001; Hume et al. 2004). Indigenous peoples have experienced the contamination of clam beds and other resource-harvesting areas in the vicinity of fish farms. Sea lice fostered in fish farms have transferred to migrating salmon smolts and decimated entire salmon runs in the Broughton Archipelago (Morton et al. 2004). Despite mounting evidence of the destructiveness of fish farm operations, federal and provincial governments disregard the traditional ecological knowledge of indigenous peoples as well as the advice of independent scientists who have called for the elimination of open-netcage fish farming. They continue to allow the salmon aquaculture industry to grow.

The above cases are only a few of many compelling examples of the multiple disruptions of indigenous peoples' cultures, traditions, and economies that have resulted from water misuse and pollution across Canada. As indigenous peoples have suffered the effects of colonization, of being displaced by newcomers, their waters continue to suffer a similar effect. Invasive and foreign species introduced into the oceans, lakes, rivers, and streams displace indigenous fish, plants, and other aquatic organisms. Foreign species are introduced through many different sources, including water released from the ballasts of ships, milfoil carried on the carriages of recreational boats, and the deliberate release of non-indigenous species for sport or other purposes. In Lake Okanagan, in a misguided attempt to provide feed for the Kokanee (landlocked sockeye salmon), the BC government introduced mysis shrimp. These shrimp have become a competitor for food sources, and the lake now produces thousands of tonnes of them per year, while the Kokanee population has become even more threatened.

Indigenous experiences and ecological knowledge are clear: a fundamental change is needed in the way that Canadian society makes and measures decisions that affect water. Canadian policy decisions regarding water systematically devalue indigenous peoples' right to continue to thrive upon our lands and waters. As a result, all life that depends upon water suffers. Environmental justice for the waters, for indigenous peoples, and for all life requires fundamentally rethinking the way that we, as humans, interact with the waters that give us life.

BOX 15.1

**Indigenous peoples' attitudes toward water**

*Planning for many generations into the future (Seventh Generation Principle)*

While newcomer society assesses decisions within a relatively short time frame (i.e., their impact on the present generation and, possibly, on one's grandchildren), indigenous decision making incorporates a broader span of living generations. It acknowledges the equal role and responsibilities of youth and elders and looks far into the future (hundreds of years). Decisions are measured according to the impact that they will have seven generations into the future. And we accept the fact that we have a collective responsibility to pass to those that will come after us the same abundance that was passed on to us. We also accept that we have a responsibility to guard the future ability of the lands and waters to live and to sustain life.

*Recognizing that all living things are connected (kincentric ecology)*

Indigenous peoples understand that the natural world is related to us. Oral traditions record the interrelationships between indigenous peoples and ancestral supernatural spirits/beings and the compacts entered into by indigenous peoples, plants, animals, and waters. When indigenous peoples say that the waters, lands, and resources are related to them, these are not mere words but, rather, deeply felt and acted upon beliefs. A completely different course of action is called for when waters, forests, and fish are viewed as our relatives than that which is called for when they are viewed merely as things available for our benefit. A kincentric notion of ecology "is an awareness that life in any environment is viable only when humans view the life surrounding them as kin" (Salmon 2000, 1332), and it calls upon humans to honour these relationships. Developing a notion of kincentric ecology could fundamentally alter the way that Canadian society makes decisions pertaining to governing waters. A kincentric approach to water would prohibit the narrow view of water as a limited or individual "human right"; instead, water use decisions would be grounded in the knowledge that water is sacred and connects all living things, that all beings have an equal right to the water necessary to sustain their own life.

*Valuing the wealth all around us*

A new way of assessing and valuing wealth is required – one that honours the inherent value in the waters and lands for their own sake rather than merely for what they can give to humans. Water and land use decisions are often made on a strictly economic basis and are often measured according to the profits (in terms of money/employment) that they will produce. This way of thinking is impoverished and fails to account for the abundance of the waters and lands or for the losses suffered by indigenous economies. Providing short-term jobs for some people may result in the long-term loss of the wildlife that is essential to indigenous peoples. Yet such losses are never counted as costs when the Canadian government assesses the costs/benefits of proposed developments.

*Environmental justice: Learning to say "no"*

Indigenous peoples recognize that they have a responsibility to consider how their actions will affect other life forms. We must not simply make decisions on the basis of our own immediate self-interest; rather, we have an obligation to preserve the integrity and intrinsic value of the waters that gave us and our ancestors life. We cannot cast them aside for our immediate benefit. It is fairly uncommon (at least from an indigenous point of view) for newcomer societies to say "no" to developments, even when their environmental impacts will be great. If we are to protect water into the future – for all life – then we need to recognize that we have a responsibility to say "no," that we need to speak for the waters, the land, and for all those life forms with whom we share ecosystems. Environmental justice requires a commitment to honouring all of our living relations and measuring the impacts of our decisions not only on humans or immediate generations but also on the present and future generations of all life forms.

## Returning Water to the Land

The example of the Siska watershed highlights important differences between the philosophies and worldviews that guide decision making in indigenous and Canadian societies. When discussing the BC government's decision to allow the logging of the Siska watershed (then the last untouched watershed in the Nlaka'pamux territory on the eastern side of the Fraser River and a place of tremendous spiritual and cultural importance), a Nlaka'pamux elder once told me that the problem with the newcomers was that they were famished. She explained that newcomers never stop eating away at the waters, at the land, at the trees, at the fish. Newcomers would log, mine, build subdivisions and highways, and fish to the point of extinction and still never feel full or satisfied. This observation is key to understanding both the problem and a way to chart a new course in our relationship with water – a way to end the famine, to end the incessant hunger for land and water, and to find a way to be full and to live on the land in peace. Instead of newcomer notions of predictability and control, of wealth based on accumulation and exclusion, what is needed is a greater appreciation of indigenous peoples' ecological knowledge and laws. Currently, decision making regarding water and land is impoverished and constrained, taking into account only a limited number of people and measuring impacts over an infinitesimal period of time. We need to shift the way we view water and our relationship with the world. It is not better and improved water management that is needed but, rather, a real and lasting commitment to managing our own actions in a way that respects all life forms that share the water and depend upon it for survival. Preserving water into the future will require a fundamental shift in how Canadians live upon the land and assess the impacts of their actions. The development of an ethic of hope and responsibility – an ethic that recognizes that change – is both possible and necessary (Box 15.1). Recognition (meaningful and not merely superficial) of indigenous laws provides hope for the future survival not only of indigenous peoples and cultures but also of the water and everything that depends upon it. Indigenous peoples' laws have taught us how to live in a way that sustains the water for future generations of all life forms.

## Awe and Reverence

Indigenous cultures recognize the spiritual nature of water and have developed complex traditions to remind people to honour it:

> There are stories that tell of Supernatural beings that live beneath the oceans, and beneath the seemingly calm surfaces of mountain lakes.
>
> There are dances that celebrate the coming of Water to the dry and parched desert lands, bringing sustenance yet again for the people.
>
> There are prayers that recognize Water as the first living thing on this earth, calling forth all other life.
>
> There are songs that celebrate the sharing of the wealth of the Water to bring life for the people.
>
> There are stories that remind us that our ancestors live in, and through, Water and that Water connects us with our past and our futures, flowing through time, sustaining us today as it sustained our great-great grandmothers. (Walkem 2003)

Honouring water as a separate being means honouring the fact that water has a life of its own, that it flows at particular places and for particular purposes. Decisions relating to water must treat it with awe and reverence rather than as merely one more resource to be managed, controlled, exploited, and used. Recovering respect for the sacred nature of water could fundamentally transform the ways in which Canadian society makes decisions regarding it. Similarly, a commitment to reversing the deliberate suppression of indigenous laws and territorial rights offers hope of restoring environmental justice and of returning water to the land for all peoples.

## NOTES

1 *Winters v. United States*, 207 U.S. 564 (1908); *United States v. Walker River Irrigation District*, 104 F. 2d 334 (9th cir. 1939).
2 *Canada (Department of Indian and Northern Affairs) and Westbank First Nation v. British Columbia (Ministry of Environment, Lands and Parks)*, [1997] B.C.E.A. No. 40. (B.C.E.A.B.).
3 *Delgamuukw v. British Columbia*, [1997] 3 S.C.R. 1010.
4 *R. v. Powley*, [2003] 2 S.C.R. 207.
5 *R. v. Van der Peet*, [1996] 2 S.C.R. 507.
6 *Claxton v. Saanichton Marina Ltd.* (1989) 57 D.L.R. (4th) 161 (BCCA).
7 *Haida Nation v. B.C. (Ministry of Forests)*, [2004] 3 S.C.R. 511. *Taku River Tlingit First Nation v. B.C. (Project Assessment Director)*, [2004] 3 S.C.R. 550.
8 The Indigenous Peoples' Declaration against Racism (Third World Conference against Racism, South Africa, 2001) defined environmental racism as follows: "Environmental racism – an historical form of racial discrimination – has led to and continues to lead to the ruination of indigenous lands, waters and environments by the implementation of unsustainable schemes, such as mining, biopiracy, deforestation, the dumping of contaminated waste, oil and gas drilling and other land use practices that do not respect indigenous ceremonies, spiritual beliefs, traditional medicines and lifeways, the biodiversity of indigenous lands, indigenous economies and means of subsistence, and the right to health."

## REFERENCES

Alfred, Taiaiake. 1999. *Peace, Power and Righteousness: An Indigenous Manifesto*. Toronto: Oxford University Press.

Assembly of First Nations. 2001. *Factsheet: Safety of First Nations Drinking Water* (August 2001).

Barsh, Russel Lawrence, and James Youngblood Henderson. 1997. The Supreme Court's Van der Peet Trivology: Naïve Imperialism and Ropes of Sand. *McGill Law Journal* 43: 993.

Bartlett, Richard. 1986. *Aboriginal Water Rights in Canada: A Study of Aboriginal Title to Water and Indian Water Rights.* Calgary: Canadian Institute of Resources Law, University of Calgary.

Borrows, John. 1997. Living between Water and Rocks: First Nations, Environmental Planning and Democracy. *University of Toronto Law Journal* 47: 417.

Brown, L. 2001. Toxic Tainted Arctic Passing Poisons on to Inuit. *Washington Post*, 24 March.

Bruce, Halie, and Ardith Walkem. 2001. Fish Farms: An Indigenous Perspective. Native Brotherhood, *Native Voice* (September).

Getches, David. 1997. *Waterlaw in a Nutshell.* 3rd ed. St. Paul, MN: West Publishing Company.

Glenn, Jack. 1999. *Once upon an Oldman: Special Interest Politics and the Oldman River Dam*. Vancouver: UBC Press.

Harris, Douglas. 2004. *Land, Fish, and Law: The Legal Geography of Indian Reserves and Native Fisheries in British Columbia, 1850-1927*. PhD diss., Osgoode Hall Law School, York University.

—. 2005. Indian Reserves, Aboriginal Fisheries and the Public Right to Fish in British Columbia, 1876-1882. In *Despotic Dominion: Property Rights in British Settler Societies,* ed. J. McLaren, 266-93. Vancouver: UBC Press.

Hume, Stephen, Alexandra Morton, Betty Keller, Rosella M. Leslie, Otto Langer, and Don Staniford. 2004. *A Stain upon the Sea: Westcoast Salmon Farming*. Vancouver: Harbour Publishing.

Kempton, Kate. 2005. *Bridge over Troubled Waters: Canadian Law on Aboriginal and Treaty "Water" Rights.* Toronto: Olthius Kleer Townsend.

Morton, A., R. Routledge, C. Peet, and A. Ladwig. 2004. Sea Lice (*Lepeophtheirus salmonis*) Infection Rates on Juvenile Pink (*Oncorhynchus gorbuscha*) and Chum (*Oncorhynchus keta*) Salmon in the Nearshore Marine Environment of British Columbia, Canada. *Canadian Journal of Fisheries and Aquatic Sciences* 61 (2): 147-57.

Moses, Ted. 1999. Water and First Nations. Conference materials for the National Symposium on Water Law, Environmental Law CLE Programme.

Notzke, Claudia. 1994. *Aboriginal Peoples and Natural Resources in Canada.* North York, ON: Captus University Publications.

Quinn, Frank. 1991. As Long as the Rivers Run: The Impacts of Corporate Water Development on Native Communities in Canada. *Canadian Journal of Native Studies* 11 (1): 137-54.

Salmon, Enrique. 2000. Kincentric Ecology: Indigenous Perceptions of the Human-Nature Relationship. *Ecological Applications* 10 (5): 1327-32.

Sherry, Erin, and Heather Myers. 2002. Traditional Environmental Knowledge in Practice. *Society and Natural Resources* 15: 345-58.

Smith, Donald. 1985. *Water Rights on Indian Reserves in British Columbia.* Report written for the Department of Indian Affairs and Northern Development (September).

Starr, Vina. 1985. *Indian Title to Foreshore on Coastal Reserves in British Columbia.* Prepared for the Department of Indian and Northern Affairs, Vancouver, BC.

Turner, Nancy. 2005. *The Earth's Blanket: Traditional Teachings for Sustainable Living.* Vancouver: Douglas and McIntyre.

Union of BC Indian Chiefs. 1991. *Indian Water Rights in British Columbia*. Vancouver: UBCIC.

Walkem, Ardith. 2003. *Lifeblood of the Land: Aboriginal People's Water Rights in British Columbia.* Vancouver: Environmental-Aboriginal Guardianship through Law and Education (EAGLE).

Walkem, Ardith, and Nicole Schabus. 2004. Our Waters, Our Responsibility: Indigenous Water Rights. Briefing paper prepared for Our Waters, Our Responsibility conference on water rights, Pinawa, MA.

Ware, Rueben. 1983. *Five Issues, Five Battlegrounds: An Introduction to the History of Indian Fishing in British Columbia, 1850-1930*. Chilliwack, BC: Coqualeetza Education Training Centre.

# 16
# Half-Empty or Half-Full? Water Politics and the Canadian National Imaginary

*Andrew Biro*

O Canada ... you are the empty chair at the end of an empty dock.

– Billy Collins, "Canada"

## Nature in Canadian Culture

All human beings, indeed virtually all forms of life on earth, are dependent on water. Water is a biophysical necessity, but, of course, it is also much more than this. Its uses and values go well beyond mere health and economics, and extend deeply into politics and culture. What cultural meanings are specific to water in Canada – a country that is often seen as "empty" and as containing an abundance of water? What political possibilities arise from (or are foreclosed by) Canada's hydrological culture? We can begin to answer these questions by looking at water, and nature, in Canadian culture.

A recent CBC radio program canvassed listeners, seeking to discover fifty "essential Canadian popular songs." When the voting ended, at the top of the list was Ian and Sylvia Tyson's folk classic, "Four Strong Winds." In the song's lyrics, the constancy of natural processes provide a contrast with the dynamic of human migration in pursuit of shifting employment opportunities. This is perhaps the core of the song's "essentially Canadian" character: rather than bending nature to suit human purposes, human conventions, such as fixed settlement or geographically stable employment, are bent to accord with the dictates of nature. (Of course, we might add that the song is not entirely upbeat about this, so it may also be heard as a lament of the failure to provide such social stability.) And if the song is indeed a distinctively *Canadian* one, then these dynamic movements, and the bending of social life to irresistible natural forces, can also be understood as taking place within the political space of the nation-state.

Indeed, defining the boundaries of "Canada" has always been a project that worked both with and against natural processes and contours. The broad expanses of salt water in the Atlantic, Pacific, and Arctic oceans isolated Canada from the colonial powers in Europe on the one hand, and Asia on the other (or "Old World man" and "Third World man," in the words of the iconic Canadian prog-rock band Rush). At the same time, the exercise of Canadian national sovereignty against American "manifest destiny" has necessitated constantly working *against* the grain of continentally unifying natural features. As generations of Canadian schoolchildren have learned, technologies like the railroad were seen as being crucial to the project of Canadian nation building precisely because they created an east-west axis that could provide an alternative to the more "natural" north-south continental flows. From the Rocky Mountains and the temperate rainforests of the west coast, to the Gulf of Maine Basin in the east, natural features tend to straddle rather than mark the political boundary between Canada and the United States. Even the Great Lakes (North America's "inland seas"), which form a long stretch of that border, are as much a point of intense connection as they are a divisive boundary.

And yet, in spite of a more or less seamless flow of nature between Canada and the United States, definitions of the Canadian nation (notoriously difficult as these may be to capture) frequently refer to the country's geography and its natural context: northern, vast and open, rugged, wild, and, of course, cold and wet. As Prime Minister William Lyon Mackenzie King famously observed (in 1936): "If some countries have too much history, we have too much geography." King's quip suggests a sort of geographic determinism that appears inherent to the question of identity in Canada (a point to which we shall return): as Canadians, *who* we are is profoundly shaped by the *where* in which we find ourselves.

To see evidence of this view, we could go back to some of the earliest English Canadian literature – Susanna Moodie's *Roughing It in the Bush* (1997), originally published in 1852, for example – or even to the material culture of the French voyageurs. But it persists even today, for example on Environment Canada's website on "Water and Culture."[1] Here, the art of the Group of Seven (a group of early twentieth-century Canadian landscape painters) is foregrounded. The group explicitly claimed that their painting was a form of politics: artistic representations of Canadian space helped to constitute Canadian identity. According to the group: "The great purpose of landscape art is to make us at home in our own country." Moreover, it was Canada's northern wilderness that was seen as "a symbol of the Canadian identity, representative of the rugged determination that had made the country great."

The website goes on to note that, by the mid-twentieth century, there was a marked decline in interest in landscape art (the Group of Seven disbanded in 1931). This was also a time of intense economic and technological modernization in Canada, and it saw the development of numerous significant hydrological engineering projects. "Canadians in general, were looking elsewhere for meaning in their country. Unlimited growth and spiraling wealth became paramount. Manipulation of rivers meant power and profit." More recently, however, "Landscape art has undergone a surge of renewed popularity in Canada ... This reflects the continuing trend toward environmental awareness. Canadians are becoming increasingly concerned about the predicament of natural systems around the world."

What is most significant about this historical trajectory from the early twentieth century to the present is the connection drawn between the popularity of landscape art and environmental awareness. More generally, the devotion of a significant portion of Environment Canada's water website to issues of *culture* reflects the view that sustainable living requires more than just the knowledge that can be captured by scientific and technical measures. If an appreciation of the cultural dimension is missing, significant ecological and social problems ensue. During that mid-century period of declining interest in landscape art, the Environment Canada website tells us, "enormous reservoirs drowned thousands of square kilometers of some of Canada's most scenic and productive regions. The people who lived at the rich boundary of water and land suffered from this development." But this period in Canadian history – where Canadian society tries to manipulate or dominate nature rather than accommodate itself to it – is presented as exceptional, squeezed between the time of the Group of Seven and the current resurgence of landscape art. Indeed, the website's discussion of Canadian culture begins with the observation that "Canada, with an abundance of water, spawns a culture particularly rich in water imagery." With all that water – and natural beauty more generally – "increasing concern" about ecological issues is presented as coming to Canadians *naturally*. And this view fits conveniently with a Canadian nationalism defined against the modern, hydrologically engineered (nature-dominating, imperialist) society to the south.

## "Hydrological Nationalism": Canada in a Comparative Context

Canadians often rely on a distinction between national cultures that are shaped by their environment, such as Canada, and national cultures that actively shape their environment. However, this distinction, although widely held, does not hold up under scrutiny. Certainly, some countries have wrought enormous changes in their hydrological landscapes – for explicitly political purposes. In China, for example, the number of large dams increased from twenty-two in 1949 to 22,000 in 2000 (the comparable world figures are 5,000 to 45,000) (WCD 2000, 8-9). China's South-North Water Transfer Scheme – a project designed to move water from the Yangtze to the Yellow River Basin – which broke ground in late 2002, is projected to cost US$58 billion over the construction project's fifty-year span. But, as a recent analysis suggests, the high costs of moving the water far outweigh any economic gains that would be made by sustaining agriculture in the North China Plain. Thus, rather than an infrastructure project that is designed to produce direct economic returns, "[the] state's endorsement of this project could imply a commitment to keep China the way it is: China can feed itself; most peasants remain on the land; urbanization and industrialization can proceed at the current pace" (Webber et al. 2005, 13). Similarly, modernization in Spain since the late nineteenth century involved the construction of many large hydraulic projects. These served not only to deal with Spain's geographical shortcomings (poor agricultural production owing to regional drought and poor soil quality) but also as a solution to the political problems posed by the loss of the last vestiges of the Spanish Empire and domestic disintegration (Swyngedouw 2003, 98-100). The reshaping of Spain's waterscape, thus, served not only an economic end but also as an important symbolic resource in the political struggle to reconstitute the modern Spanish nation.

For Spain and China, the massive transformation of waterscapes – or the domination of nature – thus seems primarily to serve a political rather than an economic purpose: shoring up people's belief in the power of a particular modern nation. The natural landscape is literally restructured to conform to the logic imposed by political boundaries. In both cases, the power of the nation is projected onto these transformations of the natural environment: tamed nature (rivers bent to human needs through the construction of dams, canals, reservoirs, and so on) functions as a symbol of the strength of the national community that does this taming.

In the Canadian case, however, such a dramatic reworking of hydrological nature would likely signify something quite different. This different symbolism is attributable to the fact that Canada shares a long border with the world's sole superpower – a superpower, moreover, whose own internal dynamics are shifting people and economic activity south and west, to its driest regions. Donald Worster describes the American west (and the Great Valley of California in particular) as a region that has been monumentally transformed by hydrological engineering: already by the early twentieth century, in the Great Valley, "the flora and fauna went through an upheaval comparable only to the cataclysmic postglacial extinctions" (Worster 1992, 10). And the rest of the twentieth century only intensified this reshaping of the landscape, taming hydrological nature to make an extremely dry region suitable for extensive and intensive human habitation. The attitude is perhaps best summed up by the succinct statement of William Mulholland, chief engineer of the Los Angeles Bureau of Water Works and Supply, at the opening of the Owens River aqueduct, which brought water to Los Angeles: "There it is. Take it." The American west thus represents what Worster calls the "modern hydraulic society" par excellence. The ecological consequences of the development of a modern hydraulic society include an enormous loss of biodiversity as a result of a massive re-engineering of the natural environment. The social consequences, meanwhile, include the formation of an "increasingly coercive, monolithic, and hierarchical system" (Worster 1992, 7), as the American west has become – thanks in part to the technologies of water management – "a principal seat of the world-circling American Empire" (15).

In the case of Spain, as Swyngedouw (2003, 98) notes, "not a single river basin has not been altered, managed, engineered, and transformed." This has been in accordance with a national ideology that stated that "not a single drop of water should reach the Ocean without paying its obligatory tribute to the earth" (cited in Swyngedouw 2003, 101). But, in Canada, such a massive system of hydraulic transfers can be (and indeed has been) imagined only for the purpose of diverting Canada's water resources south to the United States. Here, a transformation of the national waterscape would serve to indicate not Canadian national strength but, rather, its subservience to American interests, or the relative *weakness* of the Canadian nation state.

Such a focus on the Canadian nation's particular position within a continental or global order may help us to see better why symbols, discourses, or projects that provide suitable material for nationalist ideology in other places don't resonate so well here. If the domination of nature in Spain or China can serve as an index of domestic state power in those countries, in Canada, it can be read as a marker of foreign (American) state power and, hence, Canadian state weakness. For this reason, Canadian nationalism may, in some sense, be hinged not only with a reshaping or domination of nature but also with the preservation of wildness, or untamed nature.[2]

At the same time, bringing the global order – or, more specifically, "American Empire" – into view also shows the problem with saying that the transformation of the natural environment serves a political *rather than* an economic purpose. The creation or maintenance of a national ideology (perhaps most visibly in the United States but equally in Spain, China, or, indeed, Canada) is itself an economic as well as a political project. As we shall see, the idea of Canada as a place of untamed nature is also an ideological construct. It is a "myth" – not a false representation of reality but, rather, a strategic misrepresentation that serves to politically unify what might otherwise be antagonistic groups within society. In other words, it doesn't just mask the underlying reality that nature (and water) in Canada is, in fact, remarkably transformed by human action (Chapter 7, this volume), it also mystifies the political-economic character of those transformations. This mystification, in turn, suggests that the benefits and costs of these transformations might be unequally distributed across the Canadian social landscape.

In reality, Canadian culture seems to hover between seeing the Canadian nation as a product of the natural environment and seeing the Canadian nation as an agent that actively reshapes the natural environment. The tension can be seen, for example, in Environment Canada's narrative of the bases of Canadian-ness, which has two centres – water itself on the one hand, and a culture "rich in water *imagery*" on the other. At the level of the words themselves, there is perhaps a contradiction between a site devoted to *water* issues and its cultural emphasis on *land*scape painting (a contradiction to some extent resolved by reference to human settlement "where land and water meet"). But, even beneath this surface level, there is the contradiction between the emphasis on landscape painting (a static image) and what it here tries to represent: water (a fluid and dynamic substance).

As Erin Manning emphasizes, in spite of the fact that the Group of Seven represented a break from European conceptions of the proper *content* of landscape art (their depiction of distinctively Canadian "rugged" natural settings), they, nevertheless, reproduced certain ideas about the *form* (and *purpose*) of landscape art that were distinctive products of European modernity. "Landscape," newly conceived in seventeenth-century Europe as a "unified whole" rather than as "an assemblage of isolated subjects," represents a literal objectification of nature. At this point, nature became something represented in order to be "consumed, identified, and ruled" (Manning 2000, 22-24). Rather than simply drawing attention to the awesome or sublime powers of nature, the nationalist landscapes of the Group of Seven – which, as noted earlier, are "representative of the rugged determination that had made the country great" – are intended to facilitate a human imitation of the powers of nature. Canadians should *embody* the rugged determination of nature rather than, say, construct a society that shelters people from it.

It is this attempt to reproduce nature's ruggedness within Canadian society that leads to both economic and technological development *and* to drowned regions and suffering people. Moreover, Manning notes a connection between landscape painting as a specifically modern form of art and nationalism as a modern ideology. In seeking to "make us at home in our own country" the Group of Seven's Canadian nationalism fits firmly (if somewhat belatedly) into the trajectory of European imperialism. In this sense, "The Great White North" can be read as a racial, as well as a climatological, description of the country (Manning 2000, 31).

With this in mind, we can return to Mackenzie King's claim that Canada has too much geography and not enough history. The lack of Canadian "history" can be understood to refer to a lack of long-standing European settlement, or "civilized" society, while the abundant "geography" is not just the sheer physical size of the Canadian space but also the obstacles it presents to human settlement, transportation, and communication. "Too much geography, not enough history" (or a fixation on the question of "where is here?" to cite literary critic Northrop Frye's [1995, 222] definition of Canadian literature), then, suggests that a Canadian nature that remains "untamed" or apparently untransformed is a problem – perhaps *the* problem – for Canadian society.

Canadian nationalism is, thus, caught in a double bind: as with all forms of nationalism, the strength of the nation is to be measured by the extent to which the national territory is effectively brought under the control of the people and/or the state. In other words, a nation is strong to the extent that the nature within its boundaries is socialized or tamed. At the same time, because of Canada's particularly close relationship with – and its role as raw material exporter to – the world's sole superpower, the taming of Canadian nature is often understood to be an American project. So the index of Canadian national strength in this context becomes the extent to which American power is resisted or, in other words, the extent to which nature within Canada's boundaries remains untamed.

In 1898, Prime Minister Wilfrid Laurier proclaimed that "the twentieth century will belong to Canada." Not long after, the Group of Seven began to lay out a programmatic vision in which the ruggedness of Canada's natural environment could serve as the inspiration for the nation to seize its destiny. Their aesthetic representation of nature brilliantly sutured the contradiction between the need to valorize untamed nature and (through the imitation of nature's ruggedness) a desire to tame nature for society's ends. However, this ideological resolution of the contradiction soon became unravelled. A century after Laurier, Canadian nationalists on the political left are more conscious of the victims of "rugged determination" (women, First Nations, the natural environment itself). Canadian nationalists on the right, meanwhile, have all but disappeared as continental integration or (less frequently) globalization more generally is seen as the only viable way forward for Canada (Laxer 2000). To the extent that it can take an ideological form, couching a political project in apparently *apolitical* terms, Canadian nationalism takes a postmodern – ironic, self-deprecating – turn. Rather than being expressed through the forms of modern high culture (such as landscape painting), Canadian nationalism now finds expression in beer commercials (such as Molson's culturally iconic "Joe Canadian" ad, which aired in 2000-1) and folk musical comedy, such as the Arrogant Worms' (1999) "Rocks and Trees":

> My country's bigger than most,
> And if asked I boast ...
>
> Although we don't have history ...
> Still what we've got's glorious.
>
> We've got rocks and trees, and trees and rocks,
> and rocks and trees, and trees and rocks,
> and rocks and trees, and trees and rocks,
> and rocks and trees, and trees and rocks,
> and water.

## Water: Strategic Commodity or Just Another Staple?

Of course, abundant rocks and trees (among other extractive natural resources) have formed the basis of the Canadian economy for about as long as it has existed as a national economy. In his pioneering work in Canadian political economy, Harold Innis (1956) developed the notion of Canada as a "staples" economy. Against notions of American "manifest destiny," or the idea that Canada's political boundaries were at odds with the continent's natural boundaries (as was argued above), Innis claimed that it was the shape of the river systems that enabled the expansion of colonialism and the fur trade and that provided a geographic logic for the contours of the Canadian nation.

If we accept Innis' claim that Canadian nationality is rooted in, rather than cut against, the continent's physical geography, we might then see natural resource abundance ("rocks and trees, and trees and rocks") as both enabling and constraining national economic development. On the one hand, natural resource abundance drives settlement and creates economic and political linkages and institutions, creating and strengthening national scale activity. On the other hand, though, that same abundance may trap the national economy in a particular (staples-exporting) niche, impeding the development of a more diversified economy.

In this light, national economic development becomes a task of managing a persistent contradiction between the stability of the nation and the change inherent in economic development. In the Canadian case, though, there is also a more specific contradiction. On the one hand, there is the idea of Canada as a fully developed modern nation. On the other hand, there is the nature of the economic processes that underlie Canadian economic development, which suggest that Canada might be better categorized as (in Margaret Atwood's words) a "rich colony" (cited in Laxer 2004, xii) or "the world's richest underdeveloped country" (Worsley 1984, 22). The very idea of "Canadian modernity," Imre Szeman (2001, 39) notes, "is constituted along a rift between enlightenment progress, in the form of emancipation and the extension of civil liberties on one hand, and a mode of production that in a country of 'hewers of wood and drawers of water' has relied on a domestic and immigrant labour force that has rarely (and certainly not in the present era of globalization) been able to access the freedoms formally guaranteed [to] them." Abundant "rocks and trees, and trees and rocks" may provide a geographical "fix" for a sense of national community, but this is not to say that the wealth generated by the export of these abundant resources will provide the full fruits of modernity for all the residents of the nation.

Are water exports then likely to be any different? For at least a third of a century, some have suggested that Canadians should exploit the opportunity provided by a global water crisis, taking advantage of national abundance in the face of increasing global demand and shortages elsewhere (see, for example, Waterfield 1970). From this perspective, Canada, along with perhaps a few other water-rich countries, is seen as being uniquely poised to control the emerging world water market and become "the OPEC of water" (or "$H_2$0PEC"). There are serious questions to be raised both about whether or in what sense there is a global-scale water crisis (Linton 2004) and also whether massive Canadian water exports would be feasible either economically (De Villiers 2000, 370) or ecologically (Chapters 7 and 8, this volume). But even putting these sorts of questions aside, and assuming it was feasible, there remains the question of whether such a strategy would be desirable. Canada's intensified involvement in global water trade would seem to be predicated on the acceptance of economic globalization: relatively frictionless global trade, with the global market as the ideal mechanism for distributing goods. But accepting the logic of the global market effectively means an end to the sort of national economic development strategy that water exports are supposed to represent. If global free trade is going to produce an "OPEC of water," then the beneficiaries are likely to be major shareholders and executives of the world's handful of multinational water corporations, few of whom are Canadian.

Canadian water abundance, thus, may be a "myth," which, again, is not so much a false statement as a strategic misrepresentation: even if Canada's water glass is relatively full, there appears to be little that can now be done to capitalize on this situation for the benefit of Canadians. On the other hand, it is perhaps not difficult to see how a myth of water scarcity – seeing Canada's water glass as relatively empty – produces some of the same problems as does a myth of abundance. "Scarcity" lays the discursive or cultural foundations for increasing water marketization, including privatization and commodification. In this sense, the further entrenching of discourses of scarcity may have the effect of making neoliberalism – or the logic of the global market – seem natural and, thus, making alternatives to US-led economic globalization more difficult to construct or even imagine.

In the early twentieth century, the landscape art of the Group of Seven responded to the double bind of Canadian nationalism with a solution that was both true and false. The injection of uniquely Canadian content (rugged wilderness) into a European cultural form (landscape painting) provided the basis for a renewed sense of Canadian nationalism. But it also created a Canadian nation that was ultimately susceptible to the same problems as was any other modern nation, from imperialist ideology to a fetishization of national economic competitiveness.

The images prevalent in popular culture today, meanwhile, show Canada as a land of abundant fresh water (lakeside cottages, for example, figure in Canadian ads selling everything from beer to financial services). The myth of Canadian water abundance remains "firmly entrenched," according to a recent report on sustainable water management in Canada (Brandes et al. 2005, ii), in spite of the material reality that "over one-quarter of Canadian municipalities already face water shortages for one reason or another" (Brooks 2005, 2). A myth of abundance strategically misrepresents the state of Canada's waters, with potentially disastrous future consequences. At the same time, an increasingly globally accepted myth of scarcity reinforces a trend toward the privatization of water resources, which may have the consequences of exacerbating inequality and reducing Canadians' control over waters within our borders.

Seeing a way out of this contemporary double bind is not easy, particularly if we aim to move beyond the realm of myth or ideology to practices that truly are socially and ecologically sustainable. It requires us to think of water not as "ours" in strictly national terms but, rather, as a fluid resource that traverses national boundaries. At the same time, it requires us to think of "globalization" in ways that generate increased solidarity rather than increased competitiveness and inequality. But how are we to resist American imperialism, yet, at the same time, think of borders in more porous terms? A strengthening of Canadian nationalism seems a crucial resource for the former, even while it may undermine the latter. This is not a uniquely Canadian problem but, rather, the description of a global conundrum: how do we respond to global scale crises of ecological sustainability when "globalization" tends to be thought of largely in terms of growth-driven economic integration? The way forward, then, must be not simply a championing of nationalism at the expense of globalization, but also a reimagining of both global and national-scale social life in terms that are not dominated by the logic of the market. This reimagining, furthermore, must understand changes in ideas or culture as intimately connected with the economic and political processes that govern socio-ecological interactions. The most pressing question is, thus, not one of scarcity or abundance, of fullness or emptiness. Nor is it one simply of perspective, seeing the Canadian water glass as half-full or half-empty. Rather, it is how that water, and the glass itself – the cultural lens, "rich in water imagery" – can be mobilized in the construction of a more ecologically and socially sustainable world.

## ACKNOWLEDGMENT

Thanks to Karen Bakker, Gordon Laxer, and Steven Hayward for comments on an earlier draft of this chapter. This research was undertaken, in part, thanks to funding from the Canada Research Chairs program.

## NOTES

1 Unless otherwise noted, all quotations in this and the following two paragraphs are taken from pages contained in Environment Canada's "Freshwater Website: Water and Culture" at http://www.ec.gc.ca/water/en/culture/e_cultur.htm.

2 Two caveats are in order here: first the "Canadian nationalism" under discussion is largely the nationalism dominant in English Canada. Whether or to what extent the argument applies to French Canadian nationalism, or to the numerous Canadian First Nations, is an issue largely beyond the scope of this chapter. Second, it should be noted that wilderness preservation is an important current in American as well as Canadian political discourse and that the wilderness preservation movement is in many ways stronger in the United States than it is in Canada (see Worster 2002). The particular ways in which wild nature is woven into nationalist ideology, however, differ in the two countries.

## REFERENCES

Arrogant Worms (Chris Patterson, Mike McCormick, and Trevor Strong). 1999. Rocks and Trees, from the cd *Dirt!*

Brandes, O.M., K. Ferguson, M. M'Gonigle, and C. Sandborn. 2005. *At a Watershed: Ecological Governance and Sustainable Water Management in Canada*. Victoria, BC: POLIS Project on Ecological Governance.

Brooks, D.B. 2005. Beyond Greater Efficiency: The Concept of Water Soft Paths. *Canadian Water Resources Journal* 30 (1): 1-10.

Collins, B. 2001. "Canada." *Sailing Alone around the Room: New and Selected Poems*, 61-62. New York: Random House.

De Villiers, M. 2000. *Water*. Toronto: Stoddart.

Frye, N. 1995. *The Bush Garden: Essays on the Canadian Imagination*. Toronto: Anansi.

Innis, H.A. 1956. *The Fur Trade in Canada: An Introduction to Canadian Economic History*. Toronto: University of Toronto Press.

Laxer, G. 2000. Surviving the Americanizing New Right. *Canadian Review of Sociology and Anthropology* 37 (1): 55-75.

—. 2004. Preface. *Governing under Stress: Middle Powers and the Challenge of Globalization*. M.G. Cohen and S. Clarkson. Halifax, NS: Fernwood.

Linton, J.I. 2004. Global Hydrology and the Construction of a Water Crisis. *Great Lakes Geographer* 11 (2): 1-13.

Manning, E. 2000. I Am Canadian: Identity, Territory, and the Canadian National Landscape. *Theory and Event* 4 (4).

Moodie, S. 1997. *Roughing It in the Bush, or, Life in Canada*. Ottawa: Tecumseh Press.

Swyngedouw, E. 2003. Modernity and the Production of the Spanish Waterscape, 1890-1930. In *Geographical Political Ecology*, ed. T. Bassett and K. Zimmerer, 94-112. New York: Guilford.

Szeman, I. 2001. Literature on the Periphery of Capitalism. *Ilha do Desterro* 40: 25-42.

Waterfield, D. 1970. *Continental Waterboy: The Columbia River Controversy*. Toronto: Clarke, Irwin and Company.

WCD. 2000. *Dams and Development: A New Framework for Decision-Making*. London: Earthscan.

Webber, M., J. Barnett, B. Finalyson, and M. Wang. 2005. The Yellow River in the Big Picture: Questions of Borders, Boundaries and Access. Proceedings of the Fresh and Salt: Water, Borders and Commons in Australia and Asia workshop and symposium, Sydney, NSW, Australia.

Worsley, P. 1984. *The Three Worlds: Culture and World Development*. Chicago: University of Chicago Press.

Worster, D. 1992. *Rivers of Empire: Water, Aridity, and the Growth of the American West*. New York: Oxford University Press.

—. 2002. Wild, Tame, and Free: Comparing Canadian and U.S. Views of Nature. In *Parallel Destinies: Canadian-American Relations West of the Rockies*, ed. J.M. Findlay and K.S. Coates, 246-73. Montreal: McGill-Queen's University Press.

# 17 Rising Waves, Old Charts, Nervous Passengers: Navigating toward a New Water Ethic

*Cushla Matthews, Robert B. Gibson, and Bruce Mitchell*

Tough decisions are ahead, and many of them centre on water. While the world's human population tripled during the twentieth century, water consumption multiplied sixfold. And it is still rising, although, in many places, even current consumption levels cannot possibly be maintained from existing sources with present efficiencies. Meanwhile, nearly 1.1 billion people lack adequate access to potable water, 2.4 billion people do not have adequate sanitation, ecosystems are under increasing stress, a host of water contamination sources remain, and water-related diseases are leading contributors to illness and death, particularly in less developed countries.

Perhaps these circumstances can be seen as openings for innovation rather than grounds for despair. But certainly, difficult choices lie ahead, as the pressures and deeply competing demands on an inevitably limited resource escalate dramatically.

Even in Canada, with our relative water wealth and persistent myth of abundance (Foster and Sewell 1981, 6-19), water conflicts are already common and likely to expand. We, too, have good reason to think carefully not just about our options but also about how to choose among them. Decision making about water – by individuals, corporate entities, governments, and other important bodies – involves psychology and culture, science and technology, economics and politics. But underneath it all, decision making is a matter of applied ethics.

Deepening global worries about how to deal with conflicting demands for water have sparked new interest in water ethics. An emerging stream of new research, much of it published by United Nations bodies, examines the grounds for more practical, cautious, and reasonable water use (Aureli and Brelet 2004; Harremoes 2002; Selborne 2001; UNESCO 2005). Rights to water, at least rights to secure access for basic needs, are more often asserted than once was the case. So are the rights of affected people to participate in water decision making. Typically, the discussions about access and participant rights focus on justice and equity for the powerless poor, for silenced women, and for disadvantaged minorities; but there are also voices for the interests of unrepresented future generations and for ecosystems, upon which humans as well as other species ultimately and unavoidably depend (Millennium Ecosystem Assessment 2005; McDonald and Jehl 2003; Postel 2000; World Water Commission 2000).

Much of this is not new. Disputes over water are probably almost as old as our species. Deliberations on water ethics, or on the application of broader ethical approaches to resolving water conflicts, stretch back into the dust of history. The ancient religions addressed it. So did the first written legal code, issued by Hammurabi in Babylon, nearly 4,000 years ago. But the present debates about water ethics have a more fully global character. While a great deal still comes down to individual behaviour and community decisions, we now also face regional scale problems and transboundary tensions. A general failure to find solutions that are practical, effective, and clearly just will imperil us all.

In these circumstances, it would be convenient to have widespread agreement on the basics of a general water ethic or on a general ethical framework that could be easily elaborated for application to water issues. A long history of nasty experiences with imposed global ideologies has taught us to be wary of universal rules. Any universal water ethic would have to be limited to broad principles, sensitive to context, and open to local adjustment. Given the global character and seriousness of the challenges to be faced, however, an exploration of ethical options is worthwhile.

## Ethical Considerations in Water Issues

Despite increased attention to water resources in the past couple of decades, there is surprisingly little published research on the convergence of water and ethics. The most noted new publications, such as UNESCO's *The Ethics of Freshwater: A Survey* (Selborne 2000) and *Water and Ethics: Overview* (Priscoli, Dooge, and Llamas 2004), have opened up discussions on *why* fresh water must be used more responsibly but have not taken the next step of comparing or selecting among differing ethical perspectives on water use. There have been international calls for a "new water ethic" but few specifics and little underlying clarity on what such an ethic must recognize, cover, and influence.

An ethic is a set or system of moral principles or values that guides the actions or decisions of an individual or group. It helps us determine what is acceptable conduct in a society and provides a basis for judging how to act rightly or justly. No set of ethics provides all the answers. Even with an elaborate ethical framework, we face dilemmas. Sometimes, all of the evident options promise attractive but different results, or all the choices threaten negative results of debatably greater or lesser magnitude or evil. More often, the consequences are only partly predictable and the available options present mixes of positive and negative possibilities, none of which clearly offends or fully satisfies the ethical prescriptions. And sometimes, different ethical principles can conflict with one another, making choices a challenge.

For the above reasons, most ethical frameworks are not limited to guidance on what to do (e.g., comfort the afflicted and afflict the comfortable) and what not to do (e.g., thou shalt not kill, steal, etc.). They also provide directions on how to wrestle with more difficult choices. The widely recognized Golden Rule – "always treat others as you would like to be treated yourself" – is a particularly elegant example for personal decision making that aims to incorporate elements of honesty, trust, fairness, equality, and respect for others in the decision-making process. Ethics for collective decisions are more complex but often also address matters of process, at least to the extent of identifying what factors must be considered, which purposes must be respected, and whose voices must be heard.

Most peoples' and most institutions' practical ethics include a mix of explicit principles and habitual moral positions, priorities, biases, and assumptions. Traditional, habitual ethics, often buried deeply in a culture, may serve adequately in well-understood, stable, or frequently repeated circumstances. But they are problematic in dynamic conditions with serious problems and pervasive uncertainties, and they are likely to be minimally helpful, if not dangerous, in cross-cultural situations. For such applications, it is better to have explicit ethical positions that are open to examination and challenge and that can be elaborated and adjusted to fit local circumstances and to reflect local choice.

Deep thinkers and practical decision makers have been attempting to define and defend suitable ethical frameworks for millennia. The old customary ethics of humans as nomadic foragers (90+ percent of the human record) were embedded in comprehensive sets of arrangements, duties, and practices involving the community and its environment. Ethics continued to rest on the maintenance of overall social arrangements through the rise of the great civilizations and the introduction of written codes, such as Hammurabi's Code of Babylonian law and Plato's exploration of the meaning of justice in *The Republic*. But with urbanization, attention to human relations with the natural environment began to fade.

Over the past 500 years or so, another major shift has brought broad acceptance of the pursuit of economic gain (which old ethical systems had universally discouraged as personally damaging and socially pernicious) and technological advance (which the old systems also treated as dangerously disruptive). Not surprisingly, ethics today typically focus on fair dealing among individuals, personal and corporate. It is the ethics of the marketplace.

Ethics are in flux again, however. The great achievements of modern progress have been accompanied by continuing injustices and deepening worries about social and ecological effects. Injustices and tensions involving water are only part of this larger story and represent only some of the challenges for new ethical prescriptions. But, certainly, it is clear from the water issues alone that any satisfactory ethics must now extend beyond human relationships at the individual level. They must involve obligations among communities and nations, between humans and non-human species and ecosystems, and between us today and the generations to come.

## Recent Approaches to a Water Ethic

So far, there is no widely accepted and applied set of principles or ethics that has been tailored specifically to the water sector (Harremoes et al. 2002). Even in government water planning efforts, the relevant decision makers have rarely attempted to identify the common goals of conflicting stakeholders or to negotiate agreement on the means required to solve conflicts over water resources and development (Gleick 2000).

At least part of the difficulty is that identification of common ethical goals is difficult. There are many differing perceptions of what is ethical as well as a wide diversity of approaches within and among nations. Moreover, this diversity is, to some extent, inevitable and appropriate in a world of markedly different ecological, economic, and cultural circumstances.

Agreement on the fundamentals is, however, widely recognized as desirable. As Gleick (2000, 128) has noted, the continuing "lack of consensus on a guiding ethic for water policy has led to fragmented policies and incremental changes that typically satisfy none of the many affected parties." In response, organizations with water-related mandates have proposed a variety of ethical frameworks, or at least lists of key considerations for water decision making. While none of these has won broad allegiance, the best known ones have certainly helped to guide some decision making on water use and allocation.

Of the many possibilities, we will consider only three – two from international bodies and one from Canadian water experts. All are post-Brundtland approaches, that is, they all follow the 1987 report of the World Commission on Environment and Development (commonly known as the

BOX 17.1

**The Dublin Statement, 1992**

The guiding principles are:

- *Fresh water is a finite and vulnerable resource, essential to sustain life, development and the environment.* Since water sustains all life, effective management of water resources demands a holistic approach, linking social and economic development with protection of natural ecosystems. Effective management links land and water uses across the whole of a catchment area or groundwater aquifer.
- *Water development and management should be based on a participatory approach, involving users, planners and policy makers at all levels.* The participatory approach involves raising awareness of the importance of water among policy makers and the general public. It means that decisions are taken at the lowest appropriate level, with full public consultation and involvement of users in the planning and implementation of water projects.
- *Women play a central part in the provision, management and safeguarding of water.* The pivotal role of women as providers and users of water and guardians of the living environment has seldom been reflected in institutional arrangements for the development and management of water resources. Acceptance and implementation of this principle requires positive policies to address women's specific needs and to equip and empower women to participate at all levels in water resources programmes, including decision making and implementation, in ways defined by them.
- *Water has an economic value in all its competing uses and should be recognized as an economic good* ... It is vital to recognize first the basic right of all human beings to have access to clean water and sanitation at an affordable price. Past failure to recognize the economic value of water has led to wasteful and environmentally damaging uses of the resource. Managing water as an economic good is an important way of achieving efficient and equitable use, and of encouraging conservation and protection of water resources.

SOURCE: United Nations (1992b).

Brundtland Commission after its chair Grø Harlem Brundtland of Norway), which set the broad foundations for most recent thinking about water ethics by popularizing the concept of sustainable development. Although the meaning and implications of sustainable development have been much debated, the essential message clearly includes recognition of the interdependence of social justice, economic well-being, and ecological integrity, as well

BOX 17.2

**Canadian Water Resources Association's sustainability principles for water management in Canada**

*Sustainability Ethic:*

Wise management of water resources must be achieved by a genuine commitment to:

- ecological integrity and biological diversity to ensure a healthy environment;
- a dynamic economy; and
- social equity for present and future generations.

*Water Management Principles:*

Accepting this Sustainability Ethic, we will:

1 Practice integrated water resource management by:
   - linking water quality, quantity and the management of other resources;
   - recognizing hydrological, ecological, social and institutional systems; and
   - recognizing the importance of watershed and aquifer boundaries.

2 Encourage water conservation and the protection of water quality by:
   - recognizing the value and limits of water resources and the cost of providing it in adequate quantity and quality;
   - acknowledging its consumptive and non-consumptive values to both humans and other species; and
   - balancing education, market forces, and regulatory systems to promote choice and recognition of the responsibility of beneficiaries to pay for use of the resource.

3 Resolve water management issues by:
   - employing planning, monitoring and research;
   - providing multidisciplinary information for decision making;
   - encouraging active consultation and participation among all affected parties and the public;
   - using negotiation and mediation to seek consensus; and
   - ensuring accountability through open communication, education and public access to information.

SOURCE: Canadian Water Resources Association (1994).

as acceptance of responsibility for future generations. These themes are now evident in most sets of principles for water initiatives proposed by national and global bodies.

The Dublin Statement was produced by the International Conference on Water and the Environment held in Dublin in 1992 and was a contribution to the Rio Conference on Environment and Development held later that year (United Nations 1992a, 1992b, 1992c; United Nations Department of Economic and Social Affairs 1992). Its four guiding principles are accompanied by a ten-point action agenda, beginning with alleviation of poverty and disease (Box 17.1). Two years later, the Canadian Water Resources Association (CWRA) published its sustainability ethic, supplemented by a set of complementary principles (Box 17.2). The ethic rests directly on the depiction of sustainable development based on three interrelated "pillars" and stresses equity for future as well as present generations. The principles, devised to guide and facilitate implementation of the ethic, focus on integrated management, conservation, water quality protection, and conflict resolution. As with other institutions, the CWRA anticipates implementation chiefly through the decisions and practices of governments and other organizations involved in managing water resources and does not target individual behaviour.

In 1999, UNESCO began to develop of its own set of nine principles to address the ethical use of water (Box 17.3). John Selborne, elaborating on these principles in a more recent UNESCO document, stresses that water management "is fundamentally a question of social and environmental justice based on three essential concepts: equity, fairness and access between and across generations" and that useful application of the ethic relies heavily on the participative nature of decision making (Selborne in Brelet 2004, 6).

Each of the water ethics packages is attractive, and, arguably, they vary more in emphasis than in substance. Because all of them are too brief to go beyond listing the biggest areas for consideration, it would be unfair to fault them for being incomplete. Still, there are some notable differences and silences.

The Dublin Statement and the CWRA ethic are considerably stronger on integrating ecological and socioeconomic considerations than are the UNESCO principles, which emphasize the need for water decisions to combat poverty and inequality. Ecosystem priorities, including maintenance of ecosystem services for human survival, get uneven treatment. Only the CWRA ethic refers directly to future generations.[1] None of the packages includes precaution in the face of uncertainty as a key consideration.[2] And while they all recognize the overlaps between water decisions and some other large concerns – especially matters of economic and political equity – none puts much emphasis on designing water decision making to deliver multiple benefits in adjacent livelihood areas such as health, education, employment, and food security.

The most serious limitation of these sets of principles, judged as practical guides for decision makers, may be that they do not have much to say about priorities and trade-off choices. The listed principles draw attention to an admirable range of factors, all of which merit careful consideration. And for the design of decision-making processes, all three packages stress, in their various ways, the importance of open, informed, equitable, and participative deliberations. Unfortunately, participation does not ensure consensus. Especially in decision making about a crucial resource under unsustainable

BOX 17.3

**UNESCO's principles for ethical water use**

- Human dignity, for there is no life without water and those to whom it is denied are denied life.
- Participation, for all individuals, especially the poor, must be involved in water planning and management with gender and poverty issues recognized in fostering this process.
- Solidarity, for water continually confronts humans with their upstream and downstream interdependence, and initiatives for integrated water management may be seen as a direct response to this realization.
- Human equality, for all persons ought to be provided with what is needed on an equitable basis.
- Common good, for by almost everyone's definition water is a common good, and without proper water management human potential and dignity diminishes.
- Stewardship, which respects wise use of water.
- Transparency and universal access to information, for if data is not accessible in a form that can be understood, there will be an opportunity for one interested party to disadvantage others.
- Inclusiveness, for water management policies must address the interests of all who live in a water catchment area. Minority interests must be protected as well as those of the poor and other disadvantaged sectors.
- Empowerment, for the requirement to facilitate participation in planning and management means much more than to allow an opportunity for consultation.

SOURCE: Selborne in Brelet (2004).

pressures from many contending interests, even the most admirably participative processes will leave unresolved conflicts and unpleasant choices.

Perhaps no practical set of ethical principles can do much about this. Perhaps the best that can be expected is consideration of all the key factors and use of fair and open processes. But, if so, it will be particularly important to ensure that decision making regarding water does address all the crucial issues, including opportunities to contribute to related livelihood objectives, and that it both uses and contributes to the spread and entrenchment of effectively participative democracy. If this is achieved, it will represent a significant advance over what has prevailed in the practice of water management decision making in most jurisdictions.

## Water Management Paradigms

We did not always try to manage water. On the contrary, for most of human experience, we were more or less nomadic hunters, gatherers, or foragers, and we were mostly satisfied to adapt our cultures and behaviours to the environment around us. Some people used fire to alter or maintain certain kinds of ecosystems (for example, to keep encroaching forests out of meadowlands). But, until the rise of the great valley agriculturalists and their ambitious irrigation systems, we were not water managers. Our knowledge about where water could be found was built from our experience in dry lands, in the places above the flood in wet lands, where there were fish, and where the water was safe to drink, cross, or navigate. Following our customs and spiritual beliefs, we tried to respect, anticipate, and cope – but not to manage (Box 17.4).

Box 17.4

**Aboriginal perspectives on water management**

Because of the close relationship we have with the natural world, we cannot have healthy communities unless we have a healthy environment. Our ancestors have always understood this. This is why our ceremonies give thanks and gratitude to the spirits of the natural world. One example is the thunder ceremony. It is held in the spring of the year when the first thunders are heard. The ceremony is held to give thanks that the grandfather thunderers have returned with the rain that will replenish the earth, that will again quench the thirst of life and that will wash the land.

Source: J.W. Ransome (1995, Supplement, 5); see also Mercredi and Turpel (1993, 16-17, 44, 155-56).

The idea of management required a fundamentally different attitude – certainly more confidence, arguably a good deal of arrogance. Water management goes beyond use and modest adjustment. It centres on invasive manipulation (both "manage" and "manipulate" are rooted in the Latin *manus*, which refers to the human hand), and it rests on a core belief that humans can control nature to reap the benefits of a more reliable and, for human purposes, more fruitful resource (FAO 2005). Attempts to exercise control may serve a variety of purposes. They may be devoted to national economic advance or demonstration of authoritative might. They may be to enrich the few or to assist the many. They may aim to extend the reach of exploitation or to enhance its efficiency or to strengthen stewardship so that the flow of benefits will continue long into the future.

In Canada, at least, the practice of water resource management has followed a rough succession of basic ideological frameworks or paradigms. The four main ones – developmental exploitation, supply management, demand management, and integrated water resource management – are all still evident today in the work of water management bodies of one kind or another. Integrated water management has become most widely accepted today. This general acceptance of integrated water management is particularly evident in pivotal government policy documents, such as the now outdated Canadian Federal Water Policy (1987, with attempted updates in 1998) (Environment Canada 1985; Pearse and Quinn 1996), or, more recently, in Ontario's Source Water Protection Act (de Loë, Kreutzwiser, and Neufeld 2005). Both emphasize the need to manage water in a holistic way, taking into consideration economic, social, and environmental priorities. In turn, however, integrated water management now seems to be giving way to, or adjusting to incorporate, approaches reflecting the new ethics discussed above.

According to Gleick (2000, 127), the emerging changes in water management thinking have four key components: A shift away from sole, or even primary, reliance on finding new sources of water supply to address perceived new demands; a growing emphasis on incorporating ecological values into water policy; a re-emphasis on meeting basic human needs for water services; and a conscious breaking of the ties between economic growth and water use.

*Developmental exploitation* was introduced to Canada by the Europeans, who were inclined by religion and economics to see water as a resource to be used where convenient and enhanced for better use where technically and financially possible. Water was for transportation, power, irrigation, and the other fundamentals of development, and development both demanded and facilitated even greater exploitation (Box 17.5).

Over time, the projects became more ambitious in physical rearrangement as well as engineering sophistication and investment demand. Little dams and mill races for village flour mills were followed by the major diversions and dams of the Columbia River and James Bay hydro projects. Local canals and locks were supplanted by the St. Lawrence Seaway. As Ralph Pentland and Adèle Hurley note in their chapter on Canada-US water relations, some of the biggest undertakings required and inspired international cooperation rooted in a shared conviction that development is a practical, if not moral, obligation and that water is a resource that is essentially wasted if we fail to make full use of it to the extent of our energy and ingenuity.

In its rawest form, developmental exploitation is less fashionable today than it used to be. Many of the convenient opportunities have already been exploited, and the social and ecological costs of the grand schemes are now more likely to be noticed. But the next phase of big hydro development in northern Quebec – the $2 billion Eastmain-Rupert project involving 4 dams and 51 dykes, 2 diversion bays, 395 square kilometres of flooded lands, 12 kilometres of diversion channels or tunnels, and 2 permanent access roads (CEAA 2003) – is now moving toward approval. And, as Pentland and Hurley note, while the proposals for absurdly costly inter-basin water transfers from Canada to the US southwest are no longer active, and may never be ecologically or economically justifiable, their advocates survive and might, in some plausible circumstances, become powerful.

BOX 17.5

**Harnessing water for human benefit: Two perspectives**

It has long been my belief that Quebec's economic strength lies in the development of its natural resources, the most outstanding of which is its rich hydroelectric potential. And further, I have always believed that to develop these resources would require conquering and taming the North. This straightforward reasoning led me to promote the concept of developing the hydroelectric potential of the James Bay Region.

– Quebec Premier Robert Bourassa (1985, 13)

His dream has always been to flood Quebec from one border to another. He's happy to live in a lake as long as he can get electricity out of it.

– National Chief, Assembly of First Nations,
Georges Erasmus (Colombo 2000, 558)

*Supply management* was the focus of most water management agencies in the more developed parts of Canada and other developed countries through much of the twentieth century. Especially in Canada, supply-focused managers assumed that water resources, particularly freshwater resources, were practically limitless. Given properly applied management techniques and interventions, existing sources could serve a multiplicity of particular purposes, including water use for hydroelectric power, irrigation, human consumption, and even ecosystem maintenance where clear human gains resulted. Examples highlighting this type of thinking include most dam construction projects across Canada and the United States, most notably the Hoover Dam, which is located on the Colorado River at the Nevada-Arizona border. Either referred to as a modern wonder and an American icon of technology and progress or as a monolithic testament to humanity's need to control nature, the Hoover Dam, and other dams like it, represents the ultimate in supply management. Later steps to protect water quality, primarily for human health reasons, were also part of the supply management agenda. And then, gradually, attention to stewardship began to supplant the focus on exploitive manipulation. As a result, the legacy of supply management includes the provision of sewage and drinking water treatment facilities as well as the global proliferation of wells, dams, and other control and distribution structures.

Like the exploitation paradigm, however, supply management proved to be overly optimistic about managerial capabilities and insufficiently respectful of ecological complexities, proliferating demands, and/or economic realities. The beginnings of widespread water scarcity (despite advancing technology), overall declines in ecological health, and growing resistance to the costs of ambitious new supply projects (displacement of species, flooding of agricultural lands, loss of human as well as ecological heritage, along with the conventional costs of land acquisition, construction, and maintenance) pushed water managers to look for other options.

*Demand management* emerged as a third paradigm in response to evident limitations of the old approaches, including rising pressures on available sources of supply (Brooks 2005; Horbulyk 2005; Shrubsole and Tate 1994; Tate 1984, 1990). It was also pushed by recognition that efforts to curb demand and to enhance extraction and distribution efficiencies could be considerably less costly than were the increasingly ambitious projects needed to provide more supply. Demand management accepts that water resources are finite and includes at least some recognition of the limited resilience of associated biophysical systems. Also, it applies a range of tools, behavioural as well as technological.

For growing urban populations, water demand management seeks to reduce per capita consumption, in part, by discouraging wasteful and frivolous use but also by increasing per capita water-use efficiency. Some of this

BOX 17.6

**Behavioural barriers in the Waterloo Region, Ontario**

During May 2005, in Waterloo Region, Ontario (a community of approximately 500,000 people living in the cities of Cambridge, Kitchener, and Waterloo and four surrounding townships), regional councillors modified an outdoor watering bylaw, causing quite an uproar among the residents. The modified watering bylaw stipulated: (1) one lawn watering per week at specified times during the day, with the designated day determined by house numbers; (2) car washing, with a hose, on Saturday and Sunday only; (3) all other outdoor watering (gardens, for example) to be done with a watering can; and (4) all this to be in effect from May to September.

Waterloo Region is 80 percent reliant on groundwater, with the other 20 percent of its supply being taken from the Grand River. The long-term weather forecast for the summer of 2005 was hot and dry, and one of the major water supply wells had been taken off line due to industrial contamination, thereby reducing supply capacity by 5 percent. From a water conservation and allocation perspective, this modification was sensible and tried, for example, to accommodate those who enjoy having green lawns and shiny cars. The bylaw simultaneously tried to communicate the idea that we must increase our awareness regarding the links between water use and water supply. A common sense approach was being applied to water resource management.

Then came the complaints. Despite the warnings of an impending hot summer, very little rain, and the fact that most residents understand the ultimate goal of water conservation, the restrictive nature of the bylaw made some people fume. Based on media reports, some felt the bylaw violated their right to a resource they had paid for through their municipal taxes. Others disputed the new regulations, arguing that government should not interfere in the quality of life of residents by restricting them to the use of a hand-held hose. Yes, it makes watering less cumbersome; however, it is also extremely water inefficient due to its high rates of evaporation compared to, say, a watering can. Still others feared that their property values could go down because of a brown lawn.

Due to public pressure, regional councillors reviewed the bylaw and decided to strike a compromise – to allow one extra day for car washing and garden watering. To some, this was not a compromise at all but merely an indication that the regional government was inflexible; to others, the change in the bylaw reduced a strong water conservation initiative to a merely minor victory.

is achieved through infrastructural improvements such as pressure management and distribution system upgrading to cut pipeline losses. But there is also widespread use of behavioural change initiatives, including awareness-raising programs, regulatory measures (e.g., lawn watering restrictions), water metering and pricing, and incentives for use of water efficient hardware (e.g., low-flush toilets and flow restricting shower heads) (Box 17.6).

For industrial uses, the aim is less water use per unit of production. In agricultural production, for example, the goal is "more crop per drop," achieved through measures including the use of more efficient irrigation systems. Here, too, educational and regulatory measures are used, but the main tools are pricing and other economic mechanisms to encourage the adoption of process efficiencies, largely through technological improvements.

Considering demand as well as supply has increased the range of response options and has expanded the set of available management tools. While these are often used in separate initiatives, they can be combined by governments, or even by broader governance bodies, through coordinated use of law, policy, and program components. The introduction of demand management has also led to more system-oriented thinking about water issues. For example, the environment has slowly gained more prominence, especially through recognition of the desirability of protecting wetlands and other aquatic ecosystems as a means of attenuating storm runoff and preserving water quality.

No less significantly, demand management's focus on efficiencies through the use of economic tools has spurred debate on neglected larger issues. In particular, it has raised questions about how to meet the needs of the economically disadvantaged, who may find secure access to clean water even more distant when water pricing is used to encourage efficiency. More generally, treatment of water as a priced resource has raised concerns about the limitations of economic valuing and has led to new thinking about how best to recognize the full importance of water and water-dependent systems for humans and other species.

*Integrated water resource management* (IWRM) rose in the late 1980s with greater awareness of the many factors to be considered in water allocation and management decisions and, consequently, the many voices to be heard. This fourth paradigm reflects the broader water agenda required in a world that is much more complex than one centred only on supply or demand. In the integrated management approach, the relevant factors include the full range of ecological, economic, social, cultural, and technological considerations and their interrelationships, plus the great diversity of interests that may be affected in some way and that may have useful information and important perspectives to share (Hartig and Zarull 1992; Heathcote 1998; Lang 1986; MacKenzie 1996; Mitchell 1990; Shrubsole 2004). As Peigi Wilson writes (see Box 17.7), IWRM also reflects the core values of some Aboriginal peoples.

The larger scope and complexity of IWRM is evident in its approach to ecological concerns. Integrated management attempts to respect the full value of ecological services delivered by whole catchments. Rather than focusing only on particular water problems and sites, integrated approaches incorporate efforts to rehabilitate and preserve linked land and water systems throughout catchments to maintain important services such as free flood protection.

This is realistic: if forest and wetland complexes are poorly managed and flood plains filled, then their free ecological services are lost, citizens face greater flood and contamination risks, and governments incur steep costs to repair or replace the necessary services. But attention to the full suite of ecosystem factors and their interrelations also increases the technical and political challenges involved.

When integrated management adds in all the other key considerations – ensuring adequate supply, moderating demand, motivating efficiency, combatting inequities, fostering more democratic decision making, and so

BOX 17.7

**Integrated water resource management and Aboriginal values**

It is noteworthy that integrated watershed management accords with traditional First Nations' philosophies. Although there is no pan-Aboriginal view, generally speaking, First Nations traditionally hold that the world is an integrated whole. All things are connected: the plants and animals; the sun and moon, the air and water; and men and women. What affects one element affects the whole. Our relationships with each other and with the environment must be built on this holistic understanding of the world. Addressing concerns about the quality of our drinking water must take into consideration not only the quality of water coming from the tap or the end of the discharge pipe, but where that water is coming from in the first place and what is being added to it along the way. We must address not only human health and well being, but the well being of the plants, animals, birds and fish that also rely on the waters. Integrated watershed management mirrors this perspective.

SOURCE: Peigi Wilson (2004, 73).

on – the practical complexities multiply (Biswas 2004; Blomquist and Schlager 2005). Greater collaboration is required among experts in different fields, as is greater acceptance of information from unconventional sources, including Aboriginal and other long-time residents. But even with all available knowledge sources at the table, the inherent uncertainties of complex systems prevail. This encourages integrated managers to adopt precautionary approaches that favour low-risk, reversible, and diverse options. The precautionary principle, endorsed at the Earth Summit in Rio de Janeiro in 1992, posits that, in the face of possible environmental damage, even without full scientific data or knowledge concerning the extent of the environmental risk, preventative measures shall not be abandoned. Simply, we are to "always err on the side of caution" when it comes to the environment – most notably, when it comes to water resources (Dearden and Mitchell 2005, 171). It also underlines the importance of participative public choice, since no profession or authority can be in a position to decide merely on the basis of "the facts."

Recognition of complexity and uncertainty also pushes the agenda of management from managing water to managing ourselves in an adaptive manner. It also begins to undermine the underlying confidence in control, suggesting consideration of how we might, in our current circumstances, reintroduce some version of the ancient wisdom that favoured coping over controlling and avoided disruptive intervention.

All of this makes water decision making a broadly comprehensive and unavoidably political process of integration and conflict resolution. Integration has been assisted by insights from concurrent deliberations about complex systems and sustainability, both of which are integrative concepts. And inclusive approaches to conflict resolution have benefited from developments in citizen participation, multistakeholder consultation, and the use of mediation and other mechanisms to resolve conflicts among relevant interests and agencies. Integration and conflict resolution remain major challenges, however, and practitioners have reason to look for additional means of overcoming some of their difficulties without sacrificing the comprehensive scope or integrative philosophy. Widespread adoption of a new water ethic with a similarly broad agenda might be one such means.

## Basic Components of a New Water Ethic

The evolution of water management paradigms signals a shift in the way water is perceived. We are at least beginning to see that water is finite, that its ecological role is crucial and beyond mere economic calculation, that conservation and efficiency are necessary and often less expensive than other options, and that serious intra- and intergenerational equity issues are involved. We are also coming to appreciate the complexity of the issues and of decision making, and to see the implications for precaution and participation.

These insights are reflected in the water management principles that have emerged from global and national deliberations about water. The Dublin, CWRA, and UNESCO lists discussed above, and similar products from the world water forums and a host of parallel gatherings, have consistently identified a set of ethical principles that can be organized under five basic imperatives.

The first imperative is to *meet basic human needs to enhance equity today and for the future.* In order to build the foundation for the health and well-being of all people, we must ensure that everyone has essential access to sufficient clean drinking water and sanitation services. This means bringing greater equity to the present distribution of water resources and opportunities and avoiding actions that would shift water-related burdens from present to future generations. Meeting human needs is not likely to be possible in the long term unless we also *safeguard ecosystems.* This imperative involves allocating sufficient water to maintain ecosystem integrity and preserve ecosystem services. While the usual focus here has been on ensuring continued human benefits from ecosystem services, non-human interests are also likely to be served and may be recognized in the underlying ethical commitment.

*Encourage efficiency and conservation* of water resources is the third imperative. This can be done by strengthening conservation motivations and capabilities through educational, regulatory, and economic innovations and applications of appropriate technology. Demand management must be considered generally preferable to exploiting new supplies, and all new supply initiatives must be designed for efficiency, flexibility, and appropriate use.

The potential success of a new water ethic depends heavily on the processes used in its elaboration and application. Recognizing this, the fourth imperative is to *establish open and participative decision-making processes* involving those affected by water management choices. Particular emphasis must be on ensuring the effective engagement of women in water decision making since women, especially in developing countries, are chiefly responsible for water use at the household level and have the most knowledge about its location and quality. Such participative processes must facilitate use of the full knowledge base for water decisions, including Aboriginal and/or traditional ecological knowledge, while ensuring timely and convenient access to information, opportunity for public deliberation, and transparency of results and justifications. Overall, the objective is to use water decision making as a means of building social and ecological understanding, stewardship capabilities, and collective capacities for protecting water as a common good.

The fifth imperative is to *respect system complexity and emphasize precaution*. Essentially, this means applying the precautionary preference for options that are reversible, diverse, and low-risk and that maintain or enhance means of responding to unexpected results. Ultimately, this will enhance the overall integrity of interdependent socioeconomic and ecological systems.

We can legitimately add one more key consideration that may not be directly acknowledged in the established ethical statements but that, in the long run, is fundamental to successful application. It is that water initiatives never stand alone. They inevitably contribute to and are affected by other undertakings, and it is the full set of our efforts and their effects that will determine how successful we are in establishing the conditions for a better world. It is therefore important to recognize and build the links between water objectives and other sustainability requirements (and, accordingly, between water ethics and the broader ethics of sustainability).

The United Nations' new Water for Life decade initiative, for example, has been designed to emphasize the connections among access to safe, clean water (and sanitation), alleviation of poverty, and improvements in health (United Nations 2005). The water objectives are linked into a set of two targets that most national governments had agreed to at the Johannesburg Summit in 2002. And much of the publicity accompanying the initiative's launch in March 2005 repeated an earlier observation by UN secretary-general Kofi Annan (2001), underlining the interrelatedness: "We shall not finally defeat AIDS, tuberculosis, malaria, or any of the other infectious diseases that plague the developing world until we have also won the battle for safe drinking-water, sanitation and basic health care." The final imperative, then, might be stated as follows: *seek multiple sustainability benefits from water-centred initiatives, including the enhancement of livelihood opportunities, health, education, and security.*

These imperatives can, of course, be phrased, ordered, and organized in many ways. No doubt, they could be supplemented usefully by various additions and elaborations. And, for practical use, they will always need to be tuned to the particulars of the context – to the ecological realities, human priorities, and associated opportunities and choices of the places of application. But

BOX 17.8
**Six imperatives for a new water ethic**

1 meet basic human needs to enhance equity today and for the future
2 safeguard ecosystems by allocating sufficient water resources
3 encourage efficiency and conservation of water resources
4 establish open and participative decision-making processes
5 respect system complexity and emphasize precaution
6 seek multiple sustainability benefits from water-centred initiatives.

the major challenge is finding ways of encouraging sufficiently wide adoption that these principles become part of conventional thinking. Like any set of ethical principles, this one will be most powerful if it becomes ubiquitous and habitual.

## Implementation

As we have seen, new thinking is emerging in water management decision making. But, as water decision making has accepted a wider agenda and become more broadly participative, integration and conflict resolution have become understandably more difficult. While support from a new water ethic might help, this depends on the ethic being incorporated into mainstream legislation and regulation, policies, programs, and project planning decisions so that, gradually, these underlying ideas become the assumed normal framework for water-related choices.

This is not now happening in any comprehensive way, although components of the new ethic are being adopted by leading jurisdictions and agencies on several continents. A good example is provided by the policy initiatives being taken in South Africa and in Australia to recognize the importance of ecological services and to ensure sufficient water allocations to maintain these services.

South Africa's 1997 *White Paper on Water Policy* adopts an integrated approach to water management on a catchment (watershed) basis. It also recognizes two key rights – the right of all citizens to meet basic water needs and the rights of the environment to enough water to meet its requirements. According to principle 9 of the water policy, "The quantity, quality and reliability of water required to maintain the ecological functions on which humans depend shall be reserved so that the human use of water does not individually or cumulatively compromise the long term sustainability of aquatic and associated ecosystems." To highlight South Africa's commitment to social developmental policy even further, the right to basic access to food and water was entrenched in its Constitution (the Constitution of the Republic of South Africa, Act 108 of 1996, section 27[1][a]), making this one of the most progressive statutes related to advancing the status of water resources.

In Victoria, Australia, water policy reform has led to acceptance of ecosystem limits and allocation of a form of legal rights to the environment as a legitimate user of water (Victoria 2004). More generally in Australia, the Council of Australian Governments (COAG) Water Reform Agenda has accepted the provision of water to the environment, in order to mitigate ecosystem degradation, as a valid use (COAG 1994a, 1994b, 1995). The new provisions do not eliminate conflicts. Indeed, they may add to immediate tensions since human uses, including irrigation demands and other purposes beyond basic needs, are also recognized. Just how much water is needed for the ecological purposes of rivers and wetlands remains open to debate (Arthington and Pusey 2003). But a key long-term consideration is now firmly on the agenda for serious discussion.

Canadian policies on ecosystem uses are less bold and, perhaps, accordingly less successful in pushing efforts to clarify requirements and priorities than are those of South Africa and Australia. British Columbia's (1999) Water Sustainability Action Plan and its Freshwater Strategy, Nova Scotia's (2000) Water Resources Protection Act, Quebec's (2002) Water Policy (Box 17.9), and Saskatchewan's (1999) Water Management Framework, for example, generally recognize the benefits of viable ecosystems and the importance of maintaining ecosystem integrity, but they do not clarify when ecosystem purposes might have priority over human demands. Similarly, Alberta's 2003 *Water for Life* document offers a clear set of principles that indicate the province's intention to maintain and enhance ecological services and linkages, but it gives no indication of the priority that environmental integrity has in comparison that of more direct human interests.

BOX 17.9

**Quebec's water policy, 2002**

Following a symposium in 1997 and public consultation that began in 2000, the Quebec Water Policy was completed in 2002. The policy stipulates that water is a collective property, provides for protection of both public and aquatic ecosystems, and prescribes participative governance.

Clearly, just establishing an ecosystem's right to water (or finding some other means of protecting ecosystem integrity) is not enough. Incorporating such initiatives into a broad policy shift that also incorporates the full suite of water ethic principles discussed here would help. It would, at least, establish a more comprehensive, coherent, and transparent framework for deliberations about water options and their implications. But even with this, difficult choices and conflicting interests will remain.

A new water ethic is no magic wand. It merely promises to make us more aware of what needs to be considered, a little more open, and, perhaps, a little more mutually respectful in our discussions and decision making. But these are no small things. Anything that encourages us – water decision participants as well as authoritative managers and mandated institutions – to consider those whose basic needs are not being met, to appreciate the value of functioning ecosystems, to recognize the interests of future generations, and to engage more fully and humbly with others in the protection of shared goods, is worthy of pursuit.

## NOTES

1 The Dublin document does not mention respect for the interests of future generations in its principles or list of action items. However, a closing note at the end of the statement reports a children's performance that stressed decision making for the future, and this is presented as a message underlying the whole agenda.

2 In the language of the 1992 Rio Declaration (United Nations 1992c, principle 15), the precautionary principle holds that: "Where there are threats of serious or irreversible damage, lack of full scientific uncertainty should not be used as a reason for postponing measures to prevent environmental degradation." This is a little narrow for water management purposes. The key is to recognize the likelihood, if not the inevitability, of surprise in complex systems and to act in anticipatory ways to maintain flexibility and adaptive capacity and to avoid or at least minimize risks.

## REFERENCES

Alberta Environment. 2003. *Water for Life: Alberta's Strategy for Sustainability*. Edmonton: Alberta Environment. http://www.waterforlife.gov.ab.ca.

Annan, K. 2001. Statement to the fifty-fourth World Health Assembly, Geneva, 17 May.

Arthington, A.H., and B.J. Pusey. 2003. Flow Restoration and Protection in Australian Rivers. *River Research and Applications* 19: 377-95.

Aureli, A., and C. Brelet. 2004. *Women and Water: An Ethical Issue.* UNESCO, France. http://unesdoc.unesco.org/images/0013/001363/136357e.pdf.

Biswas, A.K. 2004. Integrated Water Resources Management: A Re-Assessment. *Water International* 29 (2): 248-56.

Blomquist, W., and E. Schlager. 2005. Political Pitfalls of Integrated Watershed Management. *Society and Natural Resources* 18 (1): 101-17.

Bourassa, R. 1985. *Power from the North.* Toronto: Prentice-Hall Canada.

Brelet, C. 2004. *Some Examples of Best Ethical Practices in Water Use.* UNESCO, France. http://portal.unesco.org/shs/fr/file_download.php/d0cbce96acbd61fd1e1b165686824271water_use.pdf.

British Columbia. 1999. *A Freshwater Strategy for British Columbia.* Ministry of Parks and Lands. http://wlapwww.gov.bc.ca/wat/fws/fws_nov_99.html.

Brooks, D.B. 2005. Beyond Greater Efficiency: The Concept of Water Soft Paths. *Canadian Water Resources Journal* 30 (1): 83-92.

Canadian Environmental Assessment Agency (CEAA). 2003. Environmental Assessments of the Eastmain-1-A and Rupert Diversion Project. Backgrounder. http://www.ceaa-acee.gc.ca/010/0001/0001/0017/bg030509_e.htm.

Canadian Water Resources Association (CWRA). 1994. *Sustainability Principles for Water Management in Canada.* June Pamphlet. http://www.cwra.org.

Colombo, J. 2000. *Famous Lasting Words.* Toronto: Douglas and MacIntyre.

Council of Australian Governments (COAG). 1994a. Attachment A: Water Resource Policy. http://www.coag.gov.au/meetings/250294/attachment_a.htm.

—. 1994b. *Report of the Working Group on Water Resource Policy.* http://www.coag.gov.au.

—. 1995. *The Second Report of the Working Group on Water Resource Policy.* http://www.coag.gov.au.

Dearden, P., and B. Mitchell. 2005. *Environmental Change and Challenge: A Canadian Perspective.* 2nd ed. Toronto: Oxford University Press.

de Loë, R.C., R.D. Kreutzwiser, and D. Neufeld. 2005. Local Groundwater Source Protection in Ontario and the Provincial Water Protection Fund. *Canadian Water Resources Journal* 30 (2): 129-44.

Environment Canada. 1985. *Currents of Change: Final Report.* Inquiry on Federal Water Policy. Ottawa: Environment Canada.

Food and Agricultural Organization of the United Nations. 2005. http://www.fao.org.

Foster, H.D., and W.R.D. Sewell 1981. *Water: The Emerging Crisis in Canada.* Ottawa: Canadian Institute for Economic Policy.

Gleick, P.H. 2000. The Changing Water Paradigm: A Look at Twenty-First Century Water Resources Development. *Water International* 25 (1): 127-38.

Harremoes, P. 2002. Water Ethics: A Substitute for Over-Regulation of a Scarce Resource. *Water Science and Technology* 45 (8): 113-45.

Harremoes, P., G. Gee, M. MacGarvin, A. Stirling, J. Keys, and B. Wynne, eds. 2002. *The Precautionary Principle in the 20th Century: Late Lessons from Early Warnings.* London: Earthscan Publications.

Hartig, J.H., and M.A. Zarull, eds. 1992. *Under RAPs: Toward Grassroots Ecological Democracy in the Great Lakes Basin.* Ann Arbor: University of Michigan Press.

Heathcote, I.W. 1998. *Integrated Watershed Management: Principles and Practice.* New York: John Wiley and Sons.

Horbulyk, T.M. 2005. Markets, Policy and the Allocation of Water Resources among Sectors: Constraints and Opportunities. *Canadian Water Resources Journal* 30 (1): 55-64.

Lang, R., ed. 1986. *Integrated Approaches to Resource Planning and Management.* Calgary: Univeristy of Calgary Press.

MacKenzie, S.H. 1996. *Integrated Resource Planning and Management: The Ecosystem Approach in the Great Lakes Basin.* Washington, DC: Island Press.

McDonald, B., and D. Jehl, eds. 2003. *Whose Water Is It? The Unquenchable Thirst of* a *Water-Hungry World.* Washington, DC: National Geographic Society.

Mercredi, O., and M.E. Turpel. 1993. *In the Rapids: Navigating the Future of First Nations.* Toronto: Viking.

Millennium Ecosystem Assessment. 2005. *Ecosystems and Human Well-Being: Synthesis.* Washington, DC: Island Press.

Mitchell, B., ed. 1990. *Integrated Water Management: International Experiences and Perspectives.* London: Belhaven Press.

Nova Scotia. 2000. *Water Resources Protection Act.* http://www.gov.ns.ca/legi/legc/bills/58th_1st/1st_read/b032/htm.

Pearse, P.H., and F. Quinn. 1996. Recent Developments in Federal Water Policy: One Step Forward, Two Steps Back. *Canadian Water Resources Journal* 10 (4): 329-40.

Postel, S.L. 2000. Entering an Era of Water Scarcity: The Challenges Ahead. *Ecological Applications* 10 (4): 941-48.

Priscoli, J.D., J. Dooge, and M.R. Llamas. 2004. *Water and Ethics: Overview.* Series on Water and Ethics. Essay 1 UNESCO, France. http://unesdoc.unesco.org/images/0013/001363/136343e.pdf.

Ransome, J.W. 1995. Water Is Life. *Water News,* Technical Supplement to the newsletter of the Canadian Water Resources Association, 1-8 September.

Saskatchewan. 1999. *Water Management Framework, January 1999.* http://www.se.gov.sk.ca/ecosystem/water/framework.

Selborne, L. 2000. *The Ethics of Freshwater Use: A Survey.* UNESCO, France. http://unesdoc.unesco.org/images/0012/001220/122049e.pdf.

Shrubsole, D., ed. 2004. *Canadian Perspectives on Integrated Water Resources Management.* Cambridge, ON: Canadian Water Resources Association.

Shrubsole, D., and D. Tate, eds. 1994. *Every Drop Counts.* Cambridge, ON: Canadian Water Resources Association.

South Africa. 1997. *White Paper on Water Policy.* http://www.polity.org.za/html/govdocs/white_papers/water.html?rebookmark=1.

Tate, D.M. 1984. Canadian Water Management: A One-Armed Giant. *Canadian Water Resources Journal* 9 (3): 1-6.

—. 1990. *Demand Management in Canada: A State-of-the-Art Review.* Social Science Series NO. 23. Ottawa: Environment Canada, Inland Waters Directorate, Water Planning and Managment Branch.

UNESCO [United Nations Educational, Social and Cultural Organization]. 2005. *World Water Assessment Program.* 22 March. http://www.unesco.org/water/wwap/news/wwap_wwd_05.shtml.

United Nations. 1992a. *Agenda 21.* The Rio Declaration on Environment and Development, http://www.un.org/esa/sustdev/agenda21.htm.

—. 1992b. *The Dublin Statement and the Report of the Conference.* Dublin International Conference on Water and Environment: Development Issues for the Twenty-First Century, 26-31 January, Dublin.

—. 1992c. *Report of the United Nations Conference on Environment and Development,* 3-14 June, Rio de Janeiro. Annex I, Rio Declaration on Environment and Development, A/CONF.151/26 (vol. 1). http://www.un.org/documents/ga/conf151/aconf15126-1annex1.htm.

—. 2005. *Water for Life Decade: 2005-2015.* http://www.un.org/waterforlifedecade/reference.html.

United Nations Department of Economic and Social Affairs. 1992. *Report of the United Nations Conference on Environment and Development,* 3-14 June, Rio de Janeiro. http://www.un.org/documents/ga/conf151/aconf15126-1annex1.htm.

Victoria, State of, Department of Sustainability and Environment. 2004. *Securing Our Water Future Together.*

Wilson, P. 2004. First Nations Integrated Watershed Management. *Canadian Perspectives on Integrated Water Resources Management,* ed. D. Shrubsole, 69-83. Cambridge, ON: Canadian Water Resources Association.

World Water Commission. 2000. A Report of the World Water Commission on Water for the 21st Century. *Water International* 25 (2): 194-201.

# 18
# Conclusion: Governing Canada's Waters Wisely

*Karen Bakker*

The contributors to this book have all argued that Canada's approach to water management urgently needs reworking. This includes improved water conservation, fair and efficient pricing, and enforceable water quality standards. Better data on Canada's water resources is also needed, as the data that do exist are not sufficiently standardized or accessible to managers and decision makers.

Adding to the complexity of these management problems are issues of governance. Given that water is essential for life and public health, core governance issues are of acute importance. We need to address such questions as: Who should be included in (and excluded from) decision making? What types of information should decision makers employ (or discount)? And how are decision-makers to be held accountable? As the contributors to *Eau Canada* have emphasized, there is a crucial need to address water governance in several critical areas: renewing federal water policy; clarifying federal-provincial roles in water management, particularly regarding transboundary waters; establishing a much-needed cross-Canada consensus on drinking water management; strengthening weak and poorly enforced environmental water quality regulations; and developing accountable and transparent water governance mechanisms into which indigenous peoples' water rights are fully integrated.

The remainder of this chapter reconsiders questions raised by the contributors to this volume – questions that are likely to continue to be relevant to debates over water management and governance in Canada. Central to the discussion is the question of how we can overcome the stalemates and policy vacuums that have characterized key aspects of Canadian water governance. Here, it is important to note that water is a complicated resource to manage.

By its very nature, it presents water managers with three complex issues that are very difficult to resolve: (1) dealing with competition between multiple users of water resources; (2) balancing the multiple scales at which water is managed; and (3) responding to the mismatch between geopolitical and administrative boundaries, on the one hand, and hydrological boundaries on the other.

There is no one right answer to any of the questions raised here. Rather, different communities will have to adopt models that reflect their unique hydrological, cultural, economic, and political circumstances. The questions are central to any discussion of water governance; but the answers offered are meant to initiate debate on how to best address these questions.

## How Should We Handle Multiple and Competing Uses of Water?

Water does not respect administrative boundaries. It is subject to competing demands for agriculture, energy production (e.g., hydropower), industry, urban water supply, recreation, tourism, and ecosystem services; and each sector has very different needs. Allocating responsibility for different aspects of water to different government agencies has been the norm in Canada, but this can lead to bureaucratic stalemates because no one agency has responsibility for the big picture. For example, nineteen federal departments currently have some responsibility for water management.

Effective management of multiple uses (some of which make competing demands on water resources) requires coordinating and balancing multiple goals. Provincial and federal governments have set precedents for ministries devoted to single-resource issues – forestry is an example. The complexity posed by water issues does not justify overly fragmented administrative allocation of responsibilities. Manitoba's new Department of Water Stewardship (created in 2003) is an example of how responsibility for water issues might be streamlined into one government agency. Effective jurisdictional consolidation of water management, either formally (through new agencies) or informally (e.g., through coordinating committees), is much needed in all Canadian jurisdictions, and particularly overdue at the federal level.

## At What Scale Should We Manage Water?

Water is a flow resource that constantly transgresses political boundaries. Unlike most of the resources central to our livelihoods and communities, water is constantly on the move. This means that water connects communities in ways that most other resources do not. Impacts of water use – both positive and negative – are felt far downstream, in other communities' jurisdictions. Yet, water is used locally – cheap to store but expensive to transport, water is most often used close to the point of abstraction.

Water, in other words, is subject to a continual tension between local and higher scales of governance. Distribution of governance to local levels makes particular sense in a country like Canada, in which different regions have dramatically different hydrologies, topographies, and political economies. Yet some water issues are best dealt with at regional or national scales, particularly where they affect public and environmental health. These issues of jurisdiction are exacerbated in the case of a decentralized federal state such as Canada.

Historically, the Canadian response has been to devolve water governance to intermediate scales – such as provinces or conservation authorities. This has been done in varying ways in different provinces, and it has been subject to much political influence, particularly in periods of strained federal-provincial relations.

Other jurisdictions have approached the question of scale in a far more systematic fashion than has Canada. In the European Union, for example, the allocation of water management responsibility is decided on the basis of "subsidiarity" (Bermann and Pistor 2004). This concept means, simply put, that decisions should be taken and policies implemented by the smallest (or lowest) competent authority. Subsidiarity is balanced, in the European approach to federalism, with "harmonization," by which legislation and policies are selectively standardized. In some cases, this means that member states make the decisions; in others, it means that the European Union makes them. In general, the balance has tilted toward harmonization; most national water policy in member states is now determined in Brussels, and water legislation is one of the most harmonized components of European environmental legislation.

There is no one "best answer" to the question of how to allocate responsibility for water between scales of government. However, the chances of water management being well managed are greatly increased if decisions about scale are made on the basis of sound principles – such as subsidiarity and harmonization – upon which there is political consensus. Canadians and Canadian policy makers need to engage in a dialogue about the sorts of principles that should guide our approach to water governance. If we do this, then we could match authority to jurisdictions far more appropriately than we do now, and we could also better ensure that key aspects of water management are comprehensively and effectively addressed.

## What Is the Role of Communities in Watershed-Based Management?

Internationally, there is consensus on the need to manage both land use and water resources on a watershed basis. Domestically, there is increasing recognition of the importance of watershed management to source protection and multi-barrier management – two of the key policies for drinking water protection discussed in *Eau Canada*. Many provinces have well-established watershed-based management policies. The next step could be, in some cases, to develop watershed-based management agencies. Ontario's Conservation Authorities are one example: the twenty-six conservation authorities, established after the passage of the Conservation Authorities Act in 1946, are internationally recognized as pioneers in integrated watershed management.

In other instances, provinces will choose to develop citizen-led watershed-based groups. Quebec's new citizen-run participatory watershed organizations and British Columbia's Fraser Basin Council are two good examples of community-directed watershed management groups. Alberta's new "Water for Life" policy, as another example, mandates the creation of watershed planning and stewardship councils at the community and watershed levels (Alberta Environment 2003). More watershed-based management is also likely for transboundary waters, as the International Joint Commission has been exploring a transition to more citizen participation on a transboundary watershed basis. For advocates of greater citizen participation in water management, this is a welcome development. However, as geographers Dan Shrubsole and Dianne Draper point out in Chapter 3, many of these organizations and communities are attempting to fill the void created by government retrenchment and disengagement from monitoring and regulation; as such, they are not always well networked, well resourced, or well informed. The capacity of local organizations and governments to fulfill the water management functions that have been delegated to them is a real concern.

## Should Canadians Have a Human Right to Clean Water?

No explicit right to water is expressed in the most relevant United Nations treaty.[1] South Africa and Uruguay are the only countries in the world to have granted its citizens a constitutional right to clean water.[2] Should Canada's Charter of Rights and Freedoms be amended to include a right to clean water? This question is particularly apt in light of recent media coverage of persistent failures to provide safe and adequate water supplies to a significant number of First Nations reserves in Canada.

Calls for a human right to water have been made in recent years in Canada (see, for example, Boyd 2004). Internationally, the Water Supply and Sanitation Collaborative Council's "Vision 21,"[3] the Joint Declaration of the Movements in Defense of Water,[4] the Group of Lisbon's Water Manifesto,[5] and the Declaration of the P7[6] at their fourth summit in 2000 have supported incorporating water into international law as a "third generation" right (Gleick 1998; Morgan 2004).

The argument for creating a human right to water generally rests on two facts: (1) water is "essential for life," and (2) many other human rights that are explicitly recognized in UN conventions are predicated upon an assumed availability of water (e.g., the right to food). These arguments appear to have some support in international law: in 2002 the UN Committee on Economic, Social and Cultural Rights[7] issued a comment asserting that every person has a right to "sufficient, safe, acceptable, physically accessible, and affordable water" (ECOSOC 2002; Hammer 2004).

Opponents have pointed out the difficulty of implementing a "right to water." A human right to water would need to be constrained. It could, for example, imply a basic volumetric allocation per person per day: "sufficient for everyone's need, but not for everyone's greed." The difficulties of implementing such a right are well understood in South Africa, where many citizens have been promised, but not yet provided with, the minimum level of twenty-five litres per person per day. This is the level that the government established as "sufficient" – a proposition that many citizens are contesting (McDonald and Ruiters 2005). Opponents argue that establishing a human right to water does not guarantee provision at a level acceptable to citizens and, thus, is futile. The possibility that the recognition of a human right to water might compromise national sovereignty (through being used as a legal justification for requiring one country to export water to another) has also been raised.

Proponents of a human right to water point to the impact that a lack of binding regulations on water supply has on First Nations communities. Canada's Commissioner of the Environment and Sustainable Development notes that "when it comes to the safety of drinking water, residents of First Nations communities do not benefit from a level of protection comparable to that of people who live off reserves. This is partly because there are no laws and regulations governing the provision of drinking water in First Nations communities, unlike other communities" (Commissioner of the Environment and Sustainable Development 2005). The Commissioner found that responsible federal departments (Indian Affairs and Health Canada) sought to ensure the safety of drinking water through informal policies, administrative guidelines, and funding arrangements with First Nations, but that these were not implemented consistently and did not address all of the elements found in regulatory regimes for water in Canada (Commissioner of the Environment and Sustainable Development 2005). Enshrining principles of ecological sustainability, such as clean water, in legislation will place a stronger onus on governments to enact policies that ensure sustainable water supplies for all Canadians. But a human right to water would place an even stronger obligation on all levels of government to ensure the availability of basic supplies of safe drinking water for all Canadians.

## Where to from Here?

Contributors to this volume have identified many priorities for improving water governance and management in Canada: a reinvigorated federal water policy, renewed funding for water research, efficient and fair water pricing, conservation-led management strategies, recognition of First Nations water rights, robust regulation, and more transparent and better information on Canada's water resources. This list is not exhaustive, and a different set of authors would no doubt have created a slightly different set of priorities. One point of consensus, however, is the need for good governance of water. Implicit in different definitions of good governance are assumptions about the legitimacy of different stakeholders and decision makers, about robust decision-making structures, and about accepted processes of decision making.

Thus, to some degree, good governance is dependent on how members of a society interpret the practice of deliberative democracy. In debating water management, we are debating the relationship between the environment, markets, states, and one another. Accordingly, there is no one "menu" of good governance options. Yet societies that have a coherent set of governance principles (something that is currently lacking in Canada, especially at the federal level) have a better chance of managing water wisely than do those that do not have one.

The contrast between Canada and other jurisdictions, such as the European Union, is striking. In 2000, member states of the European Union reached a historic agreement. After years of negotiations, the European Parliament passed the Water Framework Directive, a legally binding policy for water management and protection in Europe (European Commission 2000). The directive sets out a comprehensive water management strategy based on integrated watershed management, including transboundary watersheds (Kaika 2003).

The European approach to implementing the directive has been to balance subsidiarity (through the creation of watershed-based management organizations) with standardization of water quality norms and water management principles at the "federal" (or European) level. Harmonization has occurred in most areas of water management. The directive defines rigorous water quality standards that strictly limit emissions of harmful substances, which, in turn, are linked to environmental quality standards (maximum pollutant levels in the environment).[8] These standards are integrated within an overarching water quality management strategy that integrates multiple uses (such as water supply and industrial use) and multiple types of water supply (both ground and surface waters). The directive also mandates the incorporation of environmental externalities into water pricing in order to encourage demand management, and it requires substantive public participation in policy making and watershed management in order to increase transparency and compliance (Kaika and Page 2003). As such, the directive is the most ambitious water legislation in the world.

The EU initiatives contrast sharply with the situation in Canada. As the contributors to this volume have repeatedly pointed out, Canadian water legislation is a patchwork of provincial and federal laws, and it has significant inconsistencies and gaps in responsibility and oversight. The Canadian approach to water governance has produced a set of stalemates and policy gaps. Rather than selective harmonization and subsidiarity, we have produced fragmentation and an ill-coordinated downshifting of responsibilities, leaving key areas in a policy vacuum. This is, of course, a problem that is not confined to the water sector; as political scientist Kathryn Harrison (1996) has noted in her study of the impact of federalism on Canadian environmental policy, "passing the buck" between federal and provincial governments is characteristic of Canadian resource management. But it is particularly disturbing in the case of water, given that this substance is essential for human and environmental health.

Analysts of governance argue that the failure to articulate a vision is often part of the reason for poor management. "Getting our house in order" means engaging in a discussion of what we value and how we wish to achieve our collective goals. Justice O'Connor's comments in his *Report of the Walkerton Inquiry* exemplify this vision. In making recommendations on the role of municipal governments in water supply management argued that public safety was paramount and identified four principles: (1) public accountability for decisions relating to the water system, (2) effective exercise of owner's oversight responsibilities, (3) competence and effectiveness in the management and operation of the system, and (4) full transparency in decision making (O'Connor 2002b, 277). The European Union's Water Framework Directive contains a different set of principles, beginning with the statement that "water is not a commercial product like any other ... but a heritage which must be protected."[9] The message is clear: good management and wise governance of water begins with a set of principles and a vision for the future.

What is Canada's vision? The contributors to *Eau Canada* have pointed out not only where we are lacking but also where we might be heading. To begin with, we would do well to revisit the 1987 Federal Water Policy, which called for "clean, safe, and secure water for people and ecosystems." This policy has yet to be implemented. As Canada's Senate recently reminded us, failure to work toward a vision for Canada's water is "unacceptable" (Senate 2005). This failure can only have dire consequences for what is arguably the most important resource of our time. A new alliance between local communities, water managers, and all levels of government is urgently needed. We hope that this book will inspire Canadians to act together for the future of our water.

## NOTES

1 The International Covenant on Economic, Social and Cultural Rights, one of the keystones of international human rights law. None of the United Nations conventions on human rights (except the Convention on the Rights of the Child) explicitly recognizes the right to water (Morgan 2004).

2 The Constitution of the Republic of South Africa guarantees the right of citizens to have access to sufficient water (Act 108 of 1996, section 7[2]). Many other countries also mention this right in specific legislation.

3 The Water Supply and Sanitation Collaborative Council, located in Geneva, is a non-profit organization that acts as an "international policy think tank" on water management.

4 The Joint Declaration of the Movements in Defense of Water was signed by numerous campaigning NGOs during the Fourth World Water Forum in Mexico City, in March 2006. See http://www.blueplanetproject.net/Movement/Declaration.html.
5 The Group of Lisbon is a group comprised of distinguished scholars from around the world who analyze globalization and call for new types of economic governance. See Petrella (2001).
6 The P7 (now P8) annual conference was convened for the first time in June 1997 by the Green Group in the European Parliament, as an alternative Summit to the G7 (now G8). Representatives from the world's poorest countries attend the conferences, which focus on the structural causes of and solutions to poverty.
7 The Committee on Economic, Social and Cultural Rights (CESCR) is the body of independent experts that monitors implementation of the International Covenant on Economic, Social and Cultural Rights by its States parties.
8 In technical terms, two existing pollution control strategies were amalgamated: environmental quality standards (EQVs) and emission limit values (ELVs).
9 "Directive 2000/60/EC of the European Parliament and of the Council establishing a framework for the Community action in the field of water policy."

## REFERENCES

Bermann, G., and Pistor, K., eds. 2004. *Law and Governance in an Enlarged Europe.* Portland: Hart.

Boyd, D. 2004. *Sustainability within a Generation: A New Vision for Canada* Vancouver: David Suzuki Foundation.

Commissioner of the Environment and Sustainable Development. 2005. *Report of the Commissioner of the Environment and Sustainable Development.* Ottawa: Office of the Auditor General.

ECOSOC. 2002. *General Comment 15.* Geneva: United Nations Committee on Economic, Social and Cultural Rights.

Environment Canada. 1987. *Federal Water Policy.* Ottawa: Environment Canada.

European Commission. 2000. Directive 2000/60/EC of the European Parliament and the Council of 23 October 2000, establishing a framework for community action in the field of water policy. *Official Journal* 22 December L 327/1. European Commission: Brussels.

Gleick, P. 1998. The Human Right to Water. *Water Policy* 1: 487-503.

Hammer, L. 2004. Indigenous Peoples as a Catalyst for Applying the Human Right to Water. *International Journal on Minority and Group Rights* 10: 131-61.

Harrison K. 1996. *Passing the Buck: Federalism and Canadian Environmental Policy.* Vancouver: UBC Press.

Kaika, M. 2003. The Water Framework Directive: A New Directive for a Changing Social, Political and Economic European Framework. *European Planning Studies* 11 (3): 299-316.

Kaika, M., and B. Page. 2003. The EU Water Framework Directive. Part 1: European Policy-Making and the Changing Topography of Lobbying. *European Environment* 13 (6): 314-27.

McDonald, D., and C. Ruiters. 2005. *Age of Commodity: Water Privatization in Southern Africa.* London: Earthscan.

Morgan, B. 2004. The Regulatory Face of the Human Right to Water. *Journal of Water Law* 15 (5): 179-86.

O'Connor, D. 2002a. *The Report of the Walkerton Inquiry.* Part 1: The Events of May 2000 and Related Issues. Toronto: Ontario Ministry of the Attorney General.

—. 2002b. *The Report of the Walkerton Inquiry.* Part 2: A Strategy for Safe Drinking Water. Toronto: Ontario Ministry of the Attorney General.

Page, B., and M. Kaika. 2003. The EU Water Framework Directive. Part 2: Policy Innovation and the Shifting Choreography of Governance. *European Environment* 13 (6): 328-43. Toronto: Ontario Ministry of the Attorney General.

Petrella, R. 2001. *The Water Manifesto.* London: Zed Books.

Senate. 2005. *Water in the West, under Pressure.* Fourth Interim Report of the Standing Senate Committee on Energy, the Environment and Natural Resources. Ottawa: Senate of Canada.

APPENDIX 1

# A Survey of Water Governance Legislation and Policies in the Provinces and Territories

*Carey Hill, Kathryn Furlong, Karen Bakker, and Alice Cohen*

This appendix provides a comparative summary of water governance mechanisms for Canada's ten provinces and three territories, focusing exclusively on the present state of *water management-related laws, regulations, and plans* for consumptive uses (i.e., removal from the environment) of water. The survey provides a "snapshot" of current formal legislation and policies pertaining to consumption of water, excluding instream uses such as cooling for hydroelectricity or fisheries.[1] Readers interested in the links between instream and consumptive uses and trends and practices in water governance are encouraged to refer to the chapters in the first part of this book.

Water is a local resource, and management approaches vary (sometimes significantly) among communities, despite shared legislative and policy frameworks. Many of these variations will not be evident in our survey of water governance. Moreover, information on many of the water management practices of communities will not be available from our survey. For example, a survey of provincial legislation pertaining to municipal drinking water might note the near absence of references to water conservation, water efficiency, or creation of incentives to reduce domestic water use. However, in practice, many municipalities have put in place innovative programs to encourage citizens to reduce their water consumption. The "Water for Tomorrow" program of the Regional Municipality of York is one longstanding example.[2]

It is difficult to capture all the nuances and details of water governance mechanisms in each of the thirteen provinces and territories within the limited scope of this survey. This challenge results, in part, from the often discretionary, non-legalistic approach to environmental policy in Canada.[3] The considerable variation among jurisdictions outlined below is illustrative of this discretionary approach. Thus, our survey, which draws largely on secondary literature, provincial legislation, and supporting documentation, as well as other written documents, is not a fully comprehensive treatment of all aspects of water governance.

A theme that emerges from this survey is the high degree of interprovincial variation with respect to standards, norms, and laws. In the three territories, where the federal government has had more responsibility for water resources, this is less so.[4] One of the few areas where legislation is broadly similar across most provinces and territories pertains to the prohibition of bulk water exports and out-of-province diversions.

Five issues are considered here: water legislation, drinking water protection, water rights, water exports, and accountability and transparency in water governance.

## Water Legislation

Provinces play an important – and constitutionally specified – role in creating water legislation and water policies relevant to questions of water supply and water resource management (Table A.1). Provinces are generally responsible for water as a natural resource, as well as for water governance. This includes licensing, environmental protection for waters under provincial jurisdiction, and ensuring water potability. In Canada, ensuring safe drinking water is the responsibility of provincial and territorial governments.[5] Exceptions include federal lands, boundary and transboundary waters, and inland fisheries, which fall within federal jurisdiction.[6] This survey focuses on water governance at the provincial/territorial level.

Within each province, there are often many provincial departments and agencies involved in water policy. This can complicate attempts by the public to ascertain where and with whom particular responsibilities lie. As such, it is important to highlight the steps that provinces have taken toward integration.

In Ontario, Quebec, and British Columbia, whole sections of their respective environment ministries are devoted to water. Ontario's conservation authorities play an important integrative role with municipalities and the province.[7] Quebec has implemented a collaborative approach within prioritized watershed organizations to develop master plans for their watersheds.[8] In Saskatchewan, three existing water-related governmental agencies were merged to better coordinate source protection province-wide. The Saskatchewan Watershed Authority is a Crown corporation that reports to the provincial minister of the environment and has responsibility for activities, legislation, and programs that were formerly housed within SaskWater, Saskatchewan Environment, and the Saskatchewan Wetland Conservation Corporation. The Manitoba Ministry of Water Stewardship is the country's only example of an entire government ministry being dedicated specifically to water.

With respect to drinking water, several provinces provide some coordinating functions among the departments involved in this policy area. For example, Newfoundland and Labrador has developed a committee of deputy ministers, including a technical working group, that coordinates drinking water protection policy among four departments. Saskatchewan also has a steering committee composed of deputy ministers for its Safe Drinking Water Strategy. In Nova Scotia, the Department of the Environment and Labour has been designated as the lead agency and coordinating ministry for water, and it regularly convenes a committee of senior managers to deal with water issues. Quebec has also developed an interministerial committee for the implementation of the Quebec Water Policy. This committee has representation from a range of diverse government departments, from sustainable development to education, sport, and leisure. Despite attempts at coordination, however, variations in water governance between water jurisdictions persist, as does fragmentation within jurisdictions. Table A.1 identifies the most relevant water legislation in each jurisdiction, revealing the multiplicity and diversity of water-related responsibilities.

As noted above, this appendix focuses on provincial and territorial responsibility for water resources management. Responsibility for the environment was not specified in the Constitution Act (1867), resulting in considerable overlap and some uncertainty regarding jurisdictional authority between the provinces and the federal government. In light of this, and because of the need to contextualize the frameworks under which the provinces are each compelled to act, it is necessary to briefly address the federal role.

Arguably, the most important pieces of federal legislation with respect to water are the Canada Water Act (1970), the Fisheries Act (1970), the Canadian Environmental Protection Act (1988), and the Great Lakes Water Quality Agreement (1972) under the International Boundary Waters Treaty. The Canada Water Act authorizes federal-provincial joint committees on water management, enables the establishment of federal water quality management programs for interjurisdictional waters, and includes provisions for regulating discharge of waste into some protected areas. While this attempt at water management relied on provincial involvement and cooperation, it

Table A.1

**Most relevant water legislation by provincial and territorial jurisdiction**

| | |
|---|---|
| NL | Water Resources Act (2003, 2004, 2005) |
| | Municipalities Act (1999) |
| | Public Health Act (1996) |
| | Environmental Protection Act (2002, 2005) |
| | Labrador Inuit Land Claims Agreement Act (2005)[1] |
| NS | Environment Act (1994-95, 1998, 2001, 2004) |
| | Water and Wastewater Facilities and Public Drinking Water Supplies Regulations (2005) |
| | Municipal Government Act (1998, 2001, 2002, 2004) |
| | Water Resources Protection Act (2000) |
| NB | Clean Water Act (1989, 1990, 1994, 2000, 2001, 2002) |
| | Municipalities Act (1973, 1981, 1995) |
| | Public Utilities Act (1973) |
| | Health Act (1988, 2005) |
| | Clean Environment Act (1982) |
| PE | Drinking Water and Wastewater Facility Operating Regulations (2004) |
| | Sewage Disposal Systems Regulation (2004) |
| | Water and Sewerage Act (1988, 2003) |
| | Environmental Protection Act (1988, 2005) |
| | Water Wells Act (1988, 2004) |
| QC | Watercourses Act (1964, 1979, 1994, 1999, 2003) |
| | Environment Quality Act (2005)[2] |
| | Public Health Act (2001) |
| | Water Resources Preservation Act (2001) |
| ON | Ontario Clean Water Act (proposed in 2005, not yet enacted) |
| | Ontario Water Resources Act (regs. 1993, 1998, 2000, 2001) |
| | Municipal Water and Sewage Transfer Act (1997) |
| | Safe Drinking Water Act (2002) |
| | Sustainable Water and Sewage Systems Act (2002) |
| | Annex Agreement to the Great Lakes (2005) |
| | Nutrient Management Act (2002) |
| | Water Transfer Control Act[3] |
| | Drainage Act (1990) |
| | Lakes and Rivers Improvement Act (1990) |
| | Environmental Bill of Rights (1993) |
| MB | Drinking Water Safety Act (2002) |
| | Water Protection Act (2005) |
| | Public Health Act (2006) |
| | – Protection of Water Sources Regulation (1988) |
| | – Sanitary Areas Regulation (1988), Water Supplies Regulation (1988) |
| | – Water Works, Sewerage, and Sewage Disposal Regulation (1988) |
| | Water Rights Act (1987, 2005) |

▶

◀ Table A.1

| | |
|---|---|
| | Water Supply Commissions Act (2005) |
| | Water Resources Conservation and Protection Act (2000) |
| | Water and Wastewater Facility Operators Regulation (2003) |
| | Ground Water and Water Well Act (2001, 2003) |
| SK | Environmental Management and Protection Act (2002) |
| | Water Regulations (2002) |
| | Conservation and Development Act (1978, 2005)[4] |
| | Saskatchewan Watershed Authority Act (2005) |
| | Groundwater Regulation 172/66 (1966, 1967, 1968, 1971) |
| | Rural Municipalities Act (1989) |
| | Saskatchewan Water Corporation Act (2002, 2004, 2005) |
| | Public Health Act (1994, 2005)[4] |
| | Health Hazard Regulations (2002) |
| AB | Water Act (2000) |
| | Public Health Act (2000, 2005)[4] |
| | Public Utilities Board Act (2000) |
| | Environmental Protection and Enhancement Act (2000, reg. 1993) |
| | Municipal Government Act (1994, 1995, 2000, 2003) |
| | Standards and Guidelines for Municipal Waterworks, Wastewater and Storm Drainage Systems (2006) |
| BC | Drinking Water Protection Act (2001) |
| | Drinking Water Protection Regulation (2003) |
| | Water Act (1996) |
| | – Water Regulation (1988) |
| | – British Columbia Dam Safety Regulation (2000) |
| | – Ground Water Protection Regulation (2004) |
| | Water Protection Act (1996) |
| | Environmental Management Act (2003) |
| | – Code of Practice for the Discharge of Produced Water from Coalbed Gas Operations (2005) |
| | Water Utility Act (1996) |
| | Environmental Assessment Act (2002) |
| | – Reviewable Projects Regulation (2002) |
| | Fish Protection Act (1997) |
| | – Sensitive Streams Designation and Licensing Regulation (2000) |
| | Dike Maintenance Act (1996) |
| | Drainage, Ditch and Dike Act (1996) |
| YT | Yukon Waters Act and Regulation (2003) |
| | Public Health and Safety Act (2002) |
| | – Bulk Delivery of Drinking Water Regulation (proposed) |
| | – Public Drinking Water Systems Regulation (proposed) |
| | Public Utilities Act (2002) |
| | Environment Act (1991, 2002) |
| | Umbrella Final Agreement – Chapter 14, Water Management |

▶

◀ Table A.1

| | |
|---|---|
| NT | Water Resources Agreements Act (1988, 1995) |
| | Northwest Territories Waters Act (1992 – federal) |
| | Arctic Waters Pollution Prevention Act (1985, 2003)[4] |
| | Public Health Act (1990, 2004) |
| | Public Water Supply Regulations (1990, 2004) – Pursuant to Public Health Act |
| | Public Utilities Act (1988, 1993, 1995, 1998, 1999, 2004) |
| | Environmental Protection Act (1988, 1991, 1998) |
| | Environmental Rights Act (1988, 1999, 2000) |
| NU | Nunavut Waters and Nunavut Surface Rights Tribunal Act (2002) |
| | Nunavut Power Utilities Act (1999, formerly NWT Power Corporation Act) |
| | Public Utilities Act (1999, formerly the NWT Public Utilities Act) |

NOTES:

1 The Labrador Inuit Land Claims Agreement Act (2005) takes precedence over the Water Resources Act. Inuit have the right to use water for personal and domestic purposes throughout the settlement area. New commercial or industrial developers on Labrador Inuit Lands must acquire a water use permit from the government of Newfoundland and Labrador, which may only be issued if also approved by the Nunatsiavut Government.

2 The regulation pursuant to the Environment Quality Act is the Regulation respecting the quality of drinking water. Additional regulations include the Regulation respecting groundwater catchment, Agricultural operations regulation, and the Regulation respecting waste water disposal systems for isolated dwellings. Readers may also be interested in the Regulation respecting bottled water (as part of the 1996 Food Products Act).

3 This act is not yet in force. It comes into force on a day to be named by proclamation of the lieutenant governor.

4 Several amendments not noted here occurred between its enactment and the most recent amendment.

neglected to consider whether the provinces would be on-side.[9] The Canada Water Act has had a slow and stalled implementation with only part 3, the section on phosphate regulation of detergents, being fully implemented.[10] Reasons identified for its lack of implementation include provincial resistance, the "timidity" of the act itself, public indifference, and the contrasting approach of, and focus on, the Fisheries Act.[11]

The Canadian Environmental Protection Act (CEPA) pertains to water and water source pollution, focusing on the regulation of toxic substances. The Fisheries Act is highly relevant for water resources management as, by definition, protection of fish means protection of the waters in which they exist. For example, the Fisheries Act affords some source water protection, with section 36 prohibiting the deposit of "deleterious substances" into or near waters frequented by fish. Finally, the Great Lakes Water Quality Agreement (GLWQA), which now has provisions to address toxic substances, had an impact on reducing phosphates in the Great Lakes through detergent bans and sewage treatment requirements.[12] Table A.2 goes beyond the most prominent pieces of legislation to show the broad range of federal legislation and some of the programs with respect to water governance in Canada.

The following is a list of the federal government departments that have an impact on or are related to water resources management, but not in a primary capacity:

Office of the Auditor General of Canada
Finance Canada
Industry Canada
Infrastructure Canada
National Defence Canada
Policy Research Initiative
Privy Council Office
Public Works and Government Services
Treasury Board Secretariat

For example, the Office of the Auditor General includes the Commissioner of the Environment and Sustainable Development. Recently, the Commissioner released a series of reports, including two on the federal role in regulating drinking water. Similarly, the Policy Research Initiative conducts research on water policy, which began in 2003 and is scheduled to wind down in 2006. The research on water is not a direct mandate of their department but falls within its broader objectives.

Currently, the federal government's role in water management and research has been described as in "retreat."[13] Two scientific witnesses to the Senate's Standing Committee on Energy, the Environment and Natural Resources (2005) reported significant decreases in federal government monitoring of water quantity and groundwater records.[14] The government made attempts to overcome interdepartmental fragmentation by convening an Interdepartmental Water Assistant Deputy Minister's Committee (IWAC), but that committee has not met for at least two years. Thus, while efforts to coordinate across departments have been attempted, the jurisdictional fragmentation of water policy is clearly a challenge at all levels of government.

Table A.2

**Most relevant federal water legislation and/or programs and policies by federal government departments with direct roles in water management**

| Department | Legislation, programs and policies |
|---|---|
| Agriculture and Agri-Food Canada | Agricultural Policy Framework (2002) |
| Canadian Environmental Assessment Agency | Canadian Environmental Assessment Act (1992) |
| Canadian International Development Agency | CIDA's policy for environmental sustainability |
| Environment Canada | Canada Water Act (1970)<br>Canadian Environmental Protection Act (1988, 1999)<br>Canadian Wildlife Act (1985, 1994)<br>Department of the Environment Act (1978, 1985)<br>Federal Water Policy (1987)<br>Federal Wetlands Policy (1991)<br>International Rivers Improvement Act (1970) |
| Fisheries and Oceans | Fisheries Act (1868, 1970, 1985) |
| Foreign Affairs and International Trade | International Boundary Waters Treaty Act (1909)<br>– Great Lakes Water Quality Agreement (1972, 1978, 1987, 1999)<br>– Great Lakes Charter (1985)<br>– Various lakewide management plans (LaMPs) |
| Health Canada | Canadian Drinking Water Guidelines (1968, 1972, 1978, 1986, 1996, 2006)[1] |
| Indian Affairs and Northern Development | Northwest Territories Waters Act (1992)<br>Yukon Waters Act (2002) |
| Natural Resources Canada | Strategic Policy Branch: information and policy advice; scientific and technological innovation to minimize impacts of natural resource sector on aquatic ecosystems |
| Parks Canada | Dominion Water Power Act (1985) |
| Transport Canada | Canada Shipping Act (1985)<br>Navigable Waters Protection Act (1985)<br>Arctic Waters Pollution Prevention Act (1985) |

NOTE:

1 The guidelines are updated on an ongoing basis. The Federal-Provincial Territorial Committee on Drinking Water meets twice annually to discuss the guidelines, once in Ottawa, and once in one of the provinces.

## Drinking Water

In Canada, ensuring safe drinking water is primarily the responsibility of provincial and territorial governments. Notwithstanding the provincial regulatory responsibility, we recognize that municipalities generally retain responsibility for the implementation of provincial drinking water policies. Our analysis, however, is concerned with the legislative frameworks and directives to which those responsible for water resources (municipal supply, source protection, etc.) are subject.

Drinking water protection legislation exemplifies well the differentiation in water management approaches among Canadian jurisdictions (see Table A.3). National guidelines exist but only two provinces, Alberta and Nova Scotia, have fully adopted them. Requirements for monitoring of water contaminants differ vastly from province to province. In 1972, Alberta was the first province to adopt the Canadian Drinking Water Standards, making these standards binding in its Municipal Plant Regulation by 1978. In 1984, Quebec was the first to introduce binding standards for drinking water at the provincial level. Initially, forty-two parameters were set for testing, but, since then, the province has increased this number, so that some of its standards go beyond those of the Canadian guidelines. Ontario has drinking water quality standards for 161 parameters and requires monitoring for fifty-seven organics, thirteen inorganics, and three microbiological contaminants.[15] Alberta also exceeds the Canadian guideline for turbidity. In contrast, in New Brunswick, the minister of health has discretion over the sampling plan, and there are no set requirements for numbers or types of contaminants. British Columbia requires the monitoring of bacteriological contaminants only and leaves additional contaminants to the discretion of the Drinking Water Officer. Different still is Newfoundland and Labrador, which is the only province in Canada where all water quality monitoring is performed by the province.[16]

Is this variation in standards a good thing? Some experts argue that variation in standards is appropriate. Others argue, however, that flexibility in regulation may lead to laxity in monitoring and enforcement, which are important aspects of the multi-barrier approach.

Variation in standards can also arise because water issues vary among regions. For example, Prince Edward Island, unlike other provinces or territories, relies entirely on groundwater for its drinking water. This provides an interesting example of the potential benefits of fragmentation in serving the

Table A.3

**Drinking water protection by provincial and territorial jurisdiction**

| | Operator certification required | Treatment required[1] | Monitoring for contaminants: number | Water purveyors required to report annually to consumers |
|---|---|---|---|---|
| NL | No | Disinfection required. | 28 (3 microbiological and 25 physical and chemical). | Yes |
| NS | Yes | Disinfection required.<br>Filtration required | Fully adopts Guidelines for Canadian Drinking Water Quality. | No |
| NB | Yes | Policy is being developed. | According to sampling plan agreed upon by Minister of Health. | No |
| PE | Yes | While disinfection is not required by the province, in practice, all municipalities disinfect.[2]<br>Filtration is unnecessary since all drinking water is groundwater. | 50 (20 organic, 14 inorganic, and 16 other). | No |
| QC | Yes | Disinfection required.<br>Filtration required when turbidity exceeds 1 NTU. | 77 (4 microbiological, 17 inorganic, 41 organic, turbidity and PH). | No |
| ON | Yes | Disinfection required.<br>Filtration required. | 73 (3 microbiological, 70 chemical, no radiological). | Yes, reports must be made available free of charge to those who request them. |
| MB | Yes | Disinfection required. | To approval of Medical Health Officer. | No |
| SK | Yes | Disinfection required for groundwater-based supplies.<br>Filtration required for surface, mixed, or GUDI-based[3] supplies. | 65 (3 microbiological, turbidity and 61 chemical, radiological and other parameters). | Yes, a water quality notice must be sent annually. |

| | | | | |
|---|---|---|---|---|
| AB | Yes | Disinfection required.<br>Filtration required for surface and GUDI supplies.<br>Disinfection required for groundwater sources. | Fully adopts the Guidelines for Canadian Drinking Water Quality.[4] | Municipalities must submit monthly and annual reports to the department. These are public documents available for the public to review. |
| BC | Yes | Disinfection required. | 3 bacteriological with additional at discretion of Drinking Water Officer. | No[5] |
| YT | Proposed[6] | Under proposed regulation, disinfection and filtration required.[7] | 32 (30 chemical and physical parameters plus 2 bacteriological).[8] | No |
| NT | No | Disinfection required. | 32 (1 bacteriological, 31 physical and chemical). | No |
| NU | Yes | Disinfection required. | 24 (1 bacteriological, 23 physical and chemical). | No |

1 Unless specified, this section refers to surface water sources.

2 Disinfection is not required but water utilities must show that their water supply is free from bacteria over any six-month period or they must disinfect. As a result, all municipalities disinfect.

3 Groundwater under direct influence of surface water (GUDI).

4 The exception to this is naturally occurring fluoride, which is allowed to exceed 2.4 mg/L. It should also be noted that Alberta's turbidity requirements exceed those stated in the Guidelines.

5 A provincial annual report is required but each water utility is not required to report on water quality. It should be noted that requiring annual reporting appears to be within the discretion provided by the legislation.

6 Under the Bulk Delivery of Drinking Water regulation, operator certification is required with other regulations currently undergoing revision.

7 Trucked distribution systems require chlorination. The proposed Public Drinking Water Systems Regulation (February 22, 2004; currently under consultation) notes that treatment shall consist of filtration and disinfection "or other treatment capable of producing safe drinking water" and that the owner will "verify that the method of treatment consistently produces safe drinking water."

8 Yukon Territory is currently in the process of revising its drinking water regulations under the Public Health and Safety Act. Consultations are ongoing. The proposed regulation defines safe drinking water as meeting the health-related criteria set out on the Guidelines for Canadian Drinking Water Quality.

different needs of each province. Another reason for variation might be the types of expertise used to manage water issues and the degree of authority officials have to establish standards or requirements on an "as-needed" basis. Different water issues require different types of expertise. For example, in Newfoundland and Labrador, responsibilities are apportioned according to appropriate expertise. That is, health aspects are addressed by health experts, construction aspects are the purview of engineers, and hydrological and environmental aspects are overseen by hydrologists and water resources engineers. Similarly, in British Columbia, the Drinking Water Protection Act is administered by the Ministry of Health (through regional health authorities), while the Ministry of the Environment plays a supporting role in source water protection.

Provincial responsibility for water, particularly drinking water and water rights, may be reasonable from the viewpoint of subsidiarity. The devolution of drinking water management to the local level might work well, if supported by a well-designed and overarching coordinating mechanism and if local purveyors have sufficient resources to invest. In the case of drinking water, the Federal-Provincial-Territorial Committee on Drinking Water (CDW) carries out an important element of this coordinating role by establishing the Guidelines for Canadian Drinking Water Quality. Coordination in an environment of devolved responsibility can, however, be time-consuming and relatively slow. For example, the Commissioner of the Environment and Sustainable Development noted in her 2005 review of the committee's work that a "significant backlog" existed (of approximately ten years) in updating the guidelines, despite Health Canada's recommendation that a guideline should take two to three years to develop or review.[17] The commissioner found that many known contaminants are not even listed because of the length of time taken to update the guidelines.

Disinfection and filtration are vital aspects of the multi-barrier approach to drinking water protection. New Brunswick is currently developing a policy on disinfection and filtration of drinking water. As Table A.3 illustrates, fewer than half the jurisdictions have a policy on filtration of water supplies. In most cases, filtration that removes harmful contaminants and micro-organisms is required in Ontario, Alberta, Saskatchewan, Nova Scotia, and Quebec. In contrast, the US Safe Drinking Water Act has required the filtration of all surface waters, with few exceptions, since 1989.[18] After the Walkerton water tragedy, operator certification was required in most provinces, but it still remains optional in Newfoundland and Labrador, as well as in the

Northwest Territories. It is worth noting that, since Walkerton, Newfoundland and Labrador has significantly increased its number of certified operators by providing training rather than focusing on examinations.[19] It is also worth noting that, even though provinces may require certification, implementation of regulations often takes many years.[20]

The Guidelines for Canadian Drinking Water Quality, referred to in Table A.3, are developed by consensus of the provinces via the Federal-Provincial-Territorial Committee on Drinking Water (Mouldey 1994). The guidelines are neither binding nor enforceable. Consequently, drinking water standards vary across the country. Provinces can elect to make these guidelines binding within their jurisdictions by adopting them as part of broader legislative frameworks or license and approval agreements. Only Nova Scotia[21] and Alberta have fully adopted them.[22] Indeed, this differentiation demonstrates considerable fragmentation where one might reasonably expect harmonization, despite federal guidelines developed in concert with the provinces. However, this lack of interprovincial consensus may reflect the compromises required to establish uniform guidelines for such a geographically diverse resource as water.

It is also interesting to note the fragmentation that exists regarding policy for different types of waters. In Canada, far more attention is paid to surface water than groundwater. Existing regulations for groundwater seem to be far more about regulating specific technical standards, types of materials, and means of construction.[23] Alberta and Quebec are important exceptions to this apparent trend. Alberta has adopted the requirement for groundwater disinfection to achieve at least 4-log reduction of viruses recommended by the Canadian drinking water guidelines. Moreover, in Quebec, any groundwater with fecal contamination must achieve at least 4-log reduction of viruses.

Yet another example of fragmentation with respect to drinking water protection concerns the first aspect of the multi-barrier approach, source protection. While most jurisdictions do not have comprehensive watershed regulations, important examples of watershed management exist. Notably, in late 2005, Ontario introduced the Clean Water Act that, if passed, will require comprehensive management plans. Nova Scotia and New Brunswick currently regulate specific activities in entire watersheds while Newfoundland and Labrador and Prince Edward Island have designated buffer zones. As mentioned, Quebec has implemented a collaborative approach within prioritized watershed organizations that develop master plans for their watersheds.

Saskatchewan is also moving toward a commitment to protect a percentage of its lands and waters. While most provinces lack comprehensive regulations, there is evidence of source water protection via a diversity and range of approaches.

A second and related theme, to which this book has made reference throughout, is the need for stronger regulatory frameworks with respect to water governance. After Walkerton, environmental organizations and some politicians pointed to the strong regulatory frameworks for drinking water exhibited by the US Safe Drinking Water Act (SDWA). Since the Walkerton tragedy, Ontario has passed its own Safe Drinking Water Act with a strong enforcement and reporting regime.[24] The US SDWA provides nationally binding standards, which are enforced by both the federal and state levels.[25] Likewise, in the European Union, drinking water quality standards are enforceable; the European Commission can prosecute national governments for failing to comply with water legislation.

In addition to Ontario, the provinces of Alberta, Saskatchewan, and Nova Scotia have adopted binding standards and compliance programs. For Saskatchewan and Nova Scotia, these are currently being phased in for existing waterworks.[26] Alberta's abatement program includes compliance inspections, with plants inspected on a regular basis. The operators are given assistance and direction to operate the plant. Annual inspections for drinking water are undertaken in Ontario, Saskatchewan, and Nova Scotia. Ontario reserves the right to inspect without notice, while Saskatchewan Environment's internal waterworks inspection protocol calls for an unannounced inspection once every three years. Saskatchewan Environment also performs annual inspections on all regulated waterworks.[27] Nova Scotia currently audits every municipal system twice annually, having introduced an annual audit of systems in 2003, as part of its drinking water strategy.

## Water Rights

Variations among provinces and territories are further demonstrated by examining water property rights provisions. In Chapter 11 of this volume, Christensen and Lintner identify five approaches to water rights in Canada: riparian rights, prior allocation, civil code, public authority management, and Aboriginal rights. Riparian rights, as exemplified by Ontario and the Atlantic provinces, entitle landowners whose land borders on water sources to enjoy certain non-transferable water rights. Riparian rights also give

Table A.4

**Water property rights by provincial and territorial jurisdiction**

| | Approach to water rights | Can water rights be transferred? | Prioritization of water uses (from greatest to least) |
|---|---|---|---|
| NL | Riparian rights[1] | No | Domestic, municipal, agricultural, commercial, institutional, industrial, water and thermal power generations, other. |
| NS | Riparian rights | No | Not specified. |
| NB | Riparian rights | No | Not specified. |
| PE | Riparian rights | No | Domestic, commercial, industrial, agricultural/ irrigation. |
| QC | Civil code "common to all" | No | Not specified. |
| ON | Riparian rights | No | Not specified. |
| MB | Prior allocation | Currently prohibited but under discussion | Domestic, municipal, agricultural, industrial, irrigation, other. |
| SK | Prior allocation | Currently prohibited but under discussion | Not specified. |
| AB | Prior allocation | Yes | Domestic, agricultural then in order of requests received. |
| BC | Prior allocation | Yes | First come, first serve.[2] |
| YT | Public authority management | Yes | First come, first serve. |
| NT | Public authority management | Yes | First come, first serve. |
| NU | Public authority management, Aboriginal rights | Yes | Any existing use by Inuit has priority over other licensed uses. |

NOTES:

1 Riparian rights are historically recognized. However, beginning with the Crown Lands Act, very few Crown grants included riparian rights as the Crown reserved a minimum of 33 feet (now 15 metres). The reservation had the effect of preventing riparian ownership. Therefore, most bodies of water are 100 percent owned by the Crown. There are very few cases of riparian rights. Where these do exist, they are limited by Newfoundland's Water Resources Act.

2 The exception to this is when licenses with the same priority date exist on the same stream (in which case precedence is in the following order: domestic, waterworks, mineral trading, irrigation, mining, industrial, power, hydraulic, storage, conservation, conveying, land improvement). These occasions are rare.

landowners the right to extract groundwater from wells on their property. Public authority management is practised in the territories with local water boards deciding on the allocation of licenses that can be transferred to other users. Aboriginal rights are guaranteed in the 1982 Constitution Act as existing rights and within legislation passed in Nunavut as having priority over other uses. Quebec water is designated in the civil code as a resource "common to all," whereby the government holds responsibility for allocation, regulation, and establishing priority use in the public interest. By contrast, the western provinces use a prior allocation approach predicated on seniority; licensees acquire exclusive right to use the water from the date of the license. In Manitoba and Saskatchewan, such licenses cannot be transferred. While Alberta has the most mature water markets (see Chapter 10), water rights transfers are also allowed in British Columbia. Currently, water rights are granted for surface water but not for groundwater in BC.

## Water Exports and Transfers

One of the few areas of convergence across provinces with respect to water governance appears to be "laws, regulations or policies in place to prohibit bulk removal of freshwater" (Quinn et al. 2003). In 2002, the federal government made amendments to the International Boundary Waters Treaty Act, prohibiting bulk removal of boundary waters. This is very significant since this legislation is highly relevant to ongoing debates over diversions from the Great Lakes/St. Lawrence Basin (Bill C-15). With the exception of this act (which pertains to boundary waters), there is no federal law banning export of Canadian water from non-boundary waters, although at least three private members' bills have been introduced in Parliament on the issue.[28] Rather, given the constitutional allocation of responsibility for natural resources to the provinces, the federal government sought to establish a cross-Canada consensus on water exports. Currently, all provinces (with the exception of New Brunswick) have passed laws banning interbasin water diversions, and also removal of water from the province, mostly due to environmental concerns. Existing interbasin transfers remain permitted, having been grandfathered.

## Accountability and Transparency

Accountability and transparency in water governance can be assessed by examining the following: enforcement of water laws and associated penalties for contraventions, initiatives to achieve regulatory compliance, public access

Table A.5

**Water transfers, diversions, and exports by provincial and territorial jurisdiction**

| | Policy on large-scale transfers outside the province[1] | Policies on water diversions and exports and on interbasin transfers |
|---|---|---|
| NL | Prohibited under the Water Resources Act. | *Prohibited:* Removals of water from the province. Exceptions include water contained in containers of not more than 30 litres in volume; water used to power a vehicle or to transport food or industrial products; and, upon approval of the minister, water removed for a non-commercial (i.e., safety or humanitarian purposes).<br>*Permitted:* Interbasin transfers, particularly for hydroelectricity (EAA). |
| NS | Prohibited under the Water Resources Protection Act. | *Prohibited:* Water diversion and exports, under the Water Resources Protection Act. |
| NB | Not specified. | *Permitted:* Water Diversions and Exports and Interbasin Transfers under Watercourses and Wetland Alteration Regulation. |
| PE | Prohibited under the Environmental Protection Act. | *Permitted:* Water diversions and exports allowed for both groundwater and surface water. |
| QC | Prohibited under the Water Resources Preservation Act (2001).<br>Quebec, together with Ontario, has signed the Great Lakes-St. Lawrence River Basin Sustainable Water Resources Agreement (2005), which prohibits diversions between and within water basins (with limited exemptions). | *Permitted:* Water Diversions under Watercourses Act. |
| ON | Prohibited in the Water Taking and Transfer Regulation (Ontario Regulation 387/04) made under the Ontario Water Resources Act.<br>Ontario, together with Quebec, has signed the Great Lakes-St. Lawrence River Basin Sustainable Water Resources Agreement (2005), which prohibits diversions between and within water basins (with limited exemptions). | *Prohibited:* Out of basin transfers. The regulation divides Ontario into three water basins: the Great Lakes-St. Lawrence Basin, the Nelson Basin, and the Hudson Bay Basin.<br>*Permitted:* Water used to make products (e.g., beer). Water transfers ongoing prior to 1 January 1998. Stipulated under the OWRA. |

►

◀ Table A.5

| | Policy on large-scale transfers outside the province[1] | Policies on water diversions and exports and on interbasin transfers |
|---|---|---|
| MB | Prohibited under the Water Resources Conservation and Preservation Act. | *Prohibited:* Except where the user holds a permit issued by the Crown. |
| SK | Prohibited under the Saskatchewan Watershed Authority Act. | *Prohibited:* Diversions of all kinds and transfers unless they are between watersheds in Saskatchewan. |
| AB | Prohibited under the Water Act. | *Prohibited:* Water transfers out of the country, with the exception of processed water and municipal water. Water between basins, except where authorized by a specific act of the legislature stipulated under the Water Act. |
| BC | Prohibited under Water Protection Act. | *Prohibited:* Large-scale projects capable of transferring water from one major watershed to another. Diversions or water extractions that would result in the project's having the capability of transferring water at a peak instantaneous flow of 10,000 litres or more per second. Stipulated under the Water Protection Act (1996). |
| YT | Not specified: under Federal Water Policy aim to limit diversions and exports. | Federal Water Policy (1987) aims to limit diversions and exports. |
| NT | Not specified: under Federal Water Policy aim to limit diversions and exports. | Federal Water Policy (1987) aims to limit diversions and exports. |
| NU | Not specified: under Federal Water Policy aim to limit diversions and exports. | Federal Water Policy (1987) aims to limit diversions and exports. |

NOTE:

1 The term "large scale" is commonly used in the legislation but is not concretely defined except in British Columbia. It is commonly used to mean transfers requiring infrastructural intervention (e.g., dams, pipelines etc) and is not generally used to refer to moving small quantities of water for drinking by truck, train, or other method of transportation.

to information, and opportunities for public oversight. Transparency is linked to accountability, as it enables citizens to attempt to achieve redress for violations if governments fail to do their jobs as regulators.

There is some evidence of enforcement in all jurisdictions surveyed, but much is left to discretion, and data on enforcement is, in many cases, not easily accessible to the public. Compliance initiatives that stand out include those of Ontario and Saskatchewan. Ontario's Chief Drinking Water Inspector provides an annual report reviewing violations, compliance measures, and fines issued, which is available to the public. Saskatchewan undertook 792 waterworks inspections in 2004-5 and is making progress in achieving compliance. Moreover, inspections records will soon be made available online.

The need for greater transparency is evident from our survey. Our survey found that all provinces and territories provided information on their websites, but it was not always easy to locate and was often difficult to understand. Public reporting on an annual basis for drinking water quality is only required for public water systems in Newfoundland and Labrador, Ontario, and Saskatchewan. That leaves many citizens in Canada without knowledge of where their drinking water comes from and whether it is subject to occasional, regular, or no contamination.[29]

Many other developed countries have extensive water quality reporting requirements. In the United States, each water utility must provide consumers with an annual water quality report. Similarly, in the United Kingdom, statistics are available on water quality in every jurisdiction, easily downloadable from the Internet, and the performance of companies is publicly compared. The UK Environment Agency, for example, publishes the list of "top ten" polluters each year.

While Canadian provinces and territories offer some kind of drinking water information, be it legislation, strategies, or quality reports, frequently it is not centralized and may be associated with a variety of governing bodies. As such, information made available to the public by the provinces can be dispersed among many different ministries, which makes information difficult to locate. In navigating through the variety of legislation, regulations, and policies, what has been repealed and replaced is not always evident. In contrast, in the US, the EPA's Safewater website allows citizens to access one database on the status of nearly every regulated public water system. Each state, as well, has passed laws that address, at a minimum, the requirements of the national Safe Drinking Water Act (SDWA).[30] In Canada, there are increasing efforts to coordinate across departments via interdepartmental coordinating committees. Still, the need for greater public transparency is evident. In our survey, we found that provinces and territories provided some information on their websites, yet this was difficult to locate and, once found, was often presented in a way that was too complex to understand and assimilate.

While this appendix has provided an overview of water governance within Canada, much remains to be investigated with respect to water policy in the country. In particular, the need for databases, especially with respect to groundwater, has been noted by scientists (Senate 2005). Moreover, the fragmentation of water legislation, policies, and practices is overwhelmingly evident. Canada has considerable challenges ahead if it is to develop an integrated approach to water resources management. Fragmentation exists within departments, cross-departmentally, and interjurisdictionally between the federal and provincial governments. While water as a resource may require variation in programs and policies, there is clearly a need for data collection and knowledge about existing approaches, as well as for enhanced transparency and accountability.

## ACKNOWLEDGMENTS

This appendix was prepared by Carey Hill, Kathryn Furlong, Karen Bakker, and Alice Cohen. Carey Hill's contribution draws, in part, on research supported by a grant from the Weyerhaeuser US Environmental and Resource Policy Research Program at the University of British Columbia.

The appendix was sent for review to representatives of each of the ten provinces and three territories. We thank the following departments and ministries for their responses and suggestions: Department of Environment (British Columbia); Department of Environment (Alberta); Saskatchewan Watershed Authority (Saskatchewan); Ministry of Water Stewardship (Manitoba); Water Policy, Department of Environment (Ontario); Ministère du Développement Durable, de l'Environnement et des Parcs (Quebec); Department of Environment and Local Government (New Brunswick); Environmental Monitoring and Compliance Water and Wastewater Branch, Department of Environment and Labour (Nova Scotia); Department of Environment, Energy and Forestry (Prince Edward Island); Water Resources Management Division (Newfoundland and Labrador); Water and Sanitation, Public Works and Services (Northwest Territories); Water Resources Environment Programs Department of Environment, Environmental Health, Department of Health and Social Services (Yukon); Nunavut Water Board (Nunavut). While acknowledging these contributions, any errors or omissions remain those of the authors alone.

NOTES

1 Well-field protection planning is also beyond the scope of this survey.
2 This program was one of the earliest water-wise programs.
3 See, for example, Harrison (1996, 1997) and Harrison and Hoberg (1994).
4 The federal government transferred responsibility for water resources to the Yukon Territory in 2003, along with land and forest resources. The federal government retains most of the responsibility for water resources in the Northwest Territories and Nunavut.
5 It should be noted that bottled water is primarily regulated at the federal level by Health Canada.
6 Federal responsibility for First Nations reserves is complex and not well-defined, and there is no federal legislation governing drinking water protection on First Nations lands. In response to a request from the Walkerton Inquiry, the federal government defined the responsibility for drinking water on reserves as "shared among First Nation Band Councils, Health Canada, and Indian and Northern Affairs Canada (INAC)." See Walkerton Inquiry, Chapter 15, Part 2, p. 5.
7 For example, the conservation authorities provide integration among different municipalities because the municipalities in the watershed are all represented on the boards of the conservation authorities. Moreover, there is integration between the municipalities and the province because the conservation authorities act as an implementation body for provincial legislation. Finally, the conservation authorities are generally funded, in part, by provincial grants as well as by local municipalities.
8 For example, see Baril, Maranda, and Baudrand (2005).
9 See Harrison (1996, 67).
10 Ibid.
11 Ibid., 100-2.
12 For a comprehensive examination of the GLWQA and its future, see Botts and Muldoon (2005).
13 See Senate (2005, 5).
14 Ibid, 12. David Schindler, Killam Memorial Professor of Ecology at the University of Alberta, explained that monitoring of groundwater records had been reduced since 1993, while John Carey of the Environment Canada's National Water Research Institute informed the committee that the number of sites for monitoring water quantity had been reduced from 4,000 to approximately 2,500.
15 The three microbiological requirements are to be reduced to two. Note that there are seventy-eight radiological standards, but no monitoring is required for them.
16 While participating in the provincial program, St. John's uses its own staff rather than that of the province to do its testing and monitoring.

17 The report noted that Health Canada had been "unable to review chemical and physical guidelines on a timely basis." Specifically, the report explained that "about 50 of the 83 chemical and physical guidelines" were older than 15 years and needed to be updated to "reflect current science and protect the health of Canadians." See Commissioner of the Environment and Sustainable Development (2005, 13).

18 In the US, filtration was required for most surface water sources under the Surface Water Treatment Rule. Provisions for this rule existed in the 1986 amendments to the SDWA, but the Rule was not promulgated until 1989.

19 We thank the Newfoundland Department of Environment for providing this helpful information.

20 See Oborne (2005, 19).

21 Nova Scotia adopted the Guidelines for Canadian Drinking Water Quality in 2000 in its Water and Wastewater Facility Regulations.

22 Alberta requires that all health-based parameters of the guidelines be met. It allows exceedances for naturally occurring fluoride, and its standards surpass those for turbidity set out in the guidelines (see Table A.2).

23 Two-thirds of Canadians are served by surface water. Frequently, municipalities that use groundwater employ a mix of groundwater and surface water. Only one province, Prince Edward Island, is entirely dependent on groundwater.

24 It should be noted that, in 1997, Alberta was the first province in the country to adopt the USEPA's stringent Enhanced Surface Water Treatment Rule. In the US, this rule resulted in filtration being required for most US public water systems.

25 The US Safe Drinking Water Act (SDWA 1974, 1986, 1996) involves a model of delegation where the federal Environmental Protection Agency delegates responsibility for enforcement to the states if they ensure they will implement the provisions of the SDWA. The EPA can act if states fail to do this or if states request assistance.

26 Saskatchewan's standards will be phased in over a five-year time- frame (2006-10).

27 See Saskatchewan's Safe Drinking Water Strategy Performance Plan.

28 Most recently, on 18 October 2004, by MP Pat Martin (NDP) from Winnipeg Centre, Bill C-221, An Act to Prohibit the Export of Water by Interbasin Transfers.

29 In some cases, however, water suppliers do voluntarily supply drinking water quality information (see, e.g., the Greater Vancouver Regional District website at http://www.gvrd.bc.ca/water/quality-and-treatment.htm).

30 See http://www.epa.gov/safewater/dwinfo/index.html. There are two exceptions: the State of Wyoming and the District of Columbia. The SDWA includes requirements for public reporting, public right-to-know, operator certification, minimum requirements for disinfection and filtration, source water assessments, citizen suits, formal review, and monitoring for approximately ninety contaminants (National Primary Drinking Water Regulations), plus several rules (addressing surface water treatment, disinfectants and disinfection byproducts, total coliforms, lead and copper, and information collection).

REFERENCES

Baril, P., Y. Maranda, and J. Baudrand. 2005. Integrated Watershed Management in Quebec: A Participatory Approach Centred on Local Solidarity. Abstract for the International Meeting on the Implementation of the European Water Framework Directive: EURO INBO, 2005. International Network of Basin Organizations.

Botts, Lee, and Paul Muldoon. 2005. *Evolution of the Great Lakes Water Quality Agreement.* Michigan: Michigan State University Press.

Commissioner of the Environment and Sustainable Development. 2005. *Report of the Commissioner of the Environment and Sustainable Development to the House of Commons.* Chapter 4: Safety of Drinking Water – Federal Responsibilities. Ottawa: Office of the Auditor General of Canada.

Environment Canada. Freshwater Website. http://www.ec.gc.ca/water/e_main.html.

Environnement Québec. 2002. *Politique nationale de l'eau.* Québec.

—. 2003. *Bilan de la qualité de l'eau potable au Québec,* Janvier 1995–Juin 2002. Québec.

Federal-Provincial-Territorial Committee on Drinking Water of the Federal-Provincial-Territorial Committee on Health and the Environment. 2004. *Summary of Guidelines for Canadian Drinking Water Quality.* April.

Federal-Provincial-Territorial Committee on Drinking Water and the CCME Water Quality Task Group. 2002. *From Source to Tap: Guidance on the Multi-Barrier Approach to Safe Drinking Water.* http://www.ccme.ca/assets/pdf/mba_guidance_doc_e.pdf.

Fleury, Marc-Antoine. 2003. Unearthing Montreal's Municipal Water System: Amalgamating and Harmonizing Urban Water Services. *Faculty of Environmental Studies Outstanding Graduate Student Paper Series* 8 (3). York University. June.

Harrison, Kathryn. 1996. *Passing the Buck: Federalism and Environmental Policy.* Vancouver: UBC Press.

—. 1997. The Origins of National Standards: Comparing Federal Government Involvement in Environmental Policy in Canada and the United States. In Patrick Fafard and Kathryn Harrison, eds., *Managing the Environmental Union: Intergovernmental Relations and Environmental Policy in Canada.* Montreal: McGill-Queen's University Press.

Harrison, Kathryn, and George Hoberg. 1994. *Risk, Science and Politics: Regulating Toxic Substances in Canada and the United States.* Montreal: McGill-Queen's University Press.

Hill, C. 2006. Two Models of Multi-Level Governance, One Model of Multi-Level Accountability: Comparing Drinking Water Protection in Canada and the USA. PhD dissertation, Department of Political Science, University of British Columbia.

Lindgren, Richard D. 2005. *Tapwater on Trial: Overview of Ontario's Drinking Water Regime.* Prepared for Third Annual Conference on Water and Wastewater in Ontario. Canadian Environmental Law Association website at http://www.cela.ca/publications/cardfile.shtml?x=2185.

Mouldey, Sarah Elizabeth. 1994. Managing for Safe Drinking Water: A Review of Drinking Water Management in Alberta with Specific Reference to Calgary. MA thesis, University of Calgary.

Nowlan, Linda. 2005. *Buried Treasure: Groundwater Permitting and Pricing in Canada.* Toronto: Walter and Duncan Gordon Foundation.

Oborne, Bryan. 2005. Saskatchewan Provincial Case Study Analysis of Water Strategies for the Prairie Watershed Region. Working Draft for Comment presented to the Prairie Water Policy Symposium. International Institute for Sustainable Development. September.

Ontario Ministry of Environment. 2003. *Advisory Committee on Watershed-Based Source Protection Planning: Protecting Ontario's Drinking Water – Toward a Watershed-Based Source Protection Planning Framework, Final Report.* Ontario. April.

—. 2004. *White Paper on Watershed-Based Source Protection Planning.* Ontario.

—. 2005. *Progress Report on Ontario's Municipal Residential Drinking Water Systems 2003-2004: Results and Interim Results for 2004-2005.* Chief Drinking Water Inspector. 31 May.

Quinn, Frank, Michael Healey, Richard Kellow, David Rosenberg, and J. Owen Saunders. 2003. Water Allocation, Diversion and Export. Environment Canada. http://www.nwri.ca/threat2full/ch1-1-e.html.

Senate. 2005. *Water in the West, under Pressure.* Fourth Interim Report of the Standing Senate Committee on Energy, the Environment and Natural Resources. Ottawa: Senate of Canada.

Sierra Legal Defence Fund. 2001. *Waterproof: Canada's Drinking Water Report Card.* Prepared by Randy Christensen and Ben Parlin. Toronto: Sierra Legal Defence Fund.

—. 2004. *The National Sewage Report Card: Grading the Sewage Treatment of 22 Canadian Cities.* Toronto: Sierra Legal Defence Fund.

Swain, Harry, and Water Strategy Expert Panel. 2005. *Watertight: The Case for Change in Ontario's Water and Wastewater Sector.* Toronto: Ministry of Public Infrastructure Renewal.

Walkerton Commission of Inquiry. 2002. First Nations (Chapter 15, Part 2). Toronto: Publications Ontario.

—. 2002. The Legislative, Regulatory and Policy Framework (Chapter 13, Part 1). Toronto: Publications Ontario.

APPENDIX 2

# Additional Resources and Reading

### Water Websites

Environment Canada's website has links to provincial, federal, and international governmental websites: http://www.ec.gc.ca/water/en/info/pubs/wwf/e_web.htm.

Two useful nongovernmental websites with information on water protection and sustainable water management are: http://www.thewaterhole.ca (across Canada) and http://www.waterbucket.ca (focusing on British Columbia).

University-based research groups in Canada include: University of Toronto's Program on Water Issues (http://www.powi.ca), University of Victoria's POLIS Project (http://www.polisproject.org), and University of Guelph's Guelph Water Management Group (http://www.uoguelph.ca/gwmg).

Environmental NGOs are another useful source of information on water in Canada. Two of the most active groups campaigning for sustainable water management are: Canadian Environmental Law Association (http://www.cela.ca) and Pollution Probe (http://www.pollutionprobe.org).

### Global Water Issues

The California-based Pacific Institute publishes a biennial report on the world's water, packed with useful figures and information (*The World's Water*, Earthscan). The Institute also publishes reports on critical water issues (http://www.pacinst.org).

### Water Resources and Water Quality

Environment Canada's "freshwater" website has detailed information on water quantity and quality in Canada: http://www.ec.gc.ca/water/e_main.html.

### Sustainable Infrastructure

The Federation of Canadian Municipalities has published extensive resources for local water managers (http://www.fcm.ca), including the National Guide to Sustainable Municipal Infrastructure (http://www.infraguide.ca), which includes tap water.

### Privatization and Public-Private Partnerships in Canada

For a public sector union perspective, see CUPE's WaterWatch campaign (http://ww.cupe.ca). For an NGO perspective, see the Council of Canadians Blue Planet Project (http://www.canadians.org). For a business perspective, see the Canadian Council for Public Private Partnerships (http://www.pppcouncil.ca). Pollution Probe's Elizabeth Brubaker has written a book (*Liquid Assets: Privatizing and Regulating Canada's Water Utilities*, 2002), advocating for the involvement of the private sector in water supply in Canada. Maud Barlow (Council of Canadians) and Tony Clarke (Polaris Institute) argue the case against corporate management of water supply (*Blue Gold: The Battle against Corporate Theft of the World's Water*, 2002).

### Comparisons of Water Quality and Environmental Performance in Canada and Abroad

Environmental lawyer David Boyd's comparison of Canada's environmental performance to international standards revealed that Canada has one of the worst environmental records of any OECD country (*Canada vs. the OECD: An Environmental Comparison* (http://www.environmentalindicators.com).

### Water Governance in Canada

Environment Canada publications provide a good general introduction to water governance issues. See, for example, *Water in Canada: Preserving a Legacy for People and the Environment* (http://www.ec.gc.ca/water/en/info/pubs/wwf/e_contnt.htm).

### Water Conservation

Many municipalities are moving to encourage reductions in water consumption by industrial and residential customers. The Canadian Water and Wastewater Association and Environment Canada have developed a Water Efficiency Experiences Database (http://www.cwwa.ca/WEED/Index_e.asp), and the University of Victoria's Polis Project has an extensive project on water conservation (http://www.polisproject.org).

### The Walkerton Reports

The Walkerton Inquiry, led by Justice Dennis O'Connor, published two in-depth reports following the water quality contamination incident in Walkerton. The second report contains detailed recommendations for drinking water protection and sustainable water management, which have sparked debate across Canada (http://www.attorneygeneral.jus.gov.on.ca/english/about/pubs/walkerton/).

### Get Involved

Waterkeeper and Streamkeeper groups bring together local people to safeguard freshwater and marine resources. To find out more about becoming involved in a local Waterkeeper group (or about starting your own chapter) go to http://www.waterkeepers.ca.

The Council of Canadians runs a water campaign, with activities sponsored by local chapters: http://www.canadians.org.

KAIROS (Canadian Ecumenical Justice Initiatives) is running a "Water: Life before Profit" campaign: http://www.kairoscanada.org.

Many Canadian environmental NGOs run campaigns on various aspects of sustainable water management, including Friends of the Earth (http://www.foecanada.org), the Sierra Club of Canada (http://www.sierraclub.ca), the Canadian Environmental Law Association (http://www.cela.ca), and Pollution Probe (http://www.pollutionprobe.org). Specific campaigns have also emerged, including Eau Secours (http://www.eausecours.org), the Blue Planet Project (http://www.blueplanetproject.net), and CUPE's Water Watch (http://cupe.ca/waterwatch).

APPENDIX 3

# The Waterkeeper Alliance

The Waterkeeper Alliance is an example of a local grassroots water advocacy group that is playing an increasingly important role in local water management across North America. Founded in 1966, during a series of lawsuits against polluters and industrialists on New York State's Hudson River, it now has 157 local chapters. In Canada, this includes the Petitcodiac and Ottawa Riverkeepers, Fundy Baykeeper, Lake Ontario Waterkeeper, Georgian Baykeeper, Bow Riverkeeper, Fraser Riverkeeper, Grand Riverkeeper, and the Canadian-Detroit Riverkeeper. In the United States, the growth of the Alliance has been helped by the media savvy of its founder and president, environmental lawyer and activist Robert F. Kennedy Jr.

The key role for a Waterkeeper (also known as Riverkeeper or Baykeeper) is to be a local advocate for the enforcement of environmental legislation. Waterkeepers patrol the waterways, monitor pollution, and encourage concerned citizens to come forward with tips on law-breaking polluters – both governments and private companies. In Canada, this is particularly significant, given that current legislation requires strong evidence from citizens before governments will act on claims about polluters.

Waterkeepers provide assistance to pre-existing nongovernmental organizations and citizen groups with regard to dealing with pollution and taking legal action where necessary. In Canada, issues of concern have ranged from the use of pesticides along the Fraser River to oil spills in the Great Lakes. Waterkeepers can often stimulate other agencies to act; for example, in 2003, the Lake Ontario Waterkeeper led a transboundary coalition campaign to stop occurring pollution in the St. Lawrence River, allegedly in breach of the Fisheries Act, which helped to convince the Commission for Environmental Cooperation to open an investigation in August 2004.

Criticisms have been made of the Waterkeeper Alliance. Some have criticized the media hype associated with Robert Kennedy Jr. Others worry that environmental causes are cynically mobilized by self-promoting celebrities, without changing the status quo. Another criticism of the Waterkeeper Alliance focuses on its strategy of emphasizing litigation rather than prevention of environmental harm.

In response, Waterkeepers argue that most members have no celebrity ties and are independently (usually locally) funded; local support rather than celebrity connections sustain local chapters. And Waterkeepers argue that, in the absence of government enforcement and political will, their organization plays a crucial role in halting serious polluters – both state and private – that otherwise go unchallenged.

---

All proceeds from the royalties of *Eau Canada* are being donated to the Canadian chapter of the Waterkeeper Alliance.

# Contributors

**Karen Bakker** is an associate professor and director of the Program on Water Governance in the Department of Geography, University of British Columbia. Her recent book, *An Uncooperative Commodity* (2004), examines the privatization of the British water supply industry and its effects on consumers and the environment.

**Andrew Biro** holds a Canada Research Chair in Political Ecology and is an assistant professor in the Department of Political Science at Acadia University. He is the author of *Denaturalizing Ecological Politics* (2005) as well as articles on water politics, environmental political theory, and the politics of culture.

**Oliver Brandes** is a political ecologist and a senior research associate with the University of Victoria's POLIS Project on Ecological Governance, where he directs research on urban water demand-side management and water sustainability in Canada. His most recent policy report (with Keith Ferguson, Michael M'Gonigle, and Calvin Sandborn) is "At a Watershed: Ecological Governance and Sustainable Water Management in Canada" (2005).

**David Brooks** is director of research, Friends of the Earth (Canada). He has published widely on energy and water policy. He has edited several books on resource issues and on water demand management. His latest book is *Water: Local-Level Management* (2002).

**Randy Christensen** is a staff lawyer at Sierra Legal Defence Fund in Vancouver and is head of the Waterways Team. He is the author of the SLDF 2003 report *Watered Down*, which brings together information on waterborne disease outbreaks in British Columbia, and the 2001 SLDF report *Waterproof*, a cross-Canada whistle-blowing comparison of drinking water regulations.

**Alice Cohen** is a master's student in the Resource Management and Environmental Studies Program at the University of British Columbia. She holds a bachelor's degree from McGill University and has spent time in Vietnam working on climate change adaptation issues. Her research interests include groundwater management and policy as well as transboundary water governance.

**Rob de Loë** is a professor and Canada Research Chair in Water Management at the University of Guelph's Department of Geography. He is co-principal of the Guelph Water Management Group and a past national director of the Canadian Water Resources Association.

**Dianne Draper** is a professor and head of the Department of Geography at the University of Calgary. She has authored numerous publications on environmental management, including *Our Environment: A Canadian Perspective* (2004).

**Kathryn Furlong** is a doctoral candidate in the Department of Geography, University of British Columbia. Her research examines the relationship between changing governance structures for municipal water supply and the uptake of innovative technologies and programs directed at demand management within the context of neoliberalization in Canada. Her article "Hidden Theories, Troubled Waters" is forthcoming in *Political Geography.*

**Robert B. Gibson** is a professor of environment and resource studies at the University of Waterloo, where he specializes in environmental policy issues and the integration of broad sustainability considerations in project and strategic decision making. He has been co-editor, editor, and editorial board chair of *Alternatives Journal* since 1984.

**Carey Hill** is a doctoral candidate in the Department of Political Science, University of British Columbia. Her research examines multilevel governance and multilevel accountability with respect to drinking water protection policies in Canada and the United States. In addition to a long-standing interest in Canadian politics, her research interests include health policy, environmental policy, and comparative politics.

**Theodore M. Horbulyk** is an associate professor in the University of Calgary's Department of Economics and a theme leader at the Alberta Ingenuity Centre for Water Research. His most recent publication on water markets appears in the *Canadian Water Resources Journal* (Spring 2005).

**Adèle Hurley** is director of the Program on Water Issues at the University of Toronto's Munk Centre for International Studies, and she has served as Canadian co-chair of the International Joint Commission, which oversees Canada/US boundary water issues. She is also the co-author (with Andrew Nikiforuk) of a recent *Globe and Mail* op-ed piece on water diversions ("Don't Drain on our Parade," 29 July 2005).

**Reid Kreutzwiser** is a professor of geography at the University of Guelph and co-principal of the Guelph Water Management Group. He is past Canadian co-chair of the Coastal Zone Study Group, which was affiliated with the International Joint Commission's Great Lakes Water Levels Reference Study.

**Frédéric Lasserre** is a professor of geography at the Université Laval (Québec). He is the co-author of *Eaux et territoires: Tensions, coopérations et géopolitique de l'eau* (2003) and editor of *Transferts massifs d'eau: Outils de développement ou instrument de pouvoir?* (2005).

**Anastasia M. Lintner** is a staff lawyer with the Sierra Legal Defence Fund's Toronto office. Her practice focuses on Aboriginal, forestry, water, land use planning, and environmental assessment law. She is also an adjunct professor in the Department of Economics at the University of Guelph, specializing in natural resource and environmental economics.

**Cushla Matthews** is a doctoral student at the University of Waterloo's Department of Geography and a professional planner. Her research interests include water conservation, water policy, and ethics.

**Theresa McClenaghan** is counsel at the Canadian Environmental Law Association (CELA). Along with other lawyers at CELA, she represented the Concerned Walkerton Citizens at the Walkerton Inquiry. She was CELA's representative to Ontario's Source Protection Advisory Committee and currently sits on both the Nutrient Management Advisory Committee and Source Protection Implementation Committee.

**Michael M'Gonigle** is a professor of law at the University of Victoria and holds the Eco-Research Chair in Environmental Law and Policy. A director of the POLIS Project on Ecological Governance, he co-founded Greenpeace International, Smart Growth BC, and the Sierra Legal Defence Fund. He has written on environmental and international law, wilderness and forest policy, resource management, political ecology, and legal theory. His latest book (with Justine Starke) is *Planet U: Sustaining the World, Reinventing the University* (2006).

**Bruce Mitchell** is a professor in the Department of Geography and is associate provost, Academic and Student Affairs, University of Waterloo. He is a widely recognized researcher on water management issues and, in 2005, was named a Fellow of the Royal Society of Canada and a Fellow of the International Water Resources Association, in recognition of his work on water issues.

**Suzanne Moccia** is a graduate student in the Department of Geography at the University of British Columbia. She is currently conducting research on the feasibility of cooperatives as an alternative water supply strategy for poor communities in cities in Latin America, with a case study in Buenos Aires.

**Paul Muldoon** is the executive director and counsel at the Canadian Environmental Law Association in Toronto and a former member of the Science Advisory Board to the International Joint Commission. His most recent book on water policy is *Evolution of the Great Lakes Water Quality Agreement* (2005, co-authored with Lee Bott).

**Emma Norman** is a doctoral student in the Department of Geography at the University of British Columbia. She is currently conducting research on local transboundary water governance mechanisms between Canada and the United States, focusing on western Canada. Her research and volunteer work on environmental issues has taken her to South Africa, Ecuador, and Malawi. Her most recent academic article appeared in the *Journal of Borderland Studies.*

**Linda Nowlan** is an environmental lawyer and consultant in Vancouver. She has worked with NGOs, philanthropic organizations, government and international agencies and has also practised civil litigation. In 2001, the BC government appointed her to the Independent Drinking Water Review Panel. In 2004, she was commissioned by the Walter and Duncan Gordon Foundation to write a report on groundwater regulation in Canada.

**Ralph Pentland** has co-chaired three IJC binational committees dealing with Great Lakes water use and diversions, most recently in 2002. From 1978 to 1991, he was director of Water Planning and Management at Environment Canada and was responsible for negotiating and overseeing numerous Canada-US and federal-provincial agreements. He was also the prime author of the *Federal Water Policy,* which was tabled in Parliament in 1987.

**Steven Renzetti** is a professor of economics at Brock University and has published widely on water supply utility pricing in Canada. He is the author of *The Economics of Water Demands* (2002) and the editor of *The Economics of Industrial Water Use* (2002).

**J. Owen Saunders** is executive director of the Canadian Institute of Resources Law, University of Calgary, and has acted as advisor to Canadian and foreign governments on natural resources policy. He served on the binational study team advising the International Joint Commission on its 1999-2000 Water Uses Reference, in which capacity he coordinated the preparation of the legal background paper for the commission.

**David Schindler** is Killam Memorial Professor of Ecology at the University of Alberta, Edmonton. He is internationally recognized for his research on aquatic ecology and was winner of the first Stockholm Water Prize (the "Nobel Prize of Water").

**Dan Shrubsole** is an associate professor and chair of the Department of Geography, University of Western Ontario. He has authored many publications on Canadian water management, including *Canadian Water Management: Visions for Sustainability* and *Practising Sustainable Water Management: Canadian and International Experiences* (co-authored/edited with Bruce Mitchell, 1994 and 1997).

**John B. Sprague** is a former research scientist with federal Fisheries, a professor of zoology, and a consultant, now living on Salt Spring Island, British Columbia. His specialty is water pollution and ecotoxicology. One of his early publications was designated as a Citation Classic, the most highly cited paper in limnology (freshwater science) during a third of a century.

**Ardith Walkem** is an Aboriginal lawyer and a member of the Nlaka'pamux Nation in the interior of British Columbia. Her areas of practice include Aboriginal title and rights, public legal education, legislative policy analysis, and hunting and water rights. She recently completed an LLM in the area of indigenous oral traditions.

**Michael M. Wenig** is a research associate at the Canadian Institute of Resource Law and an adjunct professor at the University of Calgary Faculty of Law. His research generally relates to the intersection of natural resources and environmental law and to legal and policy aspects of ecosystem-based management, particularly watershed management. His publications include papers on water pollution liability, watershed-based pollution control, and the legal framework for watershed management.

# Index

Page references to figures, boxes, and tables are in **bold type.**

Printed and bound in Canada by Friesens
Set in Giovanni Book and Scala Sans by Artegraphica Design
Text design: Irma Rodriguez
Copy editor: Joanne Richardson
Proofreader and indexer: Dianne Tiefensee